Landscape Series

Volume 26

Series editors
Jiquan Chen, Department of Geography, Environment, and Spatial Sciences, Michigan State University, East Lansing, USA
Janet Silbernagel, Department of Planning and Landscape Architecture, University of Wisconsin-Madison, Madison, USA

Aims and Scope

Springer's innovative Landscape Series is committed to publishing high-quality manuscripts that approach the concept of landscape from a broad range of perspectives. Encouraging contributions on theory development, as well as more applied studies, the series attracts outstanding research from the natural and social sciences, and from the humanities and the arts. It also provides a leading forum for publications from interdisciplinary and transdisciplinary teams.

Drawing on, and synthesising, this integrative approach the Springer Landscape Series aims to add new and innovative insights into the multidimensional nature of landscapes. Landscapes provide homes and livelihoods to diverse peoples; they house historic—and prehistoric—artefacts; and they comprise complex physical, chemical and biological systems. They are also shaped and governed by human societies who base their existence on the use of the natural resources; people enjoy the aesthetic qualities and recreational facilities of landscapes, and people design new landscapes.

As interested in identifying best practice as it is in progressing landscape theory, the Landscape Series particularly welcomes problem-solving approaches and contributions to landscape management and planning. The ultimate goal is to facilitate both the application of landscape research to practice, and the feedback from practice into research.

More information about this series at http://www.springer.com/series/6211

Alexander V. Khoroshev • Kirill N. Dyakonov
Editors

Landscape Patterns in a Range of Spatio-Temporal Scales

 Springer

Editors
Alexander V. Khoroshev
Department of Physical Geography &
Landscape Science
Lomonosov Moscow State University
Moscow, Russia

Kirill N. Dyakonov
Department of Physical Geography &
Landscape Science
Lomonosov Moscow State University
Moscow, Russia

ISSN 1572-7742   ISSN 1875-1210   (electronic)
Landscape Series
ISBN 978-3-030-31187-2   ISBN 978-3-030-31185-8   (eBook)
https://doi.org/10.1007/978-3-030-31185-8

This Springer imprint is published by the registered company Springer Nature Switzerland AG
The registered company address is: Gewerbestrasse 11, 6330 Cham, Switzerland

Foreword: Landscape Science – A Multidisciplinary and Border-Crossing Scientific Approach

Landscapes are considered as the unifying entity incorporating a bunch of ecological processes and functions, human interventions, and interactions between multiple scales. The theoretical concept and particularly its multidisciplinary roots are considered since decades as fundamental advances in making research relevant for society and in supporting the sustainable development of social-ecological systems. This book provides an overview and synthesis on long-term research approaches and their results conducted in Russia with the aim of understanding and taking benefit from knowledge on landscape heterogeneity, functioning, and dynamics. Its particular value consists in complementing the often only two-dimensional view on landscape pattern through approaches for an integrative three-dimensional analysis of landscapes and related implications for eco-hydrological processes. This book presents a bundle of analytical approaches and their application areas to analyze landscapes and contributes particularly to advance landscape sciences in combining small- and large-scale patterns, processes, and functions. Its particular value consists in the huge range of landscapes and eco-zones covered in Russia and the tremendous monitoring which contributed to the analytical approaches.

Human impacts and pressures are included as well as aspects on how landscape pattern can help to indicate and control even dramatic situations provoked from extreme events and from historical and current land use.

As such, the book provides a brilliant overview on the state of the art of research on land systems in Russia and on the multifaceted aspects that are covered through this research. It is particularly useful for researchers and students who wish to obtain knowledge on the landscape understanding, analytical approaches, and interpretations on how to further develop landscapes in Russia.

Past President, International Association for Landscape Ecology
Halle, Germany

Prof. Dr. habil. Christine Fürst

Institute for Geosciences and Geography Dept.
Sustainable Landscape Development
Martin Luther University Halle-Wittenberg
Halle, Germany

Preface

Long-term experience of landscape research in Russia provided rationales for the diverse theoretical approaches to descriptions of landscape heterogeneity, functioning, and dynamics. Polycentric concept of landscape developed within Russian physical geography since the early twentieth century relies on the idea of strong relationships between abiotic and biotic components paying equal attention to both vertical fluxes and lateral transfer. The central objective of this book is to demonstrate multiplicity of models and multi-scale approach to description and explanation of landscape pattern and functioning. Unlike most biota-centered publications in landscape ecology, the authors use the geosystem concept which is believed to be more general than the matrix concept, because vegetation or land use patterns, which are commonly used for delineation of patches and corridors, are in most cases controlled by abiotic template. This concept treats landscape as a three-dimensional hierarchically organized heterogeneous entity where the pattern is shaped by topography and substrate which determine the distribution of matter and energy flows. In contrast to dominant two-dimensional matrix biocentric concept, we advocate three-dimensional view of landscape with strong emphasis on abiotic drivers of pattern development, including relief, geological structures, and runoff. Delineation of landscape units can be based not only on visually detected patches but also on several mutually additional approaches based on relief and substrate genesis, unilateral flows of dissolved matter, solid matter, or surface water. Moreover, the modern trend in landscape science involves the elaboration of tool to describe and depict landscape continuality. In accordance with the general ideas of the geosystem concept, we pay much attention to self-development of landscape and sequence of its temporal states and to the external drivers (e.g., climatic) of landscape evolution.

We demonstrate a variety of geosystem-oriented landscape pattern models relying on genesis-based, basin, or catena approaches with a particular focus on opportunities and advantages for actual land use decisions and understanding of human-nature relationships in the past and their remedy for the present-day landscape structure.

This book presents a multi-scale view of landscape. The contributions deal with various hierarchical levels of landscape organization ranging from within-unit interior variability to between-units interaction at the landscape level, to the regional and supra-regional zonal patterns. Specific methodologies adapted for studying pattern-process relationship at each level are demonstrated with particular focus on interactions between abiotic environment, soils and plant cover in both radial and lateral aspects. Contributions in chapters describe the tools of pattern analysis at either local, landscape or regional scale of research. Each contribution provides a particular viewpoint on pattern-process relationships with a focus on certain pattern type and research scale.

Each part of the book treats spatial pattern in one of the following aspects: (1) indicator of actual matter and energy flows; (2) control over actual processes, including disturbance expansion and determinant of future development; (3) indicator of genesis and prerequisite for future trends; (4) driver for short-term dynamics of processes; (5) response to climatic and anthropogenic influences; (6) factor of settlement network and land use adaptation at various historical epochs; and (7) framework for actual land use spatial arrangement.

Part I provides *theoretical foundations* for studying pattern-process relationships with particular emphasis on mathematical and physical rationales. Khoroshev provides a review of landscape pattern concepts elaborated in Russia and corresponding terminology. Cherkashin provides a theoretical introduction to multiplicity of landscape pattern models, hierarchy of geosystems, and multi-scale approach. Sysuev argues that quantification of structure-forming processes and physical parameters enables us to elaborate alternative classifications of landscapes; by determining boundary conditions and the relationships between the parameters, we can relate functioning model of a geosystem to structure.

In Part II landscape pattern is analyzed as an *indicator of actual processes of matter transfer*. Victorov proposes mathematical models of coarse-scale genesis-based patterns to indicate frequency and magnitude of geomorphological processes and to predict possible detrimental effects for land use. Linnik and co-authors provide arguments for stochastic nature of heterogeneity at the finest scale level for microrelief-based units using patterns of radionuclides distribution. Puzachenko demonstrates the opportunities to use information on variance of the relief properties and multispectral reflection for evaluating fractions of the spatial variance of landscape attributes and to detect processes responsible for the emergence of patterns. Sandlersky and co-authors apply thermodynamic properties of landscape cover based on the non-extensive statistical mechanics model to estimate exergy with a focus on the degree of non-linearity of relations in a system and the degree of its organization.

Part III demonstrates a variety of scale-specific approaches to treat pattern as *a control over actual processes*. Vanteeva and Solodyankina use genesis-based and gradient landscape-scale model to relate biological production to abiotic template of various altitudinal belts in Siberian mountains. Catena-based geochemical approach for local scale was applied by Avessalomova (plain region) and Semenov (mountainous regions) to identify the location of dispersion and accumulation zones according to various neighborhoods in toposequences and geochemical mobility of dissolved matter. Khoroshavin and co-authors, Erofeev and Kopysov, apply basin approach at broader scale to explain matter migration with respect to interior mosaic of topography-based units. Zolotov and Chernykh examine the interrelations between landscape and floristic units at the regional scale on the examples of the basins of medium-sized rivers and lakes with heterogeneous physical environment.

In Part IV pattern is treated as an *indicator of paleo-processes and prerequisite for evolution* with a particular focus on sedimentation, tectonic movements, and exogenous processes. Gorbunov and co-authors apply topography-based model of landscape pattern at the regional scale to explain the intensity of matter transfer that links landscape units at different elevation levels. Petrushina shows evidence that correct delineation of paragenetic geosystems with unidirectional matter flows between altitudinal belts enables us to use plant and soil cover as a valuable source of information about the evolution of mountainous landscapes in the Caucasus. Khoroshev applies multi-scale statistical analysis of landscape structure based on gradient concept to compare contributions of physiography and self-development of soil-vegetation cover to landscape heterogeneity and also evaluates uncertainty of landscape unit delineation.

Part V is devoted to spatial pattern as a *predictor of type of reversible dynamic phenomena*. Isachenko shows a variety of dynamic trajectories depending on a physiographic unit and relates local-scale structure and dynamics to land use history. Gurevskikh and Yantser apply genesis-based representation of landscape pattern and phenological methods to explain the sequence and duration of short-term temporal states at both the local and regional scales.

Part VI deals with the *response of landscape pattern to climatic and anthropogenic changes*. Dyakonov and Bochkarev use dendrochronological indicators and zonality model to analyze shifts in boundaries on ecotones between forest and non-forest belts in plain and mountainous regions. They advocate correspondence between hierarchies of spatial pattern and temporal states. Kolomyts and co-authors predict specific responses of carbon storage at supra-regional scale depending on topographically induced mosaics of plant and soil cover. Chistyakov and co-authors evaluate probabilistic linkages between occurrences of various landscape properties in the Siberian mountains and reveal ecological gradients at the inter-zonal level to model landscape pattern response to climatic changes.

Part VII deals with *historical anthropogenic drivers of landscape pattern development*. Nizovtsev and Erman analyze the adaptation of land use to physiography-induced pattern at the supra-regional scale and compare the degree and extent of anthropogenic transformations for different historical epochs. Matasov applies genesis-based landscape model to compare contributions of natural, positional factors and socioeconomic trends with land use patterns since the eighteenth century at the regional scale. Nizovtsev applies genesis-based approach at the local scale to establish the role of landscape factor in the origin and development of Moscow City. Khromykh and Khromykh combine the advantages of genesis-based and basin approaches to show the sequence "human-induced pattern – natural process – forced change in land use."

Part VIII addresses the *actual adaptation of human activity of landscape patterns*. Kiryushin proposes to apply physiography-based model of landscape pattern to the evaluation of landscape functions and spatial planning decisions in agriculture.

Field data for studies were collected in various natural zones (tundra, taiga, mixed, and broad-leaved forests, forest steppe, steppe) and regions of Russia both in plain (East Europe, West Siberia) and mountainous (the Caucasus, the Urals, South Siberia) areas. To process the data, the authors applied a wide range of tools, including statistical, geophysical, geochemical, dendrochronological, phenological, archaeological, and historical tools. To analyze the landscape patterns and processes, we used GIS modeling based on digital elevation models and remote sensing as well as traditional landscape mapping techniques.

The research outcomes involve forecast of frequency and intensity of processes, explanation of matter migration patterns, forecast of evolution trends, reconstruction of former natural and human-induced patterns, assessment of resistance in relation to exterior factors, and framework for land use decisions.

The book is based on the materials of XII International Landscape Conference "Landscape-ecological support of land use and sustainable development," which was held by IALE-Russia chapter in Tyumen, Russia, in 2017.

The authors are grateful to Kseniya Merekalova and Liudmila Khorosheva for their kind assistance in the preparation of the manuscript for publication.

Moscow, Russia   Alexander V. Khoroshev
                                                                        Kirill N. Dyakonov

Contents

List of Figures

mountain range, (3) Priol'khon plateau, (4) north-facing
macroslope of the Khamar-Daban mountain range
Fragment of the legend (prevalent types of geosystems)
Olkhinskoe plateau
NORTH ASIA TAIGA
Class of geoms—Mountain-taiga
Geom—Mountain-taiga of coniferous forests on granites
and gneiss
Class of facies—subhydromorphic
Group of facies—1. Larch (*Larix sibirica* Ledeb.) with Siberian
pine (*Pinus sibirica* Du Tour) forest with small shrub
(*Rubus arcticus* L.), small-forb (*Maianthemum bifolium*
(L.)) and sedge in herbaceous layer on combinations
of Entic Podzols and Haplic Albeluvisols Abruptic on flat
surfaces (Primary)
a. Secondary aspen-birch forest with shrubs and herbaceous
layer formed by sedge, forb and small reed (60–190 t/ha).
Geom—Subtaiga of light-coniferous forests on granites, gneiss,
dikes of dolerites and porphyrites
Class of facies—sublithomorphic
Group of facies—2. Larch-pine (*Pinus sylvestris* L.) forest with
herbaceous layer formed by forb, sedge and graminoids on
combinations of Entic Podzols and Haplic Albeluvisols
Abruptic on slopes (Pseudo-primary) (200 t/ha)
a. Secondary birch forest with herbaceous layer formed by forb,
sedge and small reed (90–190 t/ha)
Group of facies—6. Cultural landscapes (rail way
and settlements)
Barguzin mountain range
NORTH ASIA TAIGA
Class of geoms—Mountain-taiga with a certain Pacific
influence
Geom—Mountain-taiga of larch forests on biotite granites
and diorites
Class of facies—sublithomorphic
Group of facies—4. Larch-Siberian pine and Siberian pine-larch
(*Larix gmelinii* (Rupr.) Rupr.) forest with shrubs (*Rosa acicularis*
Lindl., *Salix* spp., *Ledum palustre* L.) with small-forb
(*Maianthemum bifolium*, *Linnaea borealis* L.,
Trientalis europaea L.) and red bilberry
(*Vaccinium vitis-idaea* L.) and green mosses on Umbric
Albeluvisols Abruptic on slopes or upper slopes (Primary)
(145–287 t/ha)
a. Secondary mixed aspen-birch-larch and Siberian pine
(*Populus tremula* L., *Betula platyphylla* x *B. pubescens*)
forests with red bilberry (*Vaccinium vitis-idaea*) and forb

List of Tables

Part I
Theory of Landscape Pattern
and Hierarchy

Chapter 1
Concepts of Landscape Pattern

Alexander V. Khoroshev

Abstract The term "landscape" was introduced into Russian scientific literature in 1913. Since that time a lot of competing conceptions about landscape structure have been developed. Although traditional genesis-based approach with strong focus on abiotic factors is most widely applied, today landscape is treated as a multifaceted and multifunctional phenomenon, a complex system that requires multiplicity of projections to reveal essential relations between geocomponents and spatial elements. The chapter contains a review of approaches and concepts used in Russia to describe landscape structure: genetic–morphological, positional–dynamical, para-genetical, biocentric, biocirculation, basin, and catena approaches, the concept of chorions. Landscape map provides information on various types of landscape structures although one of them is commonly chosen as a basic frame. The choice of basic type of structure used for mapping is dictated by expected practical application.

Keywords Landscape · Component · Unit · Structure · Genesis · Hierarchy · Flows · Map

1.1 What Is Landscape?

As early as in 1913, Lev Berg (1876–1950) introduced the scientific notion "landscape" based on concepts of German geographical school (A. Hettner) and ideas of V. V. Dokuchaev (1846–1903). German term "landschaft" was accepted in Russian language. Berg proposed interrelated definitions of *landscape* and *geography*. "Natural landscapes are regions similar in dominant character of relief, climate, plant cover and soil cover. In other words, landscape is a set of regularly repeating landforms" (Berg 1913, p. 117). "Geography… is chorology of interrelated groups

A. V. Khoroshev (✉)
Department of Physical Geography & Landscape Science,
Lomonosov Moscow State University, Moscow, Russia

© Springer Nature Switzerland AG 2020
A. V. Khoroshev, K. N. Dyakonov (eds.), *Landscape Patterns in a Range of Spatio-Temporal Scales*, Landscape Series 26,
https://doi.org/10.1007/978-3-030-31185-8_1

of humans, animals, plants, landforms and so forth, on the earth... Geography is the science of landscapes... The purpose of geographic research is to reveal links and regularities that exist between separate objects of interest to the geographer, the main question being, "How do particular sets of objects and phenomena affect each other, and what are the results of this spatially?" (Berg 1915, p. 469). L. S. Berg composed first small-scaled landscape maps and regional complex descriptions of Russian nature (Berg 1913). Since early period of study landscape has been understood as both system or interacting spatial units and system of interacting *geocomponents* (namely, parent rock, soil, water, air, plants, and animals, sometimes also litter). Studies of landscape composition formed the core of the structural approach within which conceptions of morphological units and their hierarchy (see below) were elaborated in 1930–1940s by L. G. Ramensky (1938) and N. A. Solnetsev (1948). Explicit empirical concept of hierarchical organization provided methodological basis studying and mapping landscape structure at various scales. The idea was applied by N. A. Gvozdetsky (1913–1994), N. I. Mikhailov (1915–1992), F. N. Milkov (1918–1997), and A. G. Isachenko (1922–2018) at regional and global scales for the purposes of physical–geographical regionalization that served a reliable basis for the country-scale economic planning. In 1926, A. A. Grigoryev (1883–1969) formulated functional approach to landscape studies. He argued that, although the study of spatial relations is important, the crucial challenge for geographers is to explain mechanisms of these relations with regard to both space and time. Landscape is a holistic entity and its properties cannot be described separately by functioning of components (Grigoryev 1926).

In 1963, V. B. Sochava (1905–1978) proposed the notion of "*geosystem,*" which was aimed at integrating structural and functional approaches to landscape studies. He relied on system approach and theory of cybernetics developing at that period. Geosystem is an open system formed by interactions of natural and cultural components generating stable patterns (invariants) at various spatiotemporal scales. He distinguished two types of geosystems forming corresponding hierarchies. Heterogeneous geosystem was referred to as *geochore*, homogeneous geosystem— as *geomer* (see Chaps. 2, 8, and 10 for details). Almost simultaneously the term "geochore" was introduced in East Germany (Neef 1967). Geochore there was regarded as associations or mosaics of basic topic elements. The term "geochore" means a geographically defined or limited unit; "tope" or topic refers to a particular locality; the properties of choric spatial units result from the combination of topic elements, as well as their arrangement in space (Bastian 2000).

1.2 Notion of Landscape Structure

The term "landscape structure" has several interpretations. In landscape ecology, landscape structure is interpreted mainly as a spatial *pattern* formed by laterally interacting elements (Forman and Godron 1986; Urban et al. 1987). In East-European landscape science (which, in this region, is commonly considered as a part of physical geography), a similar aspect is expressed by the term "horizontal structure" or "landscape morphological structure" formed by "*morphological units*" (Solntsev 1948). The term "vertical structure" is related to the composition and interactions of "*geocomponents*" (see earlier). In analogous sense, both terms are used in German-language school of landscape ecology (Bastian and Steinhardt 2002). Vertical (or radial) topological relations between geocomponents often are used as criteria for hierarchical ecosystem models applied for classification and mapping ecosystems at various spatial scales (Klijn and de Haes 1994; Rowe 1996; Bailey 2005). In English-language science, casual linkages between natural phenomena are commonly a subject of biocentered studies in functional ecology, while in Russian-language physical geography it is the core issue for geographical landscape science. Landscape ecology focuses mainly on horizontal (lateral) interactions between spatial units and resulting emergent properties of a landscape; the term "vertical structure" is applied to relations between hierarchical levels (Wu and David 2002). According to Meentemeyer (1989), relative importance of two-dimensional (horizontal, interunit) and three-dimensional (vertical, intergeocomponent) aspects of structure is scale-dependent: the finer is the scale, the higher is relative significance of the vertical aspect. Landscape ecology has recently begun to demonstrate growing interest to the third dimension of a landscape, including contribution of relief, substrate, and genesis to spatial heterogeneity (Dorner et al. 2002; Hoechstetter et al. 2008; Drăguţ et al. 2010; Cushman and Huetmann 2010). Causal relationships between landscape pattern and landform genesis were mentioned by Forman (2006). At the same time landscape science in Russia is developing toward increased interest in chorological aspect of research at a wide range of spatial scales, far beyond traditional landscape mapping. Obviously, these trends push both branches of landscape research closer to each other.

1.3   Radial Relations Between Geocomponents

V. N. Solntsev (1977), dealing with relations between geocomponents, proposed to distinguish "*linkages-interactions*" and "*linkages-relations.*" Linkages interactions can be revealed on the basis of direct investigation of matter and energy flows including experiments in nature, for example, effects of land reclamation on productivity and effects of controlled cutting in a river basin — on runoff attributes (Likens and Bormann 1995). Linkages relations do not directly indicate cause-and-effect relationships. Statistically significant correlation between a pair of landscape attributes can be interpreted in different ways. The simplest case (and the most desirable for the researcher) is when correlation can be explained by real physical one-way effect of certain geocomponent on another one or feedback due to matter and/or energy flow. For example, if canopy closure is related to herbs cover percentage, one can conclude that canopy regulates solar energy input to the soil surface. One attribute acts as an indicator of another one's state. The opportunity to interpret linkage in such a way enables us to infer stability of one attribute state from the state of another one. If the canopy closure decreases, herbs cover will increase. This variant has the highest applicability in practice of landscape planning and forecast. The second variant of correlation between two attributes A and B is related to the third attribute C, which is correlated with A and B separately, while there is no physical interaction between A and B. One can imagine a set of landscape attributes that vary in space in a similar pattern being dependent on the same environmental control, for example, groundwater level. Multidimensional statistical tools provide opportunities to separate groundwater-sensitive and insensitive attributes. Hence, the researcher can select attributes that are informative for explanation of landscape heterogeneity at the focus hierarchical level.

Theoretically intergeocomponent interactions can occur only between natural processes and bodies having comparable time and space scales (Delcourt et al. 1983; Puzachenko 1986; Shugart 1999; Wu 2013). "Characteristic time" was defined as the span of time during which particular processes occur or, in the case of self-regulating systems, the time required for a system to return to a state of equilibrium (Armand and Targul'yan 1976). To avoid uncritically applying the famous metaphor "everything is linked to everything," the concept of partial geocomplexes or partial geosystems (Neef 1967; Sochava 1978; Solon 1999) was developed in physical geography as a tool to consider strong linkages between groups of certain properties with similar space and time scale. Each of these independent holistic groups is driven by a specific ecological factor acting in a certain scale. Landscape unit ("level 0") is hypothesized to be a holistic entity united by biological circle, water cycle, and other forms of matter and energy exchange between geocomponents. Hence, the task of crucial importance is to clarify whether it interacts with environment by the whole set of properties or by groups of properties separately. If the latter holds true, this would evidence that "level 0" geosystem is a superposition of effects generated by independent processes that can produce patterns at various scale levels. In other words, the researcher tests the hypothesis that geosystem has multistructural and multiscale organization.

1.4 Causes of Spatial Heterogeneity

The basic question for any geographical investigation is as follows: Why has this or that property different values in different areas? In a simplified way, we distinguish three options of an answer: (1) variability within the natural range of permitted states, (2) influence of the other spatially independent bodies and flows, and (3) local variability of environment.

The first option means that the physical nature of a property imposes limits for possible values. Tree height in boreal forests ranges between 0 and 30–35 m. If we find a patch with dominance of 10 m tall tree, it can be concluded that ceteris paribus the forest patch is on early stage of recovery succession. However, 10 m can be the highest possible value for pines in an oligotrophic mire.

The second option means that self-organization of a geocomponent (e.g., vegetation) generates spatial mosaics. Shifting steady-state mosaics (Likens and Bormann 1995) in an undisturbed forest results in co-occurrence of patches differing in canopy cover and, hence, in properties of herb and moss layers, rate of evaporation, humus accumulation, etc. For instance, mosaics of spruce-dominated and pine-dominated patches can be controlled by differing solar radiation availability in older and younger patches.

The third option involves spatial variability of physical properties of a geosystem, for example, relief, soils, groundwater level, and infiltration capacity of soil-forming deposits. In this case, results of competition between, say, spruce and pine can be driven, for instance, by alternation of loamy and sandy soils inherited from sedimentation in Pleistocene glacial epochs. Soils and vegetation in this case are interacting geocomponents of the "0-level" system. Suppose, we consider groundwater level as a binding factor. Two similar flat surfaces with similar loamy sediments can have different groundwater level due to different landscape neighborhoods. If the flat watershed area is rather wide and poorly drained, pine acquires advantage in competition. If the flat watershed area is narrow and drained by ravines, groundwater level is deeper, which is favorable for spruce. It follows that explanation for stand composition variability is impossible until we evaluate combination and quantitative proportion of landforms within the higher-order ("+1 level") system in which focus landscape unit is embedded (Khoroshev 2019). Groundwater level and resulting spruce/pine ratio are the emergent effects generated by the "+1 level" geosystem, including flat areas, slopes, and ravines.

Thus, we face the challenge to make distinction between contributions of effects generated by at least two hierarchical levels. In real nature, *several* higher-order systems can impose constraints on properties of the focus-level system. Groundwater level can result both from certain combination of flat interfluve areas and ravines and from water redistribution among slightly convex and slightly concave patches within the interfluve area. Hence, it is possible to imagine a spatial series of interfluves ("+2 level") differing in dissection and, consequently, by ratio of spruce and pine: the more is drainage density in the area, the larger is proportion of spruce. At the same time, one can imagine the spatial series of patches within the well-drained spruce-dominated flat interfluve area. In this series, proportion of pine in spruce-dominated forests will range between, say, 10 and 30%, depending on concavity or convexity of landforms ("+1 level"). Thus, tree layer composition in a focus patch will be interpreted as a superposition of two factors acting at different hierarchical level: spruce dominates over pine due to emergent properties in a "+2 level" system; proportion of spruce is 60% and not 80% because "+1 level" system favors higher groundwater level in a concave landform. The emergent property of a heterogeneous geosystem is indicated by the rule that links geocomponents and is uniform across the entire space of this geosystem. To prove existence of such a rule, one needs to express interdependency between geocomponents by a statistically significant quantitative model for a spatial series of landscape units. By this a researcher tests the following hypothesis: if the combination of spatial units in some neighboring area changes, the properties of the focus unit will change as well. The size of neighboring area that affects processes in a focus unit is a matter of analysis. Hence, we face the need to compare quality of statistical models designed for several hypothetic higher-order geosystems.

Spatial properties of a higher-order geosystem can be described in a number of ways. The first option is to use the combination of landforms as a surrogate for characteristic of spatial heterogeneity and describe it by morphometric values (e.g., standard deviation of elevations, summary length of thalwegs, vertical curvature, horizontal curvature). Alternatively, higher-order geosystem can be characterized by different sets of properties depending on the research focus. The greatest experience in assessing emergent effects of spatial pattern is accumulated in landscape-ecological studies of how spatial pattern of landscape cover affects viability of populations. If one concentrates efforts on studying runoff regime as an emergent property of a basin geosystem, he or she proceeds from the assumption that plant cover influences time and rate of snowmelt and ratio of surface and subsurface flows. Hence, he can evaluate proportions of forested and deforested areas, small-leaved and coniferous stands, and mature and young stands as the surrogates for plant cover pattern. If one studies microclimatic controls of agricultural productivity, it is relevant to consider share of ravines as accumulators of cold air masses as well as share and connectivity of wind belts.

Thus, the organization of geosystems is, on the one hand, a geographical process of structure emergence in space and time and, on the other hand, a result of such processes manifested in stable system orderliness as a hierarchically nested temporal states (Dyakonov 2007). Integral geographical process is realized as a triad "evolution – structure – functioning."

1.5   Multiplicity of Landscape Structures

At initial stages of its development landscape science in the USSR was ideologically forced to looking for "objective reality" in nature. Physical geography in the country with extremely vast and unexplored areas was required to elaborate explicit rules for establishing unambiguous boundaries between natural units that could be used for planning decisions. These strict requirements aroused prolonged discussions concerning the essence of the "landscape" notion aimed at "single correct" way to delineate landscapes. Most researchers came to an agreement that genesis-based approach to delineation and classification of landscapes was the most effective. This approach was developed with rigid commitments to the conception that soil and plant cover are strictly determined by topographic and geologic spatial units (Solnetsev 1948; Isachenko 1973). Note that at the same period Paffen (1953) recognized the defining role of relief in the formation of ecotopes, describing them as topographic-ecological complexes (cited from Klink et al. 2002). Classical examples of physiography-based morphological units were provided in the areas with well-pronounced erosion-shaped plain landscapes of the central European Russia. It is worth noting that approximately at the same period or later similar methodologies were elaborated in Australia (Christian and Stewart 1953), Canada (Hills 1961), Germany (Paffen 1953; Neef 1963), the Netherlands (Zonneveld 1989), and the United States (Cleland et al. 1997).

In the concept of landscape morphological structure, each hierarchical level is controlled by a peculiar principal factor (Solnetsev 1948).

Facies is the smallest natural unit with uniform lithology of surficial sediments, character of relief, humidity, microclimate, soil, and biocoenosis occupying usually part of relief microform (Annenskaya et al. 1963, p. 14). The term is similar to *ecotope* in German literature (Neef 1967; Bastian and Steinhardt 2002), to *site* (Christian and Stewart 1964).

Podurochishche is a natural unit composed by group of facies densely linked by genesis and dynamic processes due to common position on element of relief mesoform with the same solar exposure (Annenskaya et al. 1963, p. 15–16). The examples are facies systems on a ravine slope, on a top of a hill, and on a floodplain of one level.

Urochishche is a system of facies interrelated by genesis, dynamic processes, and territory occupying one relief mesoform (Annenskaya et al. 1963, p. 16–17). The examples are ravine, moraine hill, river terrace, and flat interfluve area. The term is quite close to German *nanogeochore* and Australian *land unit* (Christian and Stewart 1964).

Mestnost is a territory within landscape with a particular combination of main urochishches (Annenskaya et al. 1963, p. 20). The term is close to German *microgeochore* in which the pattern of topes reflects primarily the landscape-genetic conditions (history) of their development and succession (Haase 1989).

Geographical landscape is genetically uniform territory with regular and typical occurrence of interrelated combinations of geological composition, landforms, surface and ground waters, microclimates, soil types, phytocoenoses, and zoocoenoses (Solntsev 1948, p. 258). The term is close to German *macrogeochore* or Australian *complex land system* composed of recurrent patterns of genetically related land units (Christian and Stewart 1964).

The main significance for landscape pattern is assigned to relief. Within this hierarchy, *landscape* is considered to be a quite peculiar unit with unified geological basis. Its position in hierarchy is, on the one hand, on the top level of units that are subject to typological classification. Landscape and smaller morphological units (*facies, podurochishche, urochishche, mestnost*) are not unique. On the other hand, landscape is the smallest individual unit that ensures full insight into structure of territory. The higher-level units are unique (i.e., not subject to classification) units of physical–geographical regionalization as districts, provinces, regions, and countries (Isachenko 1973). Landscape is more stable than morphological units. It can be treated as an object of particular type of economic activity and an object of management.

However, long-term stationary research showed evidence that boundaries of vegetation communities and soil units do not necessarily coincide with topographic or geological boundaries for a number of reasons. "Genetic-morphological" patterns are now believed to be a particular case of structure multiplicity (or "*polystructural-ism*"). Gradual transitions between landscape units often (though not everywhere) ignore topographic and geological boundaries due to strong influence of lateral flows that generate specific patterns.

The concept of structures multiplicity ("polystructuralism") has been elaborated since 1970s, first in seminal works by Latvian geographer K. G. Raman (1972). Any geographical complex has an infinite number of structural projections and can be analyzed by a number of system-forming relations (Topchiev 1988). Harvey (1969) reminds that in the early twentieth century in discussion between supporters of geographical determinism and possibilism, the philosophical credo of the latter was formulated as follows: Individual events in all their width and diversity are subject to explanation only in occasional intersections of mutually independent series of reasons (Harvey 1969). Recognition of geographical space polystructuralism resulted in understanding that research purposes dictate choice of this or that way for describing landscape structure (see Chap. 2). Within the framework of this concept, landscape is represented as a superposition of relatively independent spatial patterns. V. N. Solntsev (b. 1940), son of the main proponent of genetic-morphological approach N. A. Solnetsev (1902–1991), argues that each kind of geocomponent interactions generates its particular set of connections and, hence, a system that is holistic in relation to this kind of interactions (Solntsev 1981). He proposed to distinguish three types of landscape structures as follows: (1) geosta-tionary, (2) geocirculation, and (3) biocirculation. *Geostationary structures* are formed by radial (vertical) connections between geocomponents under the control of substrate that is responsible for nutrients supply, water percolation, air regime in soil, etc. *Geocirculation structures* emerge as a result of various intensity and directions of lateral flows depending on relief dissection. *Biocirculation structures* shape geographical space according to solar radiation input depending on elevation and slope aspect (Solntsev 1997). E. G. Kolomyts (1998) elaborated more detailed classification of structures and proposed to distinguish cellular (similar to geostationary structures), isopotential (similar to biocirculation structures), vectoral (similar to geocirculation structures), and basin structures. He emphasized that basin approach, though efficient for study of matter flows, has sufficient limitations in comparison with genesis-based one since it underestimates radial constituents of water balance, ignores legacy of past between-components interactions, and masks climatic influences such as insolation and atmospheric circulation inducing specific matter-and-energy connections.

The concept of poly systems analysis and synthesis that was elaborated in the Siberian school of landscape science provides a set of tools to stratify complex geographical objects into a set of possible system-based interpretations via reflection of its properties in various subject fields and establishing relations between these properties (Cherkashin 2005). Processes responsible for the emergence of these relatively independent structures differ in characteristic spatial and temporal scales and, hence, can superimpose or be in top-down relations without direct interaction. However, landscape as a complex system is not a simple sum of independent structures. Synergetic effects resulting from superpositions of multiple independent structures are believed to be one of the critical issues for landscape science.

Ukrainian researcher M. D. Grodzinsky (2005, 2014) distinguished five types of superimposing coexisting landscape structures as follows: genetic–morphological, paragenetical, positional–dynamical, basins, and biocentric-network. By so doing, he generalized a set of ideas elaborated in the former USSR and abroad:

The relationships between geotopes in common characteristics of their origin and evolution form *genetic–morphological configuration*.

The connection between geotopes with surface flows of matter and energy and their relationship with lines of change in direction and the intensity of these flows form *positional–dynamical configuration*.

The connection of geotopes with lines of concentration of the horizontal flows form *paragenetical configuration*.

The similarity of geotopes with respect to their hydrofunctioning and their relationship to the basins of surface flows form *basins configuration*.

The biotic migrations in a landscape form its *biocentric network configuration* (Grodzinsky 2005, p. 495).

This classification of configurations (referred to as structures in more recent work by Grodzinsky (2014)) has both advantages and shortcomings in comparison with V. N. Solntsev's one. It provides more details to division between geostationary and geocirculation structures.

Positional–dynamical and paragenetic configurations have much in common with geocirculation, or vector, structure (see Chaps. 3, 5, and 6). The particular case is *catena* geosystem with its center at the topmost position in relief (autonomous element) and lateral unidirectional flows that connect the center with the low-lying heteronomous elements at the slope, toeslope, floodplain, and waterbody. The catena concept initially introduced in 1930s by Milne was in 1950s developed to the theory of landscape geochemistry by B. B. Polynov (1877–1950), A. I. Perelman (1916–1998), and M. A. Glazovskaya (1912–2016). Similar concept of ecological catena, more ecologically oriented, was elaborated by G. Haase and H.-J. Klink in Germany in 1960s (Klink et al. 2002). The notion of geochemical landscape is a particular case of positional–dynamic configuration with strong focus of "history of atoms in a landscape" according to informal definition of landscape geochemistry by Perelman (1972) (see Chaps. 5, 9, and 10). Besides that, studies of positional–dynamic configuration involve plant and soil cover patterns generated by gravity-driven flows of particulate matter and water.

Basins configuration unites a set of catenas within the catchment. This type of geosystems is characterized by emergent properties depending on a number of basin attributes: proportions of spatial units, catchment area, stream orders, etc. Similar genetic–morphological units can obtain different properties depending on their position in a basin geosystem (see Chaps. 3, 11, and 12). Biocentric network configuration units can either be in compliance with a basin structure or ignore it, depending on species requirements. Since landscape science deals with emergent properties of spatial mosaics, application of basins configuration concept can be helpful in territorial planning of land units ratios aimed at regulation of runoff regime, water-dependent populations of animals, water consumption management, etc.

Paragenetic configuration is developed if spatial elements are highly contrast by their properties but do not exist without each other due to common generating matter flow. For example, snow avalanche or mudflow geosystem is composed of kar, transit zone, and fan with accumulation of debris (see Chap. 15). In this case, paragenetic geosystem can be displayed within the genetic–morphological framework as a single unit at coarse scale or as a set of units at fine scale. At the latter case the map legend needs to contain additional characteristic for this set of units in order to emphasize lateral flows connecting them. At the finest scale the map legend may reflect variation due to different intensity, period of recurring system-forming process, or age (see Chap. 15). The necessity to characterize lateral flows to describe paragenetic gravity-driven geosystems will be, in this case, in a certain contradiction with logic used to describe the other units for which the legend does not put special emphasis on lateral links between them. In the framework of the matrix concept mapping, paragenetic geosystems are problematic due to several reasons. First, their quality as potential corridor is too diverse to map them as a uniform element; hence, additional complicated classification of corridors will be needed. Second, in mountainous regions, paragenetic geosystems often extend through several altitudinal belts and, hence, through several contrast matrices with gradually changing internal properties. Third, this or that part of paragenetic geosystem may have a function of a patch (e.g., meadow at the zone of accumulation of avalanche debris surrounded by a forest) but connections with a matrix may be much weaker than connection with the upper sections of the geosystem.

At the same time, one could imagine that matter flows generating interdependent units may emerge without direct influence of gravity, unlike in Grodzinsky's examples. The example is a system of neighboring forest and overgrowing meadow linked by seeds dispersal with gradually emerging sequence of soils along the gradient of decreasing influence of forest. This highly contrast pattern may be nested within a single unit delimited according to uniform topographic position and substrate (e.g., *urochishche* within the framework of genetic–morphological configuration). Spatial gradient in values of a set of interrelated attributes testifies the action of a common factor that intensity gradually changes in space. Hence, a series of various neighboring units obeys the same rule of between-geocomponents interactions, which can be expressed mathematically, for example, by an equation. The area where the internal heterogeneity is described by a single equation comprises paragenetic geosystem, or, more precisely, geochore. Note that the equation may describe a certain set of properties but not necessarily the whole set. In this case we deal with *partial geosystem* with properly established set of elements and causal linkages between them.

The aforementioned notions can be generalized as the concept describing center-periphery relations, or concept of *nuclear systems*, or *chorions*, elaborated by A. Yu. Reteyum (1988). This author treats the geographical space as superpositions of a number of systems generated by effects of object decreasing with distance. Explicit tools for mapping chorions, unfortunately, have not been proposed. However, to some extent probabilistic techniques could be helpful. They involve calculation of probabilities that a set of attributes corresponds to typical values inherent for the core of a system. The core is treated as a site with maximum manifestation of a system-forming factor. A decrease in probability is interpreted as a decreasing control of a core over the area under its influence. If a territory consists of several highly contrast landscape units' types (e.g., nutrient-poor on sands and nutrient-rich on loams) and a series of transitional variants (e.g., on two-layered sediments with various thicknesses of sand mantle over loams), it is possible to evaluate probabilities of correspondence to each type for each unit (cell, pixel, etc.) based on the values of factor-sensitive attributes (see the example of applying discriminant analysis for this purpose in Chap. 16).

Landscape structures generated by contrast solar energy input (biocirculation structures sensu Solnetsev, or isopotential structures sensu Kolomyts) are missing in Grodzinsky's classification. However, fine-scale zonal hydrothermically driven patterns in plain regions are obviously a matter of interest for landscape studies as well as altitudinal belts and contrasts induced by slope aspects in mountains. Mapping within the genetic-morphological framework does not directly distinguish these structures since various altitudinal belts may occur within the same, say south-facing, slope with uniform geology, shape, and inclination. To some extent positional–dynamic framework can be helpful in making distinction between slopes with different input of radiation, but it concentrates mainly on differences in directions of gravity-induced flows. Since radiation contrasts cause critical differences in functioning (i.e., interactions between geocomponents) of landscape units, the tools to delineate biocirculation structures are in this or that way applied at most landscape maps.

It is worth noting that in regions with fine-scale contrasts of radiation input famous matrix concept of landscape structure can hardly be applied without corrections because the essence of the principal notions "matrix," "patch," and "corridor" changes sharply if one moves from one altitudinal belt to another or even from south-facing to north-facing slope. It is obvious that *biocentric network configuration* (Grodzinsky 2005) is the right term to express *patch-matrix-corridor concept* of landscape structure (Forman and Godron 1986), which has demonstrated its high efficiency for the purposes of biodiversity protection. Although matrix concept was elaborated as a two-dimensional representation of landscape structure, it is important to make clear distinction between two variants of visually detected heterogeneity. First, distribution of spatial elements may be the product of human activity in topographically and geologically uniform areas resulting in preservation of zonal vegetation remnants among sharply transformed areas. Second, landscape mosaics may reflect topographic and geological heterogeneity, that is visual image of genetic–morphological, positional–dynamic, paragenetic, or basins configuration. Third, anthropogenic transformation of landscape structure may be selective; that is, human-generated matrix (or patches) emerges at suitable topographic positions (e.g., well-drained interfluves) while unsuitable ones (e.g., steep slopes or nutrient-poor sandy terraces) preserve zonal vegetation.

The aforementioned concepts of landscape structure do not exclude each other but provide specific tools to solve practical issues. Landscape maps are believed to be the most useful instrument for embedding land use decisions and forecast of trends to particular spatial limits. Perfect landscape map provides information on various types of landscape structures though one of them is commonly chosen as a basic frame. The choice of basic type of structure used for mapping is dictated by expected practical application. Genetic–morphological mapping provides multifaceted information on contrasts in land quality and, hence, bioproductivity, natural hazards, etc. (see Chaps. 8, 13, 14, 17, 20, and 25). It showed high efficiency for integrated territorial planning as well as for planning particular economic activities such as forestry, agriculture, land reclamation, or settlement network (see Chaps. 22, 23, 24, and 26). Positional–dynamic structures, when depicted on landscape maps, afford to distribute technogenic loads in an environment-friendly way and to regulate matter (including pollutants) flows by means of relevant elements of ecological network in order to exclude harmful effects for nature, economy, social values, and human health (see Chaps. 5 and 9). Paragenetic structures often are depicted in combination with the other ones (commonly, genetic-morphological) with strong focus on flows intensity and diversity of development stages over the territory. By this, the maps provide opportunity to forecast probability of natural hazards and to predict the most probable rapid changes in spatial pattern and vertical composition of geosystems (see Chap. 15). Basins structures depicted in fine scale enable us to assess matter redistribution driven by soil erosion, with water redistribution resulting in diversification of forest growing conditions (see Chap. 3). At coarse scale land cover patterns within basins are responsible for runoff formation, water quality, and probability of hydrological risks (see Chaps. 11 and 12). Biocirculation structures on landscape maps are useful for depicting contrasts in biological cycle, productivity, and antierosion potential of vegetation (see Chap. 8) as well as for monitoring of responses to climate changes (see Chaps. 18 and 20). This information may be critical for planning recreational activities, grazing loads, road construction, etc. Biocentric network structures on landscape maps, despite evident emphasis on biota in its name, are preferred not only for the purposes of projecting ecological networks with focus on biodiversity. Correct display of neighborhoods, orientation, shape, and proportions of zonal and anthropogenically transformed elements allows projecting aesthetically valuable scenery, microclimate regulation measures (e.g., in agricultural, recreation, or urban areas), air purification measures in cities and industrial areas, etc. Matrix-patch representation of a landscape demonstrated its validity for the purposes of predicting hazard risks based on modeling patch distributions (see Chap. 4).

To conclude, it is obvious that it has no sense to look for the single best way for describing landscape structure. Landscape is a multifaceted and multifunctional phenomenon, a complex system that requires multiplicity of projections to reveal essential relations between geocomponents and spatial elements.

Acknowledgements This research was conducted according to the State target for Lomonosov Moscow State University "Structure, functioning and evolution of natural and natural-anthropogenic geosystems" (project no. AAAA-A16-116032810081-9).

References

Annenskaya, G. N., Vidina, A. A., Zhuchkova, V. K., Konovalenko, V. G., Mamay, I. I., Pozdneeva, M. I., Smirnova, E. D., Solnetsev, N. A., & Tseselchuk, Y. N. (1963). *Morphological structure of a geographical landscape*. Moscow: MSU Publishing House. (in Russian).

Armand, A. D., & Targul'yan, V. O. (1976). Some fundamental limitations on experimentation and model-building in geography. *Soviet Geography, 17*(3), 197–206.

Bailey, R. G. (2005). Identifying ecoregion boundaries. *Environmental Management, 34*(Suppl.1), 14–26.

Bastian, O. (2000). Landscape classification in Saxony (Germany)—A tool for holistic regional planning. *Landscape and Urban Planning, 50*, 145–155.

Bastian, O., & Steinhardt, U. (Eds.). (2002). *Development and perspectives of landscape ecology.* Boston: Kluwer Academic.

Berg, L. S. (1913). An attempt at the division of Siberia and Turkestan into landscape and morphological regions. In *Collection of papers in honor of D. N. Anuchin's seventieth birthday* (pp. 117–151). Moscow: Izdatel'stvo Imperatorskogo obshchestva Lyubitelei Estestvoznaniya, Antropologii i Etnografii pri Moskovskom universitete. (in Russian).

Berg, L. S. (1915). The objectives and tasks of geography. *Proceedings of the Imperial Russian Geographical Society, 51*(9), 463–475. (in Russian). See also in J. A. Wiens, M. R. Moss, M. G. Turner, & D. J. Mladenoff (Eds.). (2006). *Foundation papers in landscape ecology* (pp. 11–18). New York: Columbia University Press.

Cherkashin, A. K. (Ed.). (2005). *Landscape interpretative mapping.* Novosibirsk: Nauka. (in Russian).

Christian, C. S., & Stewart, G. A. (1953). *General report on survey of the Katherine-Darwin region 1946* (CSIRO Aust Land Res. Ser. No. 1). Melbourne: CSIRO.

Christian, C. S., & Stewart, G. A. (1964). Methodology of integrated survey. *Proceedings of UNESCO conference on principles and methods of integrated aerial surveys of natural resources for potential development*, Tolouse. WS/0384.15/NS.

Cleland, D. T., Avers, P. E., McNab, W. H., Jensen, M. E., Bailey, R. G., King, T., & Russell, W. E. (1997). National hierarchical framework of ecological units. In M. S. Boyce & A. Haney (Eds.), *Ecosystem management applications for sustainable forest and wildlife resources* (pp. 181–200). New Haven: Yale University Press.

Cushman, S. A., & Huettmann, F. (Eds.). (2010). *Spatial complexity, informatics, and wildlife conservation.* Tokyo: Springer.

Delcourt, H. R., Delcourt, P. A., & Webb, T. (1983). Dynamic plant ecology: The spectrum of vegetation change in space and time. *Quatenary Science Review, 1*, 153–175.

Dorner, B., Lertzman, K., & Fall, J. (2002). Landscape pattern in topographically complex landscapes: Issues and techniques for analysis. *Landscape Ecology, 17*, 729–743.

Drăguţ, L., Walz, U., & Blaschke, T. (2010). The third and fourth dimensions of landscape: Towards conceptual models of topographically complex landscapes. *Landscape Online, 22*, 1–10. https://doi.org/10.3097/LO.201022.

Dyakonov, K. N. (2007). Landscape studies in Moscow Lomonosov State University: Development of scientific domain and education. In K. N. Dyakonov, N. S. Kasimov, A. V. Khoroshev, & A. V. Kushlin (Eds.), *Landscape analysis for sustainable development. Theory and applications of landscape science in Russia* (pp. 11–20). Moscow: Alex Publishers.

Forman, R. T. T. (2006). *Land mosaics.* Cambridge: Cambridge University Press.

Forman, R. T. T., & Godron, M. (1986). *Landscape ecology*. New York: Wiley.

Grigoryev, A. A. (1926). Problems of integrated study of territory. *Priroda (Nature), 5*(6), 46–58. (in Russian).

Grodzinsky, M. (2005). *Understanding landscape. Place and space* (2 Vols.). Kiev: Kiev University. (in Ukrainian).

Grodzinsky, M. (2014). *Landscape ecology*. Kiev: Znannya. (in Ukrainian).

Haase, G. (1989). Medium scale landscape classification in the German Democratic Republic. *Landscape Ecology, 3*(1), 29–41.

Harvey, D. (1969). *Explanation in geography*. London: Hodder & Stoughton Educational.

Hills, G. A. (1961). *The ecological basis for land-use-planning* (Research report no. 26). Toronto: Ontario Department of Lands and Forests.

Hoechstetter, S., Walz, U., Dang, L. H., & Thinh, N. X. (2008). Effects of topography and surface roughness in analyses of landscape structure – A proposal to modify the existing set of landscape metrics. *Landscape Online, 3*, 1–14. https://doi.org/10.3097/LO.200803.

Isachenko, A. G. (1973). *Principles of landscape science and physical-geographic regionalization*. Carlton: Melbourne University Press.

Khoroshev, A. V. (2019). Multiscale Organization of Landscape Structure in the middle taiga of European Russia. *Landscape Online, 66*, 1–19.

Klijn, F., & de Haes, H. A. U. (1994). A hierarchical approach to ecosystems and its applications for ecological land classification. *Landscape Ecology, 9*(2), 89–104.

Klink, H.-J., Potschin, M., Tress, B., Tress, G., Volk, M., & Steinhardt, U. (2002). Landscape and landscape ecology. In O. Bastian & U. Steinhardt (Eds.), *Development and perspectives of landscape ecology* (pp. 10–24). Boston: Kluwer Academic.

Kolomyts, E. G. (1998). *Polymorphism of landscape-zonal systems*. Pushchino: ONTI PIC RAN. (in Russian).

Likens, G. E., & Bormann, F. H. (1995). *Biogeochemistry of a forested ecosystem*. New York: Springer.

Meentemeyer, V. (1989). Geographical perspectives of space, time, scale. *Landscape Ecology, 3*(3/4), 163–173.

Neef, E. (1963). Topologische und chorologische arbeitsweisen in der landschaftsforschung. *Petermanns Geographische Mitteilungen, 107*, 249–259.

Neef, E. (1967). *Die theoretischen grundlagen der landschaftslehre*. Gotha-Leipzig: Haack.

Paffen, K.-H. (1953). *Die natürlichen landschaften und ihre räumliche gliederung. Eine methodische untersuchung am beispiel der Mittel- und Niederrheinlande* (Forschung zur deutschen Landeskunde) (Vol. 68). Remagen: Verlag der Bundesanstalt für Landeskunde.

Perelman, A. I. (1972). *Landscape geochemistry*. Moscow: Vysshaya Shkola. (Translated from Russian). (Geol. Surv. Canada Trans. No. 676, Parts I and II).

Puzachenko, Y. G. (1986). Space-time hierarchy of geosystems from the standpoint of fluctuation theory. In *Issues in Geography. Vol. 127. Geosystem modelling* (pp. 96–111). Moscow: Mysl'. (in Russian).

Raman, K. G. (1972). *Spatial polystructurality of topological geocomplexes and experience of its identification in the conditions of the Latvian SSR*. Riga: Latvian State University Publishing House. (in Russian).

Ramensky, L. G. (1938). *Introduction to integrated soil and geobotanical regionalization of lands*. Moscow: Selhozgiz. (in Russian).

Reteyum, A. Y. (1988). *The terrestrial worlds (on holistic studying of geosystems)*. Moscow: Mysl'. (in Russian).

Rowe, J. S. (1996). Land classification and ecosystem classification. *Environmental Monitoring and Assessment, 39*, 11–20.

Shugart, H. H. (1999). Equilibrium versus non-equilibrium landscapes. In J. A. Wiens & M. R. Moss (Eds.), *Issues in landscape ecology* (pp. 18–21). Snowmass Village: International Association for Landscape Ecology.

Sochava, V. B. (1978). *Introduction to the theory of geosystems*. Novosibirsk: Nauka. (in Russian).

Solnetsev, N. A. (1948). The natural geographic landscape and some of its general rules. In *Proceedings of the second all-union geographical congress* (Vol. 1, pp. 258–269). Moscow: State Publishing House for Geographic Literature. (in Russian). See also in: In J. A. Wiens, M. R. Moss, M. G. Turner, & D. J. Mladenoff (Eds.). (2006). *Foundation papers in landscape ecology* (pp. 19–27). New York: Columbia University Press.

Solntsev, V. N. (1977). On the difficulties of instilling system approach into physical geography. In K. V. Zvorykin & A. Y. Reteyum (Eds.), *Issues in geography. Vol. 104. System studies of nature* (pp. 20–36). Moscow: Mysl'. (in Russian).

Solntsev, V. N. (1981). *System organization of landscapes*. Moscow: Mysl'. (in Russian).

Solntsev, V. N. (1997). *Structural landscape science*. Moscow: Mysl'. (in Russian).

Solon, J. (1999). Integrating ecological and geographical (biophysical) principles in studies of landscape systems. In J. A. Wiens & M. R. Moss (Eds.), *Issues in landscape ecology. 5th IALE-world congress* (pp. 22–27). Snowmass Village: International Association for Landscape Ecology.

Topchiev, A. G. (1988). *Spatial organization of geographical complexes and systems*. Kiev/Odessa: Vishcha Shkola. (in Russian).

Urban, D. L., O'Neill, R. V., & Shugart, H. H., Jr. (1987). Landscape ecology. A hierarchical perspective can help scientists understand spatial patterns. *Bioscience, 37*(2), 119–127.

Wu, J. (2013). Hierarchy theory: An overview. In R. Rozzi et al. (Eds.), *Linking ecology and ethics for a changing world: Values, philosophy, and action* (Ecology and ethics 1) (pp. 281–301). Dordrecht: Springer.

Wu, J., & David, J. L. (2002). A spatially explicit hierarchical approach to modelling complex ecological systems: Theory and applications. *Ecological Modelling, 153*, 7–26.

Zonneveld, I. S. (1989). The land unit – A fundamental concept in landscape ecology, and its applications. *Landscape Ecology, 3*(2), 67–86.

Chapter 2
Polygeosystem Fundamentals of Landscape Science

Alexander K. Cherkashin

Abstract Long-term experience of landscape research demonstrates the diversity of theoretical descriptions of landscape structures. System analysis leads to understanding the complementarity of knowledge obtained in various directions in landscape science and to the ideas of polysystemacity, polystructuredness, polyfunctionality, and hierarchical structure. We discuss the problem from metatheoretical and intertheoretical positions aimed at the stratification of landscape realities and knowledge on its properties. This means that to describe landscapes we need various system theories reflecting in special terms of connections and changes in both natural and socioeconomic objects on the territory. The landscape as a geographical environment for human activities is treated as the base of fibration. Landscape and landscape-typological map are treated as a fibrated space and as a spatial invariant of decision making in land use. Hence, various applications of landscape knowledge and maps are possible: interpreting mapping, that is, transformation of landscape maps to the maps with special thematic content; assessment mapping aimed at land evaluation; system mapping that is the application of the system concepts and models to describe trends in the territory; and polysystem mapping that is multidimensional modeling of geographical space for various planning tasks and forecasting. Methodology and technology of polygeosystem mapping are described.

Keywords Polysystem methodology · Theory · Differentiation · Fibration · Mapping

A. K. Cherkashin (✉)
Sochava Institute of Geography, Russian Academy of Sciences, Irkutsk, Russia
e-mail: cherk@mail.icc.ru

© Springer Nature Switzerland AG 2020
A. V. Khoroshev, K. N. Dyakonov (eds.), *Landscape Patterns in a Range of Spatio-Temporal Scales*, Landscape Series 26,
https://doi.org/10.1007/978-3-030-31185-8_2

19

2.1 Introduction

Numerous scientific challenges facing the modern landscape science require searching and defining fundamental premises for an explanation of observed processes and phenomena from the common positions, anyway connected with the general scientific level of theoretical knowledge and effective application of formal mathematical tools. This is a *complicated* problem that requires the professional work at all levels of information (data, knowledge, models, theories, metatheories), with application of the adequate mathematical tools at each level of generalization. Besides, it is necessary to consider the earlier methodological developments determining the place of different types of information in cognitive process as well as interrelations between study objects and subjects. The higher is the level of metainformation generalization, the more clearly these interrelations are understood with new models and methods aimed at explanation of the interaction. This holds true for geographical science, its branches (geomorphology, hydrology, geoecology), and for integrating disciplines (Earth science, regional studies, landscape science) in connection with emergence of sciences like landscape ecology with various interdisciplinary topics (Turner et al. 2001; Gergel and Turner 2002; Wu and Hobbs 2007) and, in general, in the context of unity of all scientific knowledge.

Long-term experience of landscape studies showed multidimensionality in the theoretical description of observed processes and phenomena by the different geographical schools, competing in scientific aspects (Harvey 1969; Haggett 1979; Isachenko 2003, 2004). The system analysis naturally leads to a conclusion about complementarity of knowledge from the different directions in landscape science and, in general, to polysystem knowledge, hence, to multistructural character, polyfunctionality, and multilevel hierarchy of knowledge (Haggett et al. 1965; Krönert et al. 2001; Wu et al. 2006; Khoroshev 2016). The problem is discussed from metatheoretical and intertheoretical positions aimed at stratifying landscape reality and knowledge about its properties. It is implied that to understand landscape the researcher needs to apply various systems theories that describe in special terms relations and change of natural, social, and economic objects at the territory.

In the intertheoretic aspect of overarching geographical research, polysystem differentiation is performed for the purpose of developing systems theories with various contents aimed at the uniform description of the nature and society from different system viewpoints in the corresponding terms (Cherkashin 1997, 2005a, b, 2009). The polysystem methodology of studying terrestrial objects is developed by systematic and mathematical means relying upon metatheoretical constructions, concretizing mathematical formulas with natural restrictions in the form of the principles of existence and change of reality. Knowledge is formalized using relevant notions of stratification (*fibration* in mathematical terms) of sets and various attribute spaces on diversity patterns as bases of stratification.

Landscape as a base of fibration is an intraregional geographical environment for realization of a different kind of activity. The geographical environment is treated as a diversity of connections between the factorial conditions that influence processes and peculiarities of their manifestation in the particular environment, corresponding to typological knowledge on this environment. For this reason, environmental approach is the dominating subject of geographical research, distinguishing geography from other spatial sciences (Cherkashin 2015; Cherkashin and Sklyanova 2016) in which knowledge relies upon the principle of environmental relativity, that is, obligatory consideration for the influence of local environment and geographical location on the final result. It allows considering the landscape and landscape typological map as the fibered space of diversity of the geographical environment and as an invariant for decision making in land use. Therefore, there are different directions of applying landscape knowledge and maps: interpretative mapping as transferring landscape maps to maps with special thematic contents (Cherkashin 2005a); assessment mapping as the use of landscape data and knowledge for land assessment; system mapping as application of system concepts and models to description of changes within the territory; and polysystem mapping as multidimensional modeling of geographical space for the solution of various planning and forecasting problems (Cherkashin 2007).

In this chapter, we describe the metatheory and methodology of polygeosystem-based display of landscape data and knowledge as well as approaches to solution of basic and applied problems in landscape science. Such approach reflects the ideas of the classical Russian landscape science and is based on original system and mathematical concepts of modern geography.

2.2 Models and Methods

Idea of the fibration of the reality as a set of independent systems is the cornerstone of polysystem methodology. Each system reflects a certain part (aspect, layer, cutting, level) of objects in terms of the corresponding special system theory. Identification of structure and function of each layer (fiber) as a special type of systems is a subject of the polysystem analysis, or polygeosystem analysis, applied to geographical topics. Functional links between the isolated fibers, their similarity, homology, and homotopy (Cherkashin and Istomina 2009) are the subjects of polygeosystem synthesis, formation of cognitive or territorial complexes, including in the form of landscape units, unity of heterogeneous parts into spatial whole. It expresses the dialectics of geographical knowledge and, in general, philosophy of searching laws of complexity of existence in different environments. The sequence of the solutions for such task implies, first, identifying distinctions and differentiation of the geographical processes and phenomena, and, second, searching similarity of the different, and identity of the opposites.

Fibration $s = (X, \pi, B)$ is continuous mapping π of space of X on space B: $\pi : X \to B$ (Postnikov 1988) (Fig. 2.1). The space X is referred to as space (set, object) of fibration, and B is the base of fibration (fiber base) consisting of a set of the elements b_i. The return $f = \pi^{-1}$ mapping $f : B \to X$ turns space X into the fiber

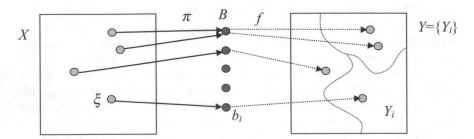

Fig. 2.1 Scheme of fibration procedure

space $Y = \{Y_i\}$. For any element of fiber base $B = \{b_i\}$ the preimage $Y_i = f(b_i)$ is called a fiber of fibration π over element $b_i \in B$.

In particular, a form $s = (X, \pi, B)$ represents the procedure π of the spatially distributed information fibration aimed at solution of geographical problems. The cornerstone is distinction between space X and fiber base B. Synthesizing them, we obtain the fiber space Y. For example, if X is the digital raster space image, B is a set of positions in landscape map legend, then Y is the landscape map constructed according to this image by means of this legend.

Other example is differentiation of scientific knowledge X on theories, theoretical layers (subjects) of knowledge $Y = \{Y_i\}$, describing objects as the systems of a certain quality. In the center of each theory, there is a set of basic system concepts and laws or axioms. The elements b_i of fiber base B are invariant qualities of the corresponding theories.

The elements b_i of base B are peculiar "information charges" with attributes that need to be identified in order to refer points of space X to one or another layer (fiber). Every element b_i is connected with coordinate x_i; therefore, fibration can be interpreted as projection of objects in different coordinate layers. Thus, there are two bases of fibration: element base $B = \{b_i\}$ and coordinate base $x = \{x_i\}$. The coordinate base is universal polysystem and forms geoinformation space of the polysystem analysis, according to features of element base B of fibration (Cherkashin 2005a, b).

The space Y is the same as space X but differentiated in parts. Layers (fibers) Y_i are called equivalence classes of space X; that is, they consist of the same points (elements) separated by the principle of mapping π. The differentiated space $Y = \{Y_i\}$ is a factor space $X|\pi$, constructed on relation π of equivalence of points X relative to elements of fiber base B.

The mathematical wording of fibration procedures allows a metatheoretical variety of interpretations of structures and functions of layers. From this the ideas of *polystructurality* and polyfunctionality of research objects follow. Formation of similar structures is the cornerstone property of the elements b_i of fiber base, which in reality at the same time belong to space X under consideration, to fiber base B and to concrete layer Y_i that is tangent to base B at point $b_i \in B$. The element b_i can be interpreted in the different ways as a reference point, equilibrium state, theory axioms, and comparison standard. For example, b_i may be considered as the original zonal type geosystem. It means that all properties of layer Y_i structure anyway result from properties of fiber base element $b_i \in B$, if simultaneously $b_i \in Y_i$, that is, object of observation is not separated from environment B.

Similar regularities may be well represented in terms of differential geometry (Postnikov 1988) (Fig. 2.2), where B corresponds to points of manifold surface, that is, a certain function $F(x)$ on indicators $x \in X$ as coordinates of vector linear space X, and layers $Y_i \in X$ are subspaces (planes), tangent to $F(x)$ in points b_i with coordinates $x_0 = \{x_{0i}\}$. In Fig. 2.2 fiber base B are points of a sphere surface $F(x)$, tangent to surfaces Y_1 and Y_2, which are layers of fibration of comprehensive space of attributes X. At the tangent point x_{01} in each layer Y_1 the local system of coordinates $\{y_{11}, y_{12}\}$ is a set. Considering a tangent layer Y_1 as manifold, on a set of its points, a new fibration is formed with layers Y_{11} and Y_{12}; therefore, through the multilevel sequence of fibration, the hierarchy of layers is generated.

In the context of this scheme, it is necessary to organize metatheoretical landscape thinking, especially, as methods of the quantitative mathematical analysis of relations between variables, in particular, using properties of Finsler's geometry (Antonelli 2003), where each functional relation $F(x) \leftrightarrow f(y)$, $(y = x - x_0)$ is considered as a metrics or norm of the studied space $x \in X$ in local coordinates y of each layer (fiber). The Finsler norm $f(y)$ satisfies the equation

$$f(y) = \sum_{i=1}^{n} a_i y_i, \quad a_i = \frac{\partial f(y)}{\partial y_i}, \tag{2.1}$$

where $y = \{y_i\}$ is a vector of values in local coordinates system; $a = \{a_i\}$ is a covector of values of internal coordinates of a system (gradient $f(y)$) of factors influence). Values $a = \{a_i\}$ are computed, according to observation data, by methods of multiple linear regression.

Fig. 2.2 Multilevel sequence of fibration

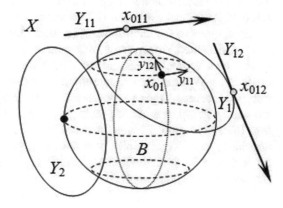

Note that in Eq. (2.1) the valuation function $f(y)$ can be any analytical homogeneous function of unit degree. It means that the uniform structure of fibration generates a set of valuation functions of the same set of variables y. This reflects the idea of polyfunctionality of the same quality systems.

Formula (2.1) is the universal equation of connection between variables $y = \{y_i\}$, which is concretized with coordinates of fiber tangent point x_0: $f(y) \rightarrow f(x_0, y)$. In this sense, Eq. (2.1) informationally includes *three general categories of environmental heterogeneity* in landscape ecology (see Lovett et al. 2005), such as patches (cores) $y = x - x_0$, gradients $a = \{a_i\}$, and *networks* (connections) $f(y)$.

Obviously, for the linear equations there is a hierarchy of attributes (Wu et al. 2006). The general Eq. (2.1) also is preserved in hierarchy of attributes. For example, let there be dependence of the variables $y_i = y_i(z)$ on indicators $z = \{z_j\}$ of system state at lower hierarchical level. In this case, if the functions $f(y)$ and $x_i = x_i(z)$ satisfy the basic Eq. (2.1), then the dependence of $f(y(z))$ on z also satisfies (2.1):

$$f(y) = \sum_{i=1}^{n} \frac{\partial f(y)}{\partial y_i} y_i = \sum_{i=1}^{n} \frac{\partial f(y)}{\partial y_i} \sum_{j=1}^{m} \frac{\partial y_i(z)}{\partial z_j} z_j = \sum_{i=1}^{n} \sum_{j=1}^{m} \frac{\partial f(y)}{\partial y_i} \frac{\partial y_i(z)}{\partial z_j} z_j = \sum_{j=1}^{m} \frac{\partial f[y(z)]}{\partial z_j} z_j. \quad (2.2)$$

Thus, deviations $y = \{y_i\} = \{x_i - x_{0i}\}$ give a way for parameterization of elements of structure and function $F(x) \rightarrow f(x_0, y)$ in local coordinates y in each layer. As a result, these structures and functions $f(x_0, y)$ become comparable $Y_i \leftrightarrow Y_j$ by comparison of their standards x_0 that allows logical transition from any layer to another one (interpretation). At fibration $F(x) \rightarrow \{f(x_0, y)\}$ on different points x_0 the doubling (complication) of data space and knowledge space presentation occurs, because each tangent layer (subspace, line, plane, type) is considered as the independent phenomenon; that is, tangent layer is orthogonal to surfaces of base $F(x)$. Each subspace with the corresponding functions of interrelation $f(x_0, y)$ among elements y can be also fibered. This results in emergence of the hierarchy of various systems (see Fig. 2.2).

This holds true, for example, when, in general, the geographical environment $F(x)$ is considered as some characteristic type of terrestrial phenomena at the stage of geological evolution of the planet. Furthermore, this type is fibered on different types of the environment; in particular, the zonal type of taiga environment is identified. According to this scheme, the territory is divided into environment types (types of natural zones), followed by the natural classification orderliness of these types in hierarchical systems. In this case hierarchical fibration is performed, when to each level of hierarchy, the discrete element of order structure corresponds, more precisely—significant indicators of objects distinction from top to lower levels of complexity. Hierarchical fibration can be based not on the general principles of typological hierarchy only, but also on various criteria, for example, fibration of *geomer* (homogeneous geosystems) on the constituent natural components.

Fibration as a division of set into the system of nonintersecting, independent subsets in geography is implemented in the form of division into districts, mapping types, development stages, levels of geosystems hierarchy, geosystem components, coordinate space of independent attributes and factors of influence, etc. The pattern of the heterogeneous territory is represented by a share (weight) of occurrence of properties of this or that system layer in a landscape. The examples of pattern representation are the share of an area occupied by different type of *facies* or the location of geosystem in attribute space of component properties coordinates (temperature, humidity, etc.).

Since elements of heterogeneous systems are defined differently in each theoretical interpretation, the hierarchy of analyzed objects in geographical space is formed in various ways; that is, the idea of hierarchy holds true for all elements and systems. In particular, at each large-scale level the systems of factorial influence of temperature condition are detected by different indicators. At lower *facial* level it is the current air or soil temperature; at landscape level it is average monthly or interannual temperature norms; and at global level it is zonal temperature regime and heat supply. It is obvious that higher levels have more stable norms with the increased variability, lower-level indicators being variants of such norms given that they are within the range of admissible variability. This domain represents the layer, or type of data, corresponding to elementary knowledge of the phenomenon under consideration $x_i = x_{0i} + y_i$, where, for example, x_{0i} is daily average air temperature and y is its standard variation.

Scale of averaging indicators is manifested in various thermal characteristics of weather and climate in time and space. Using average values x_{0i} as a first approximation and y_i as variability of parameters in each regime layer, it is possible to make judgments about local parameters of environmental heterogeneity. As the scale of the phenomenon increases, the variation of observed variables and heterogeneity of the territory increase as well. The small value of y_i indicates landscape homogeneity by ith attribute. At the same time, local temporal contrasts sometimes can be comparable in values to spatial variability at the regional and global levels (Krauklis 1979).

The structure of the layer as a manner of ordering its elements is determined by existence of a layer of group-theoretic properties in space, in particular, properties of symmetry of elements and their relations, such as temperature and moisture, productivity of vegetation cover, and heat supply in local and regional geosystems. For example, it is assumed that the system of independent coordinates $x = \{x_i\}$ of attribute space can be rotated in relation to zero coordinates so that one coordinate shifts to another (coordinates replace each other). This allows us to describe relations by the same Eq. (2.1), which is not sensitive to mutual replacement of variables. This coordination organizes elements in a system layer in accordance with the general principles of the organization of knowledge (phenomenon models). Below we investigate these regularities using the typology of land *facies* as illustration.

2.3 Facial Analysis of Landscapes

Facies is the elementary object of geographical research, which is visually detected and described in nature in space ξ and time t on a set of attributes $x = \{x_i(t, \xi)\}$. L.S. Berg (1945) has suggested to delimit in a landscape the simplest indivisible part, which he called the *facies* ("facere" is "to do, to make" in Latin). He has borrowed this concept and the term from geology (Nalivkin 1933), which assumes that land surface can be subdivided (fibered) into facies.

N. A. Solnetsev (1948, 1949) paid attention that boundaries of the *facies* were determined by geological boundaries, but not by phytocoenotic ones, since in one *facies* area several phytocoenoses can occur because of probable various stages of the same succession. Knowing natural properties of the facies, it is possible to reveal which original vegetation type (phytocoenosis) corresponds to these properties and to forecast the most probable sequence of phytocoenoses change. In this sense, the *facies* corresponds to a certain mode of vegetation restoration, as well as to the variability modes of the other geosystem components. As a result, the landscape map shows spatial location of various habitats and characterizes their major peculiarities. The corresponding spatial pattern can be treated as an invariant of landscape organization that is stable despite possible changes of component attributes.

This concept is in accord with transition from landscape morphology to interpretation of the geographical facies within the scientific direction called as structural and dynamic geosystem landscape science (Sochava 1962, 1978; Krauklis 1979; Mikheyev 1987).

According to the principles of a structural and dynamic landscape science (Mikheyev and Ryashin 1970) developed on the basis of long-time stationary research, the facies is understood as the elementary subdivision of the geographical environment, which is characterized by almost uniform relations between natural components. Homogeneity is indicated by the absence of the essential natural boundaries crossing *facies* space. The *facies* is the most short-living geosystem in comparison with physiographic complexes of higher rank orders. Besides relatively steady *original and quasi-original facies*, there are *facies* more changeable in space and time. The *facies* responding to anthropogenic influences of various intensity and character experience different modifications (*derivative and long-term derivative facies*).

The term "*facies*" is used to get insight into a set of uniform cells (layers), each of which represents an individual unit; that is, here the *facies* is a both typological and topological concept. In this sense, in Siberian school of landscape science "the *facies* represents type of an elementary physiographic (landscape) complex, ... expression '*facies* type' is redundant" (Sochava 1962, p. 20). The *facies* are identified on a full set of inherent parameters from which landforms and material substrate are the most significant. At the same time, it would be misleading to understand "attributes" as quantitative and qualitative variables only. Various regularities of *facies* development and manifestation of its properties in the area also have to be considered.

In the Siberian geographical school, the facies doctrine is a part of *geosystems doctrine* in which special attention is paid to structural and dynamical analysis of landscapes (Sochava 1978). The term "*geosystem*" has been widely used for a long time in scientific literature. A.G. Isachenko (1990) believes that the geosystems doctrine has a huge potential as a common methodological platform for physical and economic geography that inspires hopes for termination of state of degradation in geography.

In the V. B. Sochava's geosystems doctrine, landscape cover is considered as a hierarchically organized set of geosystems—"terrestrial spaces of all dimensions" (Sochava 1978, p. 292), "this is whole, consisting from the interconnected components of the nature, which are controlled by the regularities, operating in geographical *surrounding* or the landscape sphere" (Sochava 1974b, p. 4). Depending on the scale of interest, local facial-, landscape-, regional-, and planetary-level geosystems are distinguished.

Among the main objectives of the geosystems doctrine, V. B. Sochava (1978) put emphasis on the issues as follows: analysis of axioms and other propositions of the special geosystem theory, geosystems modeling with particular focus on their spontaneous and anthropogenic dynamics and corresponding natural regimes, elaborating relevant tools for quantitative assessment of geosystems and landscape-forming processes, getting insight into *spatiotemporal* regularities and the analysis of geosystem states, etc. The directions for solving these problems on an axiomatic basis were discussed in the context of a research and modeling of geosystems in special publication (Cherkashin 1985).

The geosystem doctrine as any scientific theory relies upon several basic statements with status of peculiar axioms (Sochava 1974a, b).

1. The geographical environment is organized in the form of hierarchy of managing and managed systems, that is, it includes the geosystems of various rank orders with subordination connection. The geosystem of the top level is environment for the dependent geosystem of the lower level (Sochava 1978).
2. The regularities, inherent for geosystems, are unambiguous in certain territorial boundaries and spatial scales of three scale levels: planetary, regional, and topological (local) ones.
3. Geosystems are presented by both original structure and variable states subordinated to a certain original structure invariant, which changes only in course of geosystem evolution. Changes of geosystem with preservation of an invariant determine the geosystem dynamics.
4. Combination of two phenomena (homogeneity and heterogeneity) is characteristic for the natural environment. Laws of homogenization and differentiation act simultaneously in process of the natural sphere development. The geosystems with homogeneous structure (*geomer*) and heterogeneous structure (*geochore*) coexist in landscape. This fact is reflected in principle of double-row classification of geosystems: *geochore* and *geomer* hierarchies.

These fundamental tenets were developed, supplemented, and concretized while solving the problems of modeling and mapping (Mikheyev and Cherkashin 1987;

Mikheyev 2001; Cherkashin 1997). It is worth noting that all the listed postulates are directly related to basic properties of layers and existence of different types of stratifications as follows: (1) hierarchical, (2) typological, (3) dynamic, and (4) ontological. In the last option, existence of geomers and geochores can be considered as a special case of existence of the double-layer reality (Cherkashin 2005a, b) presented by informational (typological, internal) and material (chorological, external) worlds. In geosystems doctrine, the idea of an invariant has key importance (Sochava 1967, 1973, 1978; Sochava et al. 1974). Invariant is "the essence which is invariable at dynamic changes and manifested in geosystem patterns" (Krauklis 1975, p. 26). It is such essence which appears in the various forms, admissible within geosystem boundaries, without transformation of essence. In this context the invariant by the wideness of the notion corresponds to the integrated natural regime of geosystem, that is, the generalized conditions (integrating factors) that drive everything in geosystem.

In terms of the fibered cognition $x_i = x_{0i} + y_i$, the invariant x_{0i} defines a set of the phenomena, admissible at deviation y_i connected with this invariant. It is possible to define such a set of forms of system existence within the invariant as *episystem*, that is, as a system layer of the certain type. The core and periphery are delimited in episystem, the former being defined as forms and phenomena corresponding to the entity (invariant) in the best way, the latter—as related *surroundings*. The periphery possesses all properties of the invariant resulting from dynamic changes, spatial variations, and structural subdivisions. It allows investigating various alternative manifestations of geosystems using a unified approach, that is, widely using an idea on invariants in the geosystem analysis in order to systematize and to put in order geographical data and knowledge. Such representations have something in common with geographical models "center–periphery" or *chorion* model (Reteyum 1988).

V. B. Sochava (1967, 1974a, 1978) introduced the concept of *epifacies* as theoretical framework to describe a dynamic set of the facies with identical invariant properties, and also developed the idea of *epi-association* connected to *epifacies* (Sochava et al. 1974). A core of epifacies consists of the zonal type facies, that is, the *geomers* of high scale level, which are manifested perfectly on flat surfaces in uplands and plains.

Typification and classification of geosystems are aimed at identifying the general properties of geosystems and their invariants for different locations (conditions). Typification, as one of types of the fibration procedure, assumes existence of a set of elements of fiber base, each of which can be related to an invariant of geosystem type, that is, can be assigned to type and invariant as an element from typological base. The set of geosystems indicated by a concrete element of fiber base belongs to the same episystem. Hence, the episystem is the system layer, corresponding to this element of base, and to essence of the defined set of systems.

The cornerstone of geosystem analysis lies in the classical relation between integrated and differential approaches to landscapes studies, between landscape continuity and discretization, between definition of the invariant and changing properties of geosystems. This is the way to develop the general *invariant/variant model* of cognition of geosystem structure and organization when the invariant of the lower-order geosystem becomes variant of the higher-order one. In particular, local *facies* are considered as factorial variants with various degrees of deviation from zonal background conditions and various levels of factor-induced deformation of the original zonal facies. Another example can be seen in variable states or stages of recovery changes in vegetation cover which are considered as a variation of the invariant of the facies, corresponding to original biogeocoenosis.

The concept described earlier results from the assumption that geosystem in V. B. Sochava's doctrine is treated as not spatial and morphological system only, but as functional and dynamic system that is of primary importance. Geosystem passes through a number of states; this set corresponds to a certain combination of geographical processes and phenomena; therefore, in general, it is possible to assess geosystem considering inherent temporal variations only (Krauklis 1975), that is, based on the results of regime observations.

According to the invariant/variant model, we deal with the multilevel system of nested invariants, where the invariant at one level becomes variant at higher level of system organization. Each invariant forms an episystem layer of corresponding variants. Variants of all lower levels can be referred to as epifacies of a zonal invariant that means existence of a hierarchically structured episystem. For this reason, building the whole system of geosystems taxonomy and their properties, starting with local invariants, was supposed to be possible. This model works well while solving the problems of classification (local coordinates of factorial systems), the analysis of functional connections (congruences and complexes of dependencies), the spatial organization, etc. (Cherkashin 2005a, b). This conception is believed to be much wider and more effective than the idea of a hierarchical subordination of geosystems. It clarifies connections in a system of variants and invariants. It ensures more order in numerous fractal models of self-similarity of connectivity structures. The fractality means that the natural hierarchical structure and structure of the knowledge organization at each level are presented by similar schemes of an invariant/variant relations like the "core-periphery" models, which are included naturally into hierarchical structure of subordination from top to lower level. In this sense, fractal accuracy closely correlates with the idea of geographical accuracy, which depends on degree in detail (detail scale) not only while studying the territory but also while understanding features of manifestation of spatial and classification connections at the highest levels of generalization. That is, the higher is level under consideration, the more clearly the position of each geographical situation in a landscape is elucidated.

Earlier attempts to construct models of episystem, for example, in the form of peculiar "chemical" molecules were made, where each "atom" corresponds to the geosystem of underlying level (Sochava 1965). Integrative levels are identified on the basis of homogeneity of the attributes determining existence of one or another taxon. The homogeneity acts as a form of manifestation of invariancy in attribute space. Integration of elementary geosystems is realized by establishing attribute invariants of higher level that allows considering and mapping diversity of local conditions.

To get insight into dynamic connections of geosystems, it was suggested to use methods of the theory of graphs (Sochava 1974a, 1978). Dynamics is shown within "shot" in an evolutionary series of geosystem development. Evolution is a sequence of "shots" to each of which particular invariant a set of the related variable structures corresponds. In a "shot" of *epifacies* structure (Sochava 1978), the maternal core (original facies), semioriginal, serial facies, and anthropogenic facial modifications are identified.

The invariant/variant fractal structure represents a multilevel set of fiber base elements, each of which belongs to a certain geosystem type referred to as *geomer*. In this sense, such structure has to obey the main features of the geographical environment structure, to be model of the natural geoinformation environment, and to have regularities inherent to this environment.

In particular, the landscape sphere and hierarchy of geosystems were formed in the course of historical development. Landscape cover evolution is represented as a sequence of geosystem invariants. Throughout the geological periods, it followed certain direction as a result of geosystems self-development and impact of changing external conditions. This self-development repeated and reproduced those structural relations that formed the frame of the invariant/variant scheme of the geoinformation environment. This means identity, equivalence of the structural relations and temporal changes, and adequacy of classification to evolutionary changes. The invariants, replacing each other, act as stages of evolutionary process. Throughout evolution, each invariant consisted of a series of variable states, each of which is temporal transformation of the invariant. This concept demonstrates connection between classification structure and potential dynamics of geosystems, which is believed to be one of the most important regularity of the geoinformation environment.

2.4   Polysystem and Polystructural Interpretation of Knowledge

According to the general principles of the polysystem analysis, several overarching system theoretical directions can be identified in geographical research. The classical notions "element," "connections," "structure," "system," "subsystem," and "change" obtain special contents and consequently the special models of the phenomena description, expressed in language of various theories, were elaborated.

"Structure" is a primary notion of system science that means a certain way of arranging something. Structure of elements is a set of connections between elements, that is organization of systems. Each special theory has its own peculiar understanding of structure and, hence, of system and organization. From a viewpoint of fibration methodology, there is a set of systems, their structures, and organizations that determines "polystructuredness" and "polysystemacity" at cognition of reality.

The developing polysystem methodology, first of all, is aimed at creation of theoretical bases of substantial scientific knowledge to be effectively guided in mass of various pieces of information and structurally to use information for an explanation, search, and invention. A perspective problem of scientific research is to arrange available geographical knowledge according to system directions, to create models of each subject domain of research and to use them at the solution of practical problems on the basis of the newest information.

Now, we turn to a short review of the main theoretical directions of polysystem and polystructural studies in application to studying landscape facies.

The general theory (metatheory) of fibration assumes that any object is the polysystem consisting of elements—monosystems; for example, the landscape is a set of nonintersecting monosystems *facies*. Any layer is considered as a set of types of systems, and any type is a classification element of fiber bases (classification position). Connections of monosystems are determined by similarity of their specific structure. The system of monosystems forms polysystem as a polysystem image of an object. The structure is determined by the relation of the spatial and temporal neighborhoods of monosystems in a polysystem and change of these relations.

From these positions, the facies as the complex territorial phenomenon is the polysystem, consisting of elements—monosystems. To study them, one needs to have system tools of the theoretical analysis at his disposal. On the other hand, landscape as complex territorial phenomenon is fibered on monosystems (sensu V. S. Preobrazhensky)—the facies. In both cases we deal with monosystems, but with the different aspect of fibration realized on various bases. The facies as an object is both monosystem and polysystem, but in different aspects. The geographical landscape consists of elementary areas—*facies*, that is, elementary types of the geographical environment. This knowledge concentrates on its information image (charge) as an element of typological and classification fiber base. Connections in landscapes are determined by commonalty of their facial composition (structure).

The set of *facies* areals (fiber layers) forms a polysystem image of a landscape (unit frame of a territory) and is displayed on typological landscape maps. The structure of a landscape is determined also by the spatial neighborhood of *facies*, and the temporal sequence of *facies* changes as evolutionary development of the territory. At the same time, there is a reorganization of facial structure as the polysystem space of a landscape changes. The *facies* is elementary (homogeneous in a broad sense) geomer. Compact association of facies and their parts is referred to as a heterogeneous polysystem—geochore. In a broad sense, geochore is any territory with facial structure, that is, facial polysystem. Spatially adjacent geochores unite in geochore (polysystem) of higher rank order. Association of facies in geomers of higher typological rank order is not considered in this theory. However, it becomes research object in the following system theory.

The theory of classification systems deals with elements of classifications (types, taxons). Familial relations are established through taxonomical divisions; for example, types are connected (constitute one taxonomical layer), if they belong to one taxonomical category of higher order (tangent point). All these elements are connected, but at the different taxonomical levels. The system is formed by all the related types at this associative level (layer). The structure of classification system is described with the graph of taxonomical hierarchy. Changes are considered as any displays of comparison, as symmetric image, translating hierarchical fractal structures in themselves.

The classification theory postulates discrete character of geographical space of fibration, namely existence of the typological base of fibration presented by a set of points (elements) of base (classification). Cartographic base of fibration corresponds to a map legend. The *facies* is considered as elementary geographical typological unit (type of geosystem). Any *facies* is a typological element (unit), which forms the geoinformation environment of territorial units (environment of fibration). Objective existence of the geoinformation environment is postulated. Here, an element (*facies*) corresponds to the information charge, disclosing essence and content of facial structure of taxonomical type. It is supposed that in each system and theoretical interpretation the information charge bears specific subject knowledge of the facies; therefore, typification and classification of facies are carried out in each system layer separately. However, there are natural correlations of knowledge charges at different system layers, relating to one facies. Classification emerges on a set of typological (taxonomical) units of fibration, if a structure of relations between elements is determined on this set, for example, while defining hierarchy of taxonomical units. Structures of connection between elements in the geoinformation environment model (classification) are numerous. The *facies* is at the lowest level of hierarchical classification and corresponds to a geosystem type (elementary geomer). The geoinformation environment has the fixed structure and its changes can be understood as an informative process of understanding the structural and functional analogies between various taxonomical units based on the generalized principles of symmetry of the geoinformation environment. For example, natural zones-analogs (sensu D. G. Vilensky) in different climatic zones or original facies (sensu V. B. Sochava (1978)) reveals the features of zonal regularities at the local level. Similar analogies are useful for making conclusions about properties of geographical space at both local and planetary scales. It is worth noting that association of *facies*-monosystems within the fibration theory generates geochores of various rank orders while within the classification theory—*geomers* of various levels. This is the fundamental basis of the double-row classification principle accepted in Russian geography following V. B. Sochava (1972). Here we face two different system interpretations of geographical objects, that is, structures of spatial and typological association.

The theory of complex systems is aimed at synthesizing monosystem aspects of an object to a holistic image of an object—a complex. Connections are defined through mutual reflections of monosystem layers in a complex. The structure is determined by the significance of each layer type in a polysystem and the commutative chart of image. Change is variation of the corresponding structures. It is rational to use this approach in the analysis of the complexity phenomenon in landscape ecology (Green et al. 2006; Cushman and Huettmann 2010).

It is convenient to understand the facies within the theory of complex systems with commitment to concept of Moscow landscape school, namely, following interpretation of the facies as geographical complex (Solnetsev 1949, 2001). It allows fractional fibration of facies on subject monosystems, components, variable states (biogeocoenoses), parcellas, and geohorizons, considering them as the special monosystem layers delimited with different fibration bases. The researcher has the opportunity to set the objective of synthesis of these essentially different, particular, and autonomous layers in holistic presentation of the facies. Identification of various connections between structures and functions of the monosystems is the cornerstone of synthesis, establishing similarity, correlation, interface, and balance between layers. In a landscape, for example, one component (biota) can be treated as natural model (analog) of a soil cover.

Landscape indication is based also on this kind of synthesis. Such structural and functional similarity is realized in fact, unlike virtual similarity (analogy) of elements of the classifications having other information basis. Different facial subdivisions of a landscape are also similar, which ensures unity of a landscape as geocomplex. The structural and functional similarities of systems mathematically are interpreted, respectively, as their homological and homotopic equivalence. The structural and functional similarity of components is related to continuous transformation of their homotopic coefficients, which are unambiguously characterizing each component of a geocomplex. The property of unambiguity connects homotopic coefficient with a classification position of a component (layer). *Facies* are identified as complexes on the basis of studying of their component structure and the nature of functional links of components (integrity).

The theory of rank distributions studies structures of ordering specific layers by decreasing rank value in accordance with the degree of their manifestation in a polysystem (weight, area, number, frequency, abundance). Connections are set by order of types in the ranged row and by the equations of rank distributions (degree of manifestation as a function of rank number of a position). Any ranged row with the corresponding function of distribution forms rank system. Its change is expressed in shift of types placed in a row as a result of change of its rank value.

The theory of rank distributions studies quantitative regularities of complex systems structure. Rank distributions determine an order of layers (types) resulting from a decrease of their importance and occurrence. The *facies* has rank structure in general and for components. A set of *facies* forms facial rank structure of landscape, for example, in the form of distribution of landscape facies on the land area. The facial structure of landscape is a basis for regionalization (method of filling individual units with typological units of mapping) and generalization of cartographic images (by dominant type). The biotic structure of facies is a basis for studying biological diversity at the levels of species and ecosystem, and facial structure of the territory for that at landscape level. The equations of rank distributions are equivalent homotopically; that is, equations are interrelated by changing a single coefficient. Rank distributions are system basis of ordinal approach in economic estimation of lands and resources. Facies are delineated by structure and texture of geo-images (heterogeneity of composition) and by dominating landscape-forming components.

The theory of discrete spatial systems considers the land surface as a fractal multilevel structure of the basic (nodal) points (elements) connected in the different ways by coordinate lines. As a result, the rectangular and triangular systems of coordination determine lineament orderliness of space and time. The examples are Christaller's model in the theory of central places and network of crust faults with well-expressed hierarchical structure. It is supposed that overarching centers (points, cores) of space and time unambiguously determine a set of situations and events as a structure of spatial and time series of attributes.

In this theory, space and time of the facies are filled up (covered) regularly with a multilevel set of the nodal points connected with lines in the different ways. In the general sense, facies is assumed to be closed spatiotemporal surroundings (unit and stage) of one or several such points, which are unambiguously corresponding to a certain classification position in the geoinformation environment. These points control position of the facies in space and by the hierarchical level of manifestation. This model relates classification and polysystem approaches, referring situations and events of different type to the land surface. The hierarchical numerical code of location and homotopic coefficients of system parameters evolve to a spatial and temporal series of geographical characteristics. As a result, on the Earth's surface the multilevel coordinate system (grid-structure) of reference points (epicenters) with fractal properties of self-similarity will be defined. It is a natural geographic information system that allows us to individualize the facies while performing estimating and forecasting calculations. In space, this approach is used to delimit facies of invariant landscape structure linked with lineaments, mainly with water divides, thalwegs, and geological faults. Such structures determine an ecological network of a landscape.

In the theory of geospatial field elements are the values of coordinates of physical and evolutionary space of the planet that define location (point on a surface) and intensity of geological development in this point. System is the vector field of rotation in space of these coordinates. The structure is determined by coordinates of a point of rotation (epicenter) and velocity of evolution in this point. Changes of structure are expressed in variation of these parameters.

In this theory the relief development is the factor, determining spatial properties of landscape. Here each reference point is taken as an epicenter of a multilayered cone of rotation of evolutionary space characteristics. As a result, the circles with cyclic structure (concenter) are designed on the Earth's surface. Their superposition generates a spatial grid of facies boundaries (epigenetic structures). The spatial rotation model describes process of facies boundaries emergence, which are subject to changes during geological and biological evolution of the Earth. The invariant basis of boundaries is formed that is common for all other system interpretations of the facies. The intensity of evolutionary processes correlates with a numerical code of epicenter and its classification position. *Facies* are delineated with boundaries, which follow special lines of relief. To draw boundaries, large-scale topographic maps, satellite imagery, and methods of relief plasticity are used.

Totally, two latter theories can form a basis for the known chorological principles of landscape ecology postulating spatiotemporal definiteness of landscapes, heterogeneity and connectivity, observed processes, and the phenomena (Turner 1987; Farina 2006).

The theory of dynamic systems studies a set of the similar elements (particles, bodies, units), shifting from one state to another. Subsystems unite elements in the same state and are connected by flows of elements. The system structure is determined by distribution of elements by states and the scheme (graph) of flows of elements between states, and intensity of flows. Change of structure is dynamic processes of change of states and the reproduction of elements. Similar flow models are used to explain landscape heterogeneity by structure and functions (Lovett et al. 2005).

The theory of dynamic systems is a basis of the geosystem theory that studies transitions of the similar elements from one state to another. The *facies* is considered as an elementary spatial cell of uniform physiographic dynamic process (flow) regulated by the integrated intrinsic natural regime, that is, in compliance with position of the facies in classification. The facies is manifested in the form of various spatial and temporal variable states of diurnal, seasonal, long-term, and century-long changes. All elements of the *facies* are characterized by this or that variable state. In fact, the *facies* is studied as a set of the variable states, which are ordered in the dynamic sequences and cycles. A bright example is the model of recovery series (successions) of a vegetation cover when recovery stages are described as variable states (Shugart 1984). The restored (potential) facies pattern is unambiguously determined by its natural regimes. Typification of geographical facies within structural and dynamic geosystem approach is carried out on the basis of identification of dynamic series for different facies components and the corresponding natural regimes.

The theory of functional systems considers the factors as elements connected through functions of their influence on each other (cause–effect relationships). System is the complex of factors (integrated factor) that is an analytical function of particular factors. The systems of integrated factors are functionally assembled in higher order factors (functional systems). The structure of such system is set by "weight" of factors' influence and the graphs of their cause–effect functional connections.

In this theory, the facies is either position in space of factors, that is, function of many factors, or an integrated factor. Change of this function leads to the change of position in factorial space of facies (ordination space) in which each facies and the corresponding natural regimes are considered as modification of normal zonal facies (i.e., functional core) on one or several integrated factors.

Depending on degree of factorial deformation, the original, semioriginal, and serial (transformed) facies are distinguished. In serial facies, influence of factor is hypertrophied. As a result, most facies in a landscape of natural zone are represented by factor-induced deviation from zonal norm. Totally, they form a set of facies belonging to one type of environment. Part of facies in landscapes may be relicts. Some facies may be generated by invasion of facies from adjacent natural zones; they are referred to as *extrazonal facies*. With this approach, the facies are identified on the basis of structure and intensity of factors' influence, manifestation of serial character, and the nature of connections among particular factors and their displays in remote sensing images.

Functional approach involves an important concept of *geosystem niche*, which, analogously to ecological niche notion, means position and role of *geomer* in physical space and space of environmental factors. Structurally the niche is set by frequency distribution of *facies* in the system of a factorial ordination or as estimated functions, for example, the functions of bioproductivity of communities, depending on the size of the influencing factors. Such approach is important for studying natural functions of geosystems and assessment of the ecological services provided (Groot et al. 2002). Another field of application is determination of type spatial distribution under the habitat conditions (Guisan and Zimmermann 2000).

The theory of potential systems considers objects as the systems of elements (properties) or potentials (*geotropical* formations). Structural connections will organize particular potentials in the generalized potentials (the system of potentials) by means of indicative functions of usefulness. The structure of potential systems is determined by a set of extensive potentials and its sensitivity (intensive potentials) of change in the generalized potentials to change in particular ones. Change of systems is expressed in change of the generalized potentials at change of their information capacity (variety, entropy). Examples of extensive (summable) potentials are the areas, volumes, stocks, and fluxes, and functions are the organization, energy, working capacity, and usefulness. Intensive potentials are temperature and pressure, which are understood in a broad (not only physical) sense as a measure of organization (significance).

The theory of potential systems, unlike that of functional systems, considers the *facies* not from the external (factorial) viewpoint, but from the internal position as the system of its properties, that is, indicator system. Various properties of the facies are integrated into the generalized indices (potentials) that enable us to estimate the natural and economic usefulness of the facies in various ways. Production of similar economic estimates is called *cardinalist estimation*. Potentials of external geographical environment act as a reference system. In environment with different potentiality, the system of one potential has the different final index of development. The facies differ in organizational structure of its properties and organizational uniformity (connectivity) of spatial characteristics. Degree (index) of organization determines usefulness of the facies, that is, its ability to make work, including work instead of man. The facies is defined by a set of properties and parameters (descriptive approach), by organizational uniformity (connectivity) of its spatial characteristics (analytical approach) or by position in space of the generalized potentials as integrated properties (quantitative methods, multidimensional statistical analysis).

The theory of systems of regulation mechanisms describes behavior of objects in phase space of their independent characteristics $y = \{y_i\}$. Elements are deviations y_i of values of characteristics x_i of an object from equilibrium values x_{0i}. Connections are described by the equations "in deviations" when, for example, change of any characteristic is function of deviations of values of all other characteristics from equilibrium. The structure is determined by the graph that shows the direction of influence of attributes on each other and by coefficients a_{ij} that express power of interaction.

The equations of models in this theory, as well as all the other quantitative theoretical models, are a special case of the universal Eq. (2.1) at $f_j(y) \leftrightarrow \dfrac{\mathrm{d}y_j}{\mathrm{d}t}$:

$$f_j(t) = \frac{\mathrm{d}y_j}{\mathrm{d}t} = \sum_{i=1}^{n} a_{ij} y_i, \ a_{ij} = \frac{\partial f_j}{\partial y_i}, \tag{2.3}$$

Theory of interaction (self-regulation) mechanisms can be used to describe peculiarities of facies behavior, that is, orientation, intensity, and frequency of change of the attributes. Changes occur in the surroundings of an equilibrium state x_0, corresponding to a classification position of the facies. The surrounding y includes a full set of variable states of facies, for example, of the interacting vegetable communities in a recovery series (succession) sensu F. Clements or stages of relief development in W. M. Davis geographical cycles. Parameterization is performed in deviations of current process values from equilibrium ones (e.g., biomass of original vegetation, elevation of a peneplain). The size of set of variable states is determined by the degree of spatial and temporal variability of facies. The original facies have the low variability unlike serial ones. If *facies* attributes naturally deviate from equilibrium state (norm) and are beyond a set of admissible variable states (standard), the facies behavior is considered to be in nonsteady state. If *facies* stays within the admissible range of state, it is treated in steady state. If *facies* develops toward equilibrium, it is treated as asymptotically steady. In this case, the behavior of *facies* and their components, namely restoration of the parameters, is described in terms of homeostatic regulation models (2.3). The *facies* are defined by character of their spatial and temporal variability, mainly by mosaic of spatial pattern and species composition of biota, as well as power and frequency of catastrophic exterior influences.

The theory of public systems deals with social systems, consisting of people and related objects of their activity. Public relations are shown as a result of the human relations and activity. Public structures are realized in a form of public and production organizations, and their change is an organizational development.

In the theory of public systems, the *facies* is considered as an object of land use, assuming legal zoning of the territory. The *facies* is elementary territorial unit of economic activity (land use policy cell, unit of legal zoning, areal of landscape planning). The most effective way of economic activity ideally is based on knowledge of potentials and development tendencies of facies. Hence, territorial limits for economic activity and nature protection measures are expected to coincide with facies boundaries. This association provides information for landscape planning and allows optimizing environmental management.

The aforementioned multifaceted theoretical approaches study landscape objects from the different viewpoints, that is, in various projections. This list of approaches is not limited to the given theoretical interpretations of reality. For example, we did not mention interpretation of a landscape as an aesthetic system as a natural image with the own structural composition.

The *facies* as any real object is multifaceted. It is the areas where the laws of any science partially are manifested and intertwined. The plurality of geographical theories causes a variety of definitions for facies and criteria for these definitions. Each system theory is based on its peculiar definitions and laws and has specific space of data presentation, for example, space of states, factorial space, spaces of potentials, and shifts of characteristics from equilibrium. Any model of facies correctly applied to processing geodata has to ensure equivalent spatial pattern as the landscape map with typological contents, and map of geographical environment heterogeneity.

The presented system-based interpretations of facies, as geographical research objects, clearly demonstrate a variety of possible approaches to definition and delineation of *facies* on a territory. Despite fundamental difference of approaches and methods, results of facies delineation have to be identical, that is, facies delineated by different methods (different interpretations of *facies*) are expected to have identical boundaries. In concrete landscape investigation, morphological (facies as a geocomplex), dynamic (facies as geosystem), and spatial (*facies* as special location in relief) approaches are preferred.

2.5 Conclusion

Review of polysystem methodology content in geographical studies shows how metatheoretical fibration of spaces of data and knowledge, as well as that of models and theories, provides an opportunity to formulate and explain how structure and function of geosystems develop. Fundamental insights were supplemented, developed, and concretized while solving problems of landscape modeling and mapping. Various interpretations of the landscape facies concept mean polysystem, polystructural, multifunctional, and polymodel approach to their delineation and description.

The polygeosystem methodology has a vast field of application in landscape science. First of all, it is aimed at progressive understanding landscape as polysystems and, hence, as polystructure and polyfunctional. This or that structure of a system layer (fiber) generates the particular function. Each system theory describes a certain class of processes and phenomena and can be used for the solution of corresponding problems.

To solve a geographical problem in the most complete and precise way, one needs to find the most adequate approach, its appropriate reflection in modeling and mapping, and the corresponding model of data presentation. Thus, the problem is projected to this or that subject domain, where it is being solved using laws of the special system theory with its particular understanding of landscape structure and functions. Such approach can be called as projective modeling, when the problem formulation and solution are considered as a problem projection in this subject domain. Then, if the problem remains unsolvable in one domain, it is solved in another one, and results of the decision are used in various fields of knowledge. We believe that this is the principal practical advantage of polygeosystem methodology of scientific research.

References

Antonelli, P. L. (Ed.). (2003). *Handbook of Finsler geometry*. Dordrecht: Kluwer Academic.

Berg, L. S. (1945). Facies, geographical aspects and geographical zones. *Proceedings of the All-Union Geographical Society, 77*(3), 162–164. (in Russian).

Cherkashin, A. K. (1985). Mathematical problems of the geosystems doctrine and possible ways of their decision. *Geography and Natural Resources, 2*, 34–44. (in Russian).

Cherkashin, A. K. (1997). *Polysystem analysis and synthesis. Application in geography*. Novosibirsk: Nauka. (in Russian).

Cherkashin, A. K. (2005a). *Polysystem modeling*. Novosibirsk: Nauka. (in Russian).

Cherkashin, A. K. (Ed.). (2005b). *Landscape interpretative mapping*. Novosibirsk: Nauka. (in Russian).

Cherkashin, A. K. (Ed.). (2007). *Geographical research in Siberia. V. 4. Polysystem thematic mapping*. Novosibirsk: Geo. (in Russian).

Cherkashin, A. K. (2009). Polysystem modeling of geographical processes and phenomena in nature and society. *Mathematical Modeling of Natural Phenomena, 4*(5), 4–20.

Cherkashin, A. K. (2015). Geography and non-geography. *Proceedings of the Irkutsk State University, Series Earth Sciences, 14*, 108–127. (in Russian).

Cherkashin, A. K., & Istomina, E. A. (Eds.). (2009). *Homology and homotopy of geographical systems*. Novosibirsk: Geo. (in Russian).

Cherkashin, A. K., & Sklyanova, I. P. (2016). Manifestation of the geoecological ethics principles: Environmental approach. *Geography and Natural Resources, 3*, 189–199. (in Russian).

Cushman, S. A., & Huettmann, F. (Eds.). (2010). *Spatial complexity, informatics, and wildlife conservation*. Tokyo: Springer.

Farina, A. (2006). *Principles and methods in landscape ecology: Towards a science of the landscape* (Landscape series 3) (2nd ed.). Dordrecht: Springer.

Gergel, S. E., & Turner, M. G. (Eds.). (2002). *Learning landscape ecology: A practical guide to concepts and techniques*. New York: Springer.

Green, D. G., Klomp, N., Rimmington, G., & Sadedin, S. (2006). *Complexity in landscape ecology* (Landscape series 4). Dordrecht: Springer.

Groot, R. S., Wilson, M. A., & Boumans, R. M. J. (2002). A typology for the classification, description and valuation of ecosystem functions, goods and services. *Ecological Economics, 41*, 393–408.

Guisan, A., & Zimmermann, N. E. (2000). Predictive habitat distribution models in ecology. *Ecological Modelling, 135*, 147–186.

Haggett, P. (1979). *Geography: A modern synthesis*. New York: Harper and Row International.

Haggett, P., Chorley, R. J., & Stoddart, D. R. (1965). The importance of scale standards in geographical research: A new measure of areal magnitude. *Nature, 205*, 844–847.

Harvey, D. (1969). *Explanation in geography*. London: Hodder & Stoughton Educational.

Isachenko, A. G. (1990). Geography at the crossroads: Lessons of the past and way of reorganization. *Proceedings of the All-Union Geographical Society, 122*(2), 89–96. (in Russian).

Isachenko, A. G. (2003). *Introduction to ecological geography*. St. Petersburg: St. Petersburg University Publishing House. (in Russian).

Isachenko, A. G. (2004). *Theory and methodology of geographical science*. Moscow: Academia. (in Russian).

Khoroshev, A. V. (2016). *Polyscale organization of a geographical landscape*. Moscow: KMK. (in Russian).

Krauklis, A. A. (1975). Idea of dynamics in the theory of geosystems. *Proceedings of the Institute of Geography of Siberia and Far East SB AS USSR, 48*, 24–30. (in Russian).

Krauklis, A. A. (1979). *Problems of experimental landscape science*. Novosibirsk: Nauka. (in Russian).

Krönert, R., Steinhardt, U., & Yolk, M. (Eds.). (2001). *Landscape balance and landscape assessment*. Berlin/Heidelberg: Springer.

Lovett, G. M., Jones, C., Turner, M. G., & Weathers, K. C. (Eds.). (2005). *Ecosystem function in heterogeneous landscapes*. New York: Springer.

Mikheyev, V. S. (1987). *Landscape-geographical maintenance for integrated problems of Siberia*. Novosibirsk: Nauka. (in Russian).

Mikheyev, V. S. (2001). *Landscape synthesis of geographical knowledge*. Novosibirsk: Nauka. (in Russian).

Mikheyev, V. S., & Cherkashin, A. K. (1987). Geographical fundamentals of creation and information support of mathematical models. In *Ecologo-economic systems: Models, information, experiment* (pp. 127–148). Novosibirsk: Nauka. (in Russian).

Mikheyev, V. S., & Ryashin, V. A. (1970). The principles and a technique of drawing up the map of landscapes of Transbaikalia. In *Problems of thematic mapping* (pp. 183–192). Irkutsk: Institute of Geography of Siberia and Far East SB AS USSR. (in Russian).

Nalivkin, D. V. (1933). *The doctrine of facies*. Moscow/Leningrad: Georazvedizdat. (in Russian).

Postnikov, M. M. (1988). *Differential geometry*. Moscow: Nauka.

Reteyum, A. Y. (1988). *The terrestrial worlds (on holistic studying of geosystems)*. Moscow: Mysl'. (in Russian).

Shugart, H. H. (1984). *A theory of forest dynamics. The ecological implications of forest succession models*. New York: Springer.

Sochava, V. B. (1962). Starting positions of taiga lands typification on landscape and geographical basis. *Proceedings of the Institute of Geography of Siberia and Far East SB AS USSR, 3*, 14–23. (in Russian).

Sochava, V. B. (1965). Practical sense of geographical research and concept of applied geography. *Proceedings of the Institute of Geography of Siberia and Far East SB AS USSR, 9*, 3–12.

Sochava, V. B. (1967). Structural and dynamic landscape science and geographical problems of the future. *Proceedings of the Institute of Geography of Siberia and Far East SB AS USSR, 16*, 18–31. (in Russian).

Sochava, V. B. (1972). Classification of vegetation as hierarchy of dynamic systems. In *Geobotanical mapping* (pp. 3–18). Moscow/Leningrad: Nauka. (in Russian).

Sochava, V. B. (1973). A system paradigm in geography. *Proceedings of the All-Union Geographical Society, 105*(5), 393–401. (in Russian).

Sochava, V. B. (1974a). Epilog. Problems of modern theoretical geography. In D. Harvey (Ed.), *A scientific explanation in geography* (pp. 471–481). Moscow: Progress. (in Russian).

Sochava, V. B. (1974b). Geotopology as the section of the geosystems doctrine. In V. B. Sochava (Ed.), *Topological aspects of the geosystems doctrine* (pp. 3–86). Novosibirsk: Nauka. (in Russian).

Sochava, V. B. (1978). *Introduction to the theory of geosystems*. Novosibirsk: Nauka. (in Russian).

Sochava, V. B., Krauklis, A. A., & Snytko, V. A. (1974). To unification of the concepts and terms used at complex researches of a landscape. *Proceedings of the Institute of Geography of Siberia and Far East SB AS USSR, 42*, 3–9. (in Russian).

Solnetsev, N. A. (1948). The natural geographic landscape and some of its general rules. In *Proceedings of the Second All-Union Geographical Congress* (Vol. 1, pp. 258–269). Moscow: State Publishing House for Geographic Literature. (in Russian). See also in: In J. A. Wiens, M .R. Moss, M. G. Turner, D. J. Mladenoff (Eds.) (2006), *Foundation papers in landscape ecology* (pp. 19–27). New York: Columbia University Press.

Solnetsev, N. A. (1949). On the morphology of the natural geographical landscape. *Questions of geography* (pp. 61–86). Issue 16. Moscow: Geographical Publishing House. (in Russian).

Solnetsev, N. A. (2001). *The doctrine of landscape*. Moscow: Publishing House of MSU. (in Russian).

Turner, M. G. (Ed.). (1987). *Landscape heterogeneity and disturbance*. New York: Springer.

Turner, M. G., Gardner, R. H., & O'Neill, R. V. (2001). *Landscape ecology in theory and practice: Pattern and process*. New York/Berlin/Heidelberg: Springer.

Wu, J., & Hobbs, R. (Eds.). (2007). *Key topics in landscape ecology*. Cambridge: Cambridge University Press.

Wu, J., Jones, K. B., Li, H., & Loucks, O. L. (Eds.). (2006). *Scaling and uncertainty analysis in ecology: Methods and applications*. Dordrecht: Springer.

Chapter 3
Multipattern (Polystructural) Organization of a Landscape: Geophysical Approach

Vladislav V. Sysuev

Abstract We propose the tools to describe the formation of geostructures distinguished by classical landscape analysis using parameters of force geophysical fields. The parameters of gravity and insolation fields were calculated from digital elevation models in GIS. The resulting formalized mathematical algorithms aimed at classifying the elementary and the higher order relief units acquire a fundamental geophysical meaning. In this case, the concept of a landscape multipattern organization is absolutely logical: choosing different structure-forming processes and physical parameters, one can implement different classifications of landscapes. Models of geosystem functioning are closely related to their structure through the boundary conditions and the relationship between the parameters. All models of the processes and structures are verified by field experimental data obtained under different environmental conditions.

Keywords Geosystem · Model · Morphometric parameters · Geophysical field · Flow

3.1 Common Principles of the Landscape Structure Models

The wide use of GIS technologies and measurement procedures resulted in the accumulation of a huge amount of data on the state of the Earth's surface and encouraged application of mathematical methods for their processing as well as constructing models of the phenomena under investigation. The potential for the development of classical issues in landscape studies is associated with the reliance on the physical and mathematical instruments that ensure the reliability of the results and the integration of physical geography into holistic science of nature (Dyakonov 2008).

V. V. Sysuev (✉)
Lomonosov Moscow State University, Moscow, Russia

© Springer Nature Switzerland AG 2020
A. V. Khoroshev, K. N. Dyakonov (eds.), *Landscape Patterns in a Range of Spatio-Temporal Scales*, Landscape Series 26,
https://doi.org/10.1007/978-3-030-31185-8_3

The construction of any physical and mathematical models begins with basic axioms and postulates. The principal point is the identification of elementary material objects (particles and points) forming the system and the assignment of independent variables and functions of the system states. Further, it is necessary to adopt a number of binding postulates so that it is possible to apply the physical laws of this or that degree of generality. It is important that physical laws and their parameters can be applied precisely from the point of view of describing the structure-forming landscape processes.

The unequilibrium thermodynamics shows the principles on which the classification of natural-territorial complexes (NTCs) should be based. In accordance with Onsager's bilinear equation, classifications can be carried out by: (1) system-forming flows; (2) force fields and their gradients; and (3) phenomenological coefficients (generalized conductivities).

The most common for any geosystem are the fields of gravity and insolation. The selection (as well as classification and integration) of geographical objects in terms of parameters describing geophysical fields and their gradients leads to the classification of geosystems due to the flows of matter and energy (Kleidon 2010). This is the functional approach to the identification and study of geosystems, which is consistently developing in the works of D.L. Armand, K.N. Dyakonov, A.Yu. Reteum, and other authors. The boundaries of the geosystems in this case will be determined by the magnitude and sign of the divergence of the flows. For example, if we consider the behavior of elementary volumes of water in the geopotential field, we obtain a hierarchy of catchment geosystems (river basins) corresponding to the formalized Horton–Strahler–Tokunaga schemes. The selection of a sharp change in the phenomenological coefficients leads to the classification of the NTCs by the principle of homogeneity (in accordance of N.A. Solnetsev's theory). Considering the spatial distribution of plants and animals simultaneously in the geopotential field and in other physical fields (insolation, chemical, thermodynamic, etc.), one can obtain a hierarchy of ecosystems or biogeocoenotic systems and their distribution in space. Obviously, these approaches to separating geosystems are mutually complementary and their contraposition is meaningless. For example, V.N. Solntsev (1997) considers three mechanisms of landscape structuring—geostationary, geocirculation, and biocirculation ones, either separately or simultaneously. The identification of multiscale polystructural geosystems and the boundaries between them is one of the central problems in landscape studies.

To describe geosystems, it is necessary, first, to substantiate the potentials of the main force geophysical fields that determine the structure-forming processes, and, second, to formalize the description of elementary geosystems and hierarchical invariants of geosystems. The quantitative values of the spatially distributed physical parameters of landscape state can be obtained: (1) from digital elevation models (DEM)—morphometric parameters (MPs) describing the gradients of the gravity and insolation fields; (2) from digital remote sensing data—parameters of the Earth's surface cover; (3) from measurements in nature and in laboratory conditions; and (4) in special experiments.

The space of geographical coordinates is provided by the construction of a DEM. *Pixels* of the 3D DEM *are elementary material points* (similar to the material points in theoretical mechanics), from which the NTC structure is synthesized using the formalized procedures. DEMs are built to achieve the maximum resolution of a particular hierarchical level of geosystems. For example, the construction of a DEM based on the contours from a detailed topographic map M 1:10,000 by the method of a regular grid, assume that possible pixel size can be 10 × 10 m. However, the pixel size is dictated by resolution of aerial photo or satellite image as well. So, the resolution of Landsat image 30 m allows us to identify the NTC of *urochishche*[1] level but not of the lower ones.

Differentiation of geographical space can be revealed using various mathematical methods (cluster analysis, neural networks, etc.). However, to distinguish uniform areas we need numerical parameters of the state of elementary material objects (pixels). Theoretical description of the geostructure, that is, stationary (on a certain time interval) state of the dynamic geosystem, begins with identifying MPs describing the force fields that determine the main structure-forming processes. Morphometric formalization of the Earth's surface in the gravitational field was systematized by Shary (1995). Logically meaningful association of morphometric parameters includes three groups describing: (1) the distribution of solar energy— the dose of direct solar radiation (daily and annual), aspect and illumination of slopes; (2) the distribution and accumulation of water under the influence of gravity—the specific catchment area and the specific dispersive area, depth of *B*-depressions and the height of *B*-hills, the slope gradient; and (3) the mechanisms of matter redistribution under the influence of gravity—horizontal, vertical, and mean curvature, slope gradient, and height (Sysuev 2014).

The physical meaning of the morphometric parameters is quite clear. For example, the dose of radiation characterizes the potential input of direct insolation. Slope aspect and gradient are the components of the geopotential vector gradient. Horizontal curvature is responsible for the divergence of flow lines. Vertical curvature is the derivative of the steepness factor and characterizes slope convexity/concavity. Specific catchment area shows the area from which suspended and dissolved substances can be transported to a surface element; it is a component of a number of related indices (water flow capacity index, erosion index, etc.).

The state of the Earth's surface covers (vegetation, snow, soil cover, etc.) is detected from the digital data of space spectral image and from the related indices (e.g., the normalized difference vegetation index, snow index, and humidity [water] index—NDVI, NDSI, and NDWI). The most important sources of obtaining data, as well as verifying the interpretation of the state of the covers, are field studies. In addition to the traditional integrated methods of landscape science, it is necessary to use automated complexes for recording the geophysical parameters of the near-surface layers of the atmosphere, natural waters, and soils. Hopes are inspired by the

---

[1] *Urochishche* is the term for hierarchical level of landscape morphological units (NTCs) higher than *facies* and lower than *mestnost* used in Russian landscape science. See Glossary and Chap. 1 for details.

perspective methods of applied geophysics, based on measuring the spatial distribution of gravitational, electromagnetic, and other parameters.

The choice of parameters for describing the structure is carried out in accordance with the classical approaches of landscape studies. Any formal algorithms for selecting the smallest and the higher order units of the relief surface based on parameters of the gravity and insolation fields acquire a fundamental geophysical meaning. In this case, the concept of polystructural landscapes is absolutely logical: choosing different types of physically meaningful structure-forming processes and their parameters, it is possible to implement different classifications of landscapes. Below we illustrate the approach on concrete examples.

3.2   Typological Model of the Landscape Structure

Typological approach allows obtaining a hierarchy of classical NTC (*facies—urochishche—mestnost—landscape*[2]) in accordance of N.A. Solntsev's theory (Solntsev 1948). The choice of parameters is carried out in accordance with the widely known definitions. "Elementary NTC—*facies* ... is confined to one element of mesorelief; this territory is homogeneous in terms of its three characteristics: the lithological composition of the rocks, the slope aspect and gradient. In this case, the total solar radiation and atmospheric precipitation entering the surface are the same at any part of it. Therefore, one microclimate and one water regime are formed, ... one biogeocenosis, one soil unit and a uniform complex of soil mesofauna" (Dyakonov and Puzachenko 2004, p. 22). As follows from the definition, the selection of elementary NTCs (*facies*) can be carried out by the parameters of the distribution of solar radiation and water over the surface; more precisely, by the distribution of the insolation and gravity fields gradients. Thus, the classical definition requires the description of the NTC differentiation based on the theory of field and the morphometry of the Earth's surface. The classification results essentially depend on the weight values and number of parameters. By changing the latter, it is possible to optimize the classification of landform elements by a known landscape structure. On the other hand, the change in the set of parameters and their numerical value allows one to model the change of the landscape structure under the influence of climate change, neotectonic events, etc. With this modeling, a rigorous landscape approach is needed that allows us to identify the role of the main factors of differentiation and exclude derivative or dependent variables. Automatically obtained classes of landscape cover require identification and verification of their physical content.

Verification of automatically decoded classes of vegetation cover from space spectral image was carried out based on field data (Akbari et al. 2006). Some results are given in Fig. 3.1. The investigated transect (representative of series of analogous

---

[2] *Facies, urochishche, mestnost,* and *landscape* are the terms for hierarchical levels of NTC used in Russian landscape science. See Glossary and Chap. 1 for details.

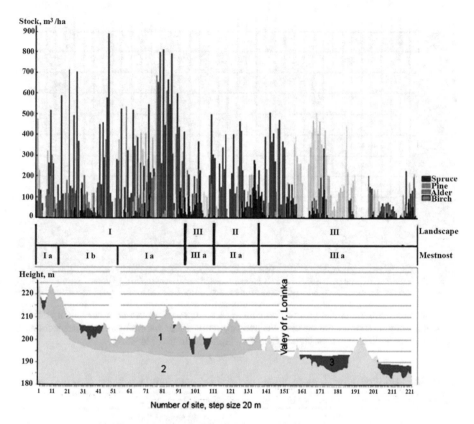

Fig. 3.1 The distribution of the stand volume (m³/ha) along transects crossing typical landscapes of the Valdai National Park: (i) landscape of the ridge-hilly moraine-kame plain on carbonate morainic loams often covered with silty-sandy loam deposits of low thickness; (ii) landscape of ridge-hilly kames-eskers plain on sandy-loam sediments; and (iii) landscape of outwash glaciofluvial plain with sand ridges. The alphabetic indices characterize the locality ("*mestnost*"). In the lower part—elevations above sea level (m) and a schematic lithological profile: (1) moraine deposits, locally overlaid by kames; (2) glaciofluvial sands; and (3) peat deposits

transects) was 5 km long; leveling was performed at 5 m interval; integrated descriptions were made at sample plots 20 m apart. Along with complex descriptions, a continuous forest inventory was carried out at 239 plots 20 × 20 m.

The investigated territory of the National Park "Valdai" is located in the central part of the Valdai Upland, which belongs to the end moraine belt of the last Valdai (Würm) glaciation in the northwestern East European Plain. The loamy moraine deposits with residual carbonate reach a thickness of 25 m in the ridges of the last Crestets glaciation stage and overlie the glaciofluvial sands of the preceding stages. Locally, the moraine is covered by kame silty-sandy loam sediments. The glaciofluvial plains are covered with peat bog sediments, the massifs of which are separated by sandy eskers. Peatland systems are connected by streams and the Loninka and

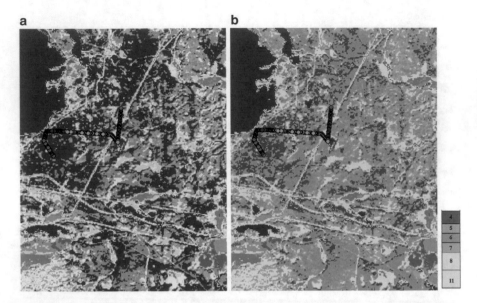

Fig. 3.2 Identification of the physical content of landscape cover classes in the Valdai National Park according to a priori information (**a**) and vegetation cover on the basis of identification by field data (**b**). See Table 3.1 for legend. The dots depict the transect location

Table 3.1 Identification of decoded classes on the continuous forest inventory along the transect

Class (Fig. 3.2a)	Number of sites	Stand
4	66	Closed spruce forests
5	59	Spruce-pine and pine forests
6	50	Mixed and spruce-pine forests
7	25	Boggy pine forests
8	22	Small-leaved and boggy pine forests
11	9	Bogs with stunted pine trees

the Chernushka rivers. Such a variety of landforms and sediments cause a high degree of biological and landscape diversity of the study area (Fig. 3.2).

The map of the automatically decoded classes (Fig. 3.2a; Table 3.1) compiled from a priori geophysical data was refined based on the field data geo-referenced to space image. Map of vegetation cover shows classes according to field-based verification (Fig. 3.2b).

The scheme of the geosystems structure (NTC at the *urochishche* level) is shown in Fig. 3.3. We used successive dichotomous grouping of landform elements (DEM pixels) based on the parameters of geophysical fields and the state of landscape cover. Independent morphometric parameters were chosen on the basis of preliminary analysis (digital modeling): the dose of direct annual solar radiation, elevation, slope gradient, horizontal, and vertical curvature, the specific catchment area, as well as numerical data of Landsat 7 spectral channels and the NDVI.

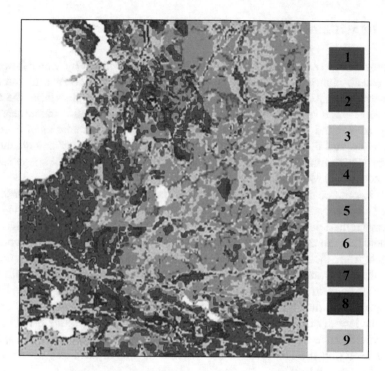

Fig. 3.3 The structure of the NTC based on the relief classification by the parameters of geophysical field gradients and the space image data Landsat-7

(1) Moraine ridges and kame hills with loamy Umbric Albeluvisols (sod-podzolic soils) under *Piceetum oxalidosum* forests; (2) summits of kame hills and ridges with sandy Umbric Podzols under *Pinetum cladinosum, Pinetum cladinoso-hylocomiosum*, and *Pinetum herbosum* forests; (3) foot slopes of the hills and flat concave hollows with Umbri-Endogleyic Albeluvisols and Umbric Albeluvisols under mixed forests; (4) river and lake terraces with Endogleic Umbrisols and Distri-Fibric Histosols under spruce forests and mixed forests; (5) dune ridges and sandy hills with Umbric Albeluvisols under pine forests; (6) flat and convex upland bogs with thick Histosols under sparse pine forests; (7) river floodplains with Endogleic Umbrisols under flooded meadows; (8) steep slopes of hills with Umbrisols under coniferous forests; and (9) anthropogenically modified and anthropogenic landscapes (roads, power lines, quarries, farmland, nurseries, and residential areas)

Revealing spatial structures in this way is a *process of synthesis*, since the material points (pixels) were integrated into elementary natural-territorial complexes (at a given scale level) according to the selected parameters of the main geophysical fields and the state of the cover. The landscape map resulting from numerical experiments reproduces the boundaries of the NTC independently obtained by the classical landscape-field method with sufficient accuracy (Sysuev and Solntsev 2006).

3.3 Functional Model of the Geosystem Structure

The functioning of low rank order geosystems is determined by water flows to a great extent. Hence, classification is aimed at the construction of a hierarchy of catchment geosystems according to the morphometric values describing the redistribution of water in the gravity field—slope gradient, the specific catchment area, and horizontal and vertical curvatures—that determine the boundaries of the divergence zones and the convergence of streamlines. The hierarchy of catchment geosystems is determined in accordance with the Horton–Strahler–Tokunaga scheme (Tarboton et al. 1991).

The automated algorithms for identifying drainage channels based on raster layers of GIS involves three main steps. First, we use the digital map of one of the abovementioned parameters to select cells with values exceeding a predetermined threshold that are treated as the potential points of the sources. At the second step, channels from the given sources are delimited, removing those sources through which the flow from higher elevation passes. At the third step, channels smaller than a certain minimum length are cut off (Fig. 3.4). The process can be adjusted easily by changing the threshold values of the drainage accumulation parameter and the minimum length of the raster layer drainage channels. The resulting array of morphometric characteristics, geo-referenced to various order catchments, is appropriate for revealing the landscape-related causes of their anomalies, which can be checked in the field.

Analysis of the catchment parameters and hydrological measurements showed a close relationship between the structure and functioning of geosystems (Chow et al. 1988). This provides opportunity to calculate the water flow discharge based exclusively on the geosystems structure and data on precipitation.

The calculation of the surface runoff from the a priori topographic DEM data can be performed in different GIS supporting hydrological procedures. For example, in

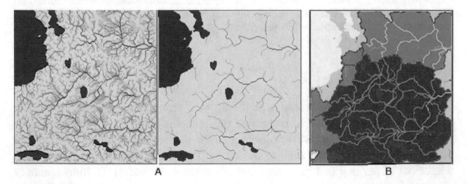

A   B

Fig. 3.4 Automated delineation of the catchment basins of the Loninka River (National Park "Valdai") in MapWindow GIS with the TauDem module (Tarboton et al. 1991). (**a**) The procedure for cutting off the drainage channels that are shorter than critical value. (**b**) Catchment areas of various order basins

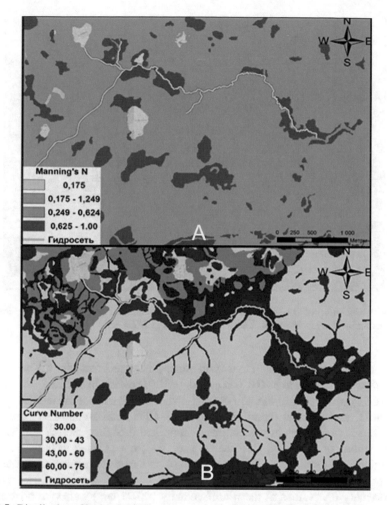

Fig. 3.5 Distribution of hydrophysical parameters in the basin of the Loninka River for modeling water flow based on the landscape structure (SAGA GIS). (**a**) Manning's n (MN); (**b**) curve number (CN)

SAGA (Olaya 2004), in order to calculate the water flow in addition to the required parameters (elevation, slope gradient, specific catchment area, etc.), one needs a sufficiently large number of individual catchment parameters such as the Shezi–Manning coefficient of surface roughness ("Manning's n"—MN) and the coefficient of soil influence on the intensity of surface runoff ("Curve number"—CN).

Values of the parameters must be assigned to each pixel of the model, which is objectively possible only with reliance upon information on the landscape structure. In Fig. 3.5, the numerical values of MN, taken from the standard Chow tables (Chow 1959), are depicted in accordance with the typological structure of the landscape (Fig. 3.3). These data form the basis for a hydrological distributed model aimed at calculating the flow velocity and discharge of the Loninka basin.

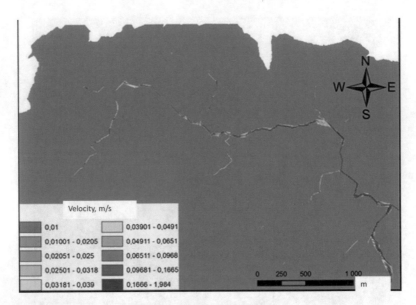

Fig. 3.6 Calculation of surface water velocity in the Loninka River basin in SAGA GIS. The precipitation intensity is 10 mm/h, CSS = 1, MFT = 180, CDT = 360

The average precipitation intensity in numerical experiments for the calculation of runoff was assumed to be 0.0, 0.66, 10.0, or 100.0 mm/h. In addition, channel site slope (CSS) parameters and runoff characteristics (slope surface, mixed flow threshold—MFT, and channel definition threshold—CDT) and some other parameters were subject to changes in calculations. Numerical modeling has shown that even tabular values of MN, CN, CSS, MFT, and CDT, not adapted to taiga wetlands, reveal significant features in the distribution of surface water runoff in various geosystems (Fig. 3.6).

Extremely low values of the runoff velocity (<0.01 m/s) were observed on the bulk of the basin. Higher velocities were observed only in the channels of streams and rivers (0.025–0.2 m/s). Velocity increases up to 2 m/s only in certain sections of the river. The pattern of velocity distribution is quite realistic, since the catchment of the Loninka River is a flat swamped hummocky sandy plain, cut by rare channels with runoff.

The results of runoff simulation using various parameters of average rain intensity (ARI) and CSS revealed some regularities. In all cases, with an increase in ARI, there was also an increase in runoff, for example, at the source point, which is located near the drainage pipe under the railway embankment (the source of the Loninka River). This site is highly modified by human activities, and as a consequence, it has low MN values (0.025), contributing to surface runoff, and high CN

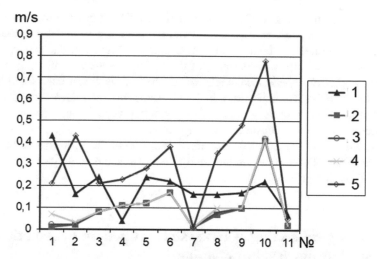

Fig. 3.7 Comparison of the measured and calculated flow velocities in the Loninka River at the gauging stations. (1) Measurement in nature; (2) calculation by model, precipitation intensity 0.0 mm/hour; (3) calculation by model, precipitation intensity 0.66 mm/h; (4) calculation by model, precipitation intensity 10.0 mm/h; (5) calculation by model, intensity of precipitation 100.0 mm/h. The abscissa shows the numbers of the gauging stations from the sources of the river

(98) interfering with moisture infiltration into the soil. In connection with this, a change in the intensity of precipitation successively leads to a change in the values of the surface runoff velocity. That is, in this section the sensitivity of the runoff to the precipitation intensity is great, although the flow rates are not very high. Most other observation points are located in natural forest and mire landscapes. These sites are characterized by high MN values (0.5–0.9), and low CN (30–40). High values of MN prevent surface runoff, and low values of CN favor active infiltration. In connection with this, the flow rates at these observation points decrease substantially and respond weakly to changes in the precipitation intensity. Thus, the differences in the landscape location of the observation points lead to significant differences in the characteristics of the runoff. Low steady flow velocity at zero precipitation intensity confirms the high capacity of the flat overhumidified catchment to accumulate water and regulate the runoff in geosystems.

Verification of model calculations was carried out in the field. Experimental measurements of velocities and discharge of the Loninka River at the gauging stations showed the following. In all cases, the predicted velocities differ from the measured ones, but the calculated values are not so far from the real ones as expected (Fig. 3.7). The closest results were obtained for a precipitation intensity of 10 mm/h, although precipitation of a lower intensity is more realistic.

More accurate simulation results can be obtained, apparently, by adjusting the values of the model coefficients (in the calculations were used tabulated values for nonwaterlogged rivers), as well as a more detailed DEM. It turned out that the channel width 1.0–1.5 m, and the pixel size 30 m does not allow us to delimit valleys, as well as the microrelief which is very important for flow from flat plains. On the other hand, errors in the measurement of flow velocities are possible in conditions of steep, boggy meandering channels, often blocked by forest debris and beaver dams. Nevertheless, in conditions of lack of information, the values obtained during modeling in GIS can serve as a basis for predicting runoff values in areas where measurements are labor consuming or impossible.

Thus, the possibility of calculating the parameters of the hydrological functioning of geosystems was shown on the basis of geophysical modeling of the structure of the NTC and runoff by a priori data.

3.4   Ecological Zoning of Geosystems Based on Information on Hydrological Functioning

The lag time for the water flow to reach the river or control gauging stations is an important environmental characteristic of processes in a catchment. To evaluate the lag time by means of the SAGA GIS we applied the calculation of the isochrone to the user-defined ending gauging stations of river (Fig. 3.8a). This makes it impossible to use isochrones to predict the time of pollutants entering from the lateral runoff to the river, which is important for taking measures to intercept pollution before entering the riverbed. We elaborated the modified cascade algorithm (Sysuev et al. 2011) in which the time for runoff for each first-order catchment to a first-order channel is calculated, and then the time of approach of each second-order channel to the merging with the next second-order channel, and so on (Fig. 3.8b).

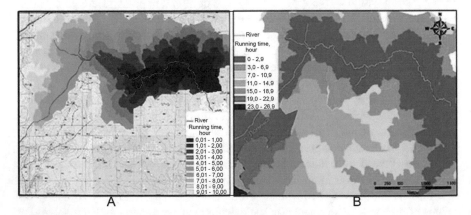

Fig. 3.8 Running time of surface runoff water, hours. (**a**) Calculation to the control section of the Loninka River, according to SAGA GIS algorithm; (**b**) Calculation to the river at each particular point of the channel, according to the cascade algorithm

3.5 Conclusion

In this chapter, we demonstrated the principle of theoretical geophysical construction of low-ranking geosystems. Landscape was assumed as a geosystem of regional dimension, consisting of genetically and functionally interconnected local geosystems formed within a uniform relief type in a local climate. Under this assumption, the deterministic linear approach of describing the structure of geosystems is fully justified. However, with increasing dimensionality (space-time scales), on the one hand, it becomes necessary to describe the complex hierarchical structure of geosystems, and, on the other hand, the nonlinearity and stochastic character of formation of specific fractal structures are increasingly manifested (Puzachenko 2014).

Acknowledgments This research was conducted according to the State target for Lomonosov Moscow State University "Structure, functioning and evolution of natural and natural-anthropogenic geosystems" (project no. AAAA-A16-116032810081-9).

References

Akbari, K. K., Bondar, Y. N., & Sysuev, V. V. (2006). Indicative properties of the stand in the landscapes of the edge zone of the Valdai glaciation. *Proceedings of Moscow University, Series 5 Geography, 6,* 59–66. (in Russian).

Chow, V. T. (1959). *Open-channel hydraulics.* New York: McGraw-Hill.

Chow, V. T., Maidment, D. R., & Mays, L. W. (1988). *Applied hydrology.* New York: McGraw-Hill.

Dyakonov, K. N. (2008). Basic concepts and intention of landscape studies. In N. S. Kasimov (Ed.), *Geographical scientific schools of Moscow University* (pp. 348–381). Moscow: Gorodets. (in Russian).

Dyakonov, K. N., & Puzachenko, Y. G. (2004). Theoretical positions and directions of modern landscape studies. In K. N. Dyakonov & E. P. Romanova (Eds.), *Geography, society and environment. Vol. II. Functioning and present-day state of landscapes* (pp. 21–36). Moscow: Gorodets. (in Russian).

Kleidon, A. (2010). Life, hierarchy, and the thermodynamic machinery of planet Earth. *Physics of Life Reviews, 7*(4), 424–460.

Olaya, V. (2004). *A gentle introduction to SAGA GIS* (Edition 1.2). Gottingen: SAGA.

Puzachenko, Y. G. (2014). Organization of a landscape. In K. N. Dyakonov, V. M. Kotlyakov, & T. I. Kharitonova (Eds.), *Issues in geography. Vol. 138. Horizons of landscape studies* (pp. 35–64). Moscow: Kodeks. (in Russian).

Shary, P. A. (1995). Land surface in gravity points classification by a complete system of curvatures. *Mathematical Geology, 27*(3), 373–390.

Solntsev, N. A. (1948). The natural geographic landscape and some of its general rules. In *Proceedings of the second all-union geographical congress* (Vol. 1, pp. 258–269). Moscow: State Publishing House for Geographic Literature. (in Russian). See also in: J. A. Wiens, M. R. Moss, M. G. Turner, & D. J. Mladenoff (Eds.). (2006). *Foundation papers in landscape ecology* (pp. 19–27). New York: Columbia University Press.

Solntsev, V. N. (1997). *Structural landscape studies.* Moscow: Faculty of Geography MSU. (in Russian).

Sysuev, V. V. (2014). Basic concepts of the physical and mathematical theory of geosystems. In K. N. Dyakonov, V. M. Kotlyakov, & T. I. Kharitonova (Eds.), *Issues in geography. Vol. 138. Horizons of landscape studies* (pp. 65–100). Moscow: Kodeks. (in Russian).

Sysuev, V. V., & Solnetsev, V. N. (2006). Landscapes of the edge zone of the Valdai glaciation: Classical and morphometric analysis. In K. N. Dyakonov (Ed.), *Landscape science: Theory, methods, regional studies, practice. Proceedings of the XI international landscape conference* (pp. 249–252). Moscow: MSU Publishing House. (in Russian).

Sysuev, V. V., Sadkov, S. A., & Erofeyev, A. A. (2011). Basin principle of functional zoning: Modeling of the structure and drainage of catchment geosystems based on a priori data. In K. N. Dyakonov (Ed.), *Landscape science: Theory, methods, regional studies, practice. Proceedings of the XI international landscape conference* (pp. 101–105). Moscow: MSU Publishing House. (in Russian).

Tarboton, D. G., Bras, R. L., & Rodriguez-Iturbe, I. (1991). On the extraction of channel networks from digital elevation data. *Hydrological Processes, 5*(1), 81–100.

Part II
How Patterns Indicate Actual Processes

Chapter 4
Representation of Process Development Laws in Morphological Pattern Laws: Approach of the Mathematical Morphology of Landscape

Alexey S. Victorov

Abstract The research is aimed at showing interrelations between quantitative parameters of the landscape morphological pattern and those of landscape forming processes based on the mathematical morphology of landscape. Landscape of thermokarst plains with fluvial erosion is a typical example of an area under a unidirectional process. We obtained the general expressions for morphological patterns describing interrelations of thermokarst and fluvial erosion processes with regularities of morphological patterns. Thermokarst lake radii obey the chi-distribution, areas obey the gamma distribution. Khasyrei radii obey the Rayleigh distribution, areas obey the exponential distribution. Landscapes with the broad development of landslides is an example of an area in a dynamic balance state. The mechanism of the relationship among dynamic parameters of geomorphological processes and the morphological pattern at any time was theoretically substantiated for a wide class of geomorphological processes which can be named the complex cyclic processes. The mathematical model of geomorphological processes, called the complex cyclic processes, was developed based on Markov chain technique. We obtained the expressions allowing us to estimate dynamic parameters of the current process with the help of quantitative characteristics of the morphological pattern from a single time slice.

Keywords Mathematical morphology · Landscape · Mathematical models · Morphological patterns · Thermokarst plains · Fluvial erosion · Complex cyclic geomorphological processes · Landslide

A. S. Victorov (✉)
Institute of Geoecology, Russian Academy of Science, Moscow, Russia
e-mail: vic_as@mail.ru

© Springer Nature Switzerland AG 2020
A. V. Khoroshev, K. N. Dyakonov (eds.), *Landscape Patterns in a Range of Spatio-Temporal Scales*, Landscape Series 26,
https://doi.org/10.1007/978-3-030-31185-8_4

57

4.1 Introduction

Numerous researchers are confronted with the need to represent process development in morphological pattern laws. The challenge is how to detect genesis as a set of processes that generate the observed morphological pattern and form a landscape itself (Nikolaev 1978; Vinogradov 1998; Victorov A.S. 2011; Abaturov 2010). On the other hand, some authors treat the problem as a matter of indication and focus on the question to what extent the up-to-date appearance of a landscape is indicative of the current process including its stages and their order (Victorov S.V. and Chikishev 1985). Both abovementioned approaches investigate quantitative peculiarities of the processes and morphological patterns. However, there is still relatively little focus on the quantitative independencies among morphological patterns and the corresponding processes is often omitted from the analysis.

This chapter deals with interrelations among quantitative parameters of the landscape morphological pattern and quantitative parameters of the corresponding processes using the concept of mathematical morphology of landscape.

4.2 Methods

The mathematical morphology of landscape is the perspective method for solving the abovementioned problem. We define it as a branch of science studying quantitative regularities (laws) of mosaics formed by land units on the Earth surface (landscape patterns) and techniques of their mathematical analysis (Victorov A.S. 1998, 2006; Kapralova 2014; Victorov et al. 2016). Hence, this field of knowledge examines quantitative aspects of landscape morphological patterns; figuratively we can call it "the landscape geometry." Mathematical models of the morphological patterns are the uniting kernel for the whole set of problems concerning the quantitative analysis of the morphological patterns. A mathematical model of a morphological pattern is a set of mathematical relationships reflecting its most essential geometric properties. The random process theory is a base for the mathematical models of morphological patterns. Our studies revealed an exciting essential independence (invariant) of the models on the local environment. In fact, the model of a morphological pattern corresponds to a genetic type of the area (Victorov A.S. 1998). The mathematical models proved to be useful for a number of land use decisions which required mathematical rationales. The main decision technique is mathematical analysis of the morphological pattern models.

4.3 Results

Let us consider two variants of situations:

1. A unidirectional process develops within the area.
2. A dynamic balance takes place within the area.

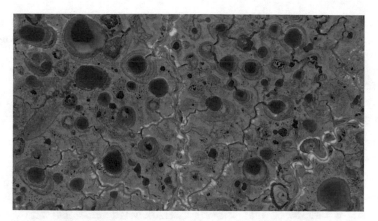

Fig. 4.1 Typical view of a thermokarst plain with fluvial erosion on a space image

The first situation results in the changing morphological pattern when some of its elements increase while other ones decrease. The second situation is a result of two opposite processes within the area when, despite specific changes of land units, the general quantitative ratio among the morphological pattern elements remains stable.

A thermokarst plain with fluvial erosion is a typical example of the first situation. It is commonly a slightly undulating subhorizontal area covered by tundra vegetation, interspersed with lakes and *khasyreis*[1] and rather rare fluvial erosion network. The lakes of isometric, often roundish, shape are randomly scattered across the plain. *Khasyreis* are also isometric flat-bottomed and flattened peaty depressions covered with meadow or bog vegetation similar to lakes randomly scattered across the plain. Most researchers expect *khasyreis* to be the result of thermokarst lake drainage usually due to fluvial erosion.

Khasyreis often have terraced stair-step topography resulted from reducing area and lowering levels of lakes (Slagoda and Ermak 2014). Some lakes can remain within a *khasyrei*; little ones around the periphery and large lakes in the central part. Within a *khasyrei* secondary permafrost and frost mounds can develop. Typical view of a thermokarst plain with fluvial erosion is given in Fig. 4.1.

Thermokarst, thermo-abrasion, and thermo-erosion processes have complex interrelations and determine the type of the area. Thermokarst depressions appear and grow independently from each other as lakes due to thermo-abrasion after being filled with water. Their growth depends on occasional factors associated with a thermic situation of a particular geological strata and soil condition. Finally, at any time, a lake can be lowered by erosion processes and transformed into *khasyrei*—the post-lake depression, overgrown with meadow, and bog vegetation with separate relict lakes. Thus, the growth of the depression stops because of lack of water. At the same time, permafrost can reappear within the *khasyrei*.

---

[1] A *khasyrei* is a drained lake.

The transition of a lake to a *khasyrei* is, of course, not instantaneous but the growth of the depression stops quickly, as far as the water body moves away from the edge of the depression.

The whole area appears to be a complex mosaic of sites, which were covered by lakes or *khasyreis* in different periods.

We consider two variants of developing morphological patterns for thermokarst plains with fluvial erosion:

* Synchronous start of thermokarst lake appearance
* Asynchronous start of thermokarst lake appearance

Our recent research (Victorov A.S. et al. 2015) has revealed that the hypothesis of the synchronous start is in better compliance with the reality. In case of synchronous start, we supposed that primary thermokarst depressions were appearing within a relatively short period compared to duration of their further development. Hence, this model of the landscape morphological pattern can be built from the following assumptions:

1. Thermokarst depressions were appearing within a relatively short period (i.e., synchronous start). This stochastic process was running independently across the different nonadjacent landscapes. At the same time, the probability of a new depression appearance within a sample plot (p_k) was dependent exclusively on the plot area (Δs):

$$p_1 = \tau_l^0 \Delta s + o(\Delta s),$$

$$p_k = o(\Delta s), \quad k = 2, 3 \ldots,$$

where τ_l^0 is an average density of lakes within the area at the starting moment of synchronous thermokarst, k is the quantity of thermokarst depressions within a sample plot.

2. Change in the radius of a primary thermokarst depression is a random process; it goes independently on other depressions (lakes), and its rate is directly proportional to the density of heat waste through the underwater side surface of the lake depression.

3. In the course of its growth, a lake can turn into a *khasyrei* after draining by the fluvial erosion network; the probability of this does not depend on other lakes. If it happens, the depression stops to grow.

4. The appearance of new fluvial erosion sources within an occasional area is a random number, which probability is directly proportional to the size of the area.

The model deals with the simplified case of relatively constant climate.

The analysis of the model allows us to obtain main expressions describing structure and dynamics of the morphological pattern of thermokarst plains with fluvial erosion in case of the synchronous start. Considering the two first assumptions of the model we get the equation for an unlimited lake growth as a radius distribution in time t (see Victorov A.S. 2007).

$$f_0\left(x,t\right) = \frac{1}{\sqrt{2\pi}\sigma x\sqrt{t}} e^{-\frac{\left(\ln x - at\right)^2}{2\sigma^2 t}} \tag{4.1}$$

A number of primary thermokarst depressions within as trial plot is described with a Poisson distribution

$$P\left(k\right) = \frac{\left(\tau_l^0 S\right)^k}{k!} e^{-\tau_l^0 S},$$

where a, σ are distribution parameters, t is time since the process has started, S is trial plot area, τ_l^0 is an average primary (at the starting moment of synchronous thermokarst) density of lakes.

Now, let us detect a distance distribution from the center of a growing depression (a focus of the process) to the closest fluvial erosion source. As we have shown earlier (Victorov and Trapeznikova 2000), from the last assumption of the model we can conclude that the distribution of the quantity of fluvial erosion sources within a random plot obeys the Poisson distribution:

$$P\left(k\right) = \frac{\left(\gamma S\right)^k}{k!} e^{-\gamma S},$$

where γ is an average density of erosion sources, S is a plot area.

We can get the distribution for a distance from the center of a growing depression (a focus of the process) to the closest fluvial erosion source, which could stop the lake growth by turning it into a *khasyrei* (so-called degenerate focus of the thermokarst process) if we take into account that probability of an event "a distance from the center of a growing *focus* to the closest erosion source more or equal x" is equal to probability of an event "a cycle of x radius from the center of the focus does not contain sources." The probability of the latter event results from the previous expression if we take that $k = 0$. Thus, coming to the complementary event, we get that the distance distribution from the center of a growing focus to the closest fluvial erosion source obeys the Rayleigh distribution

$$F\left(x\right) = 1 - e^{-\pi\gamma x^2}. \tag{4.2}$$

One of our goals is to get radii distribution for *khasyreis*. At any moment of time the distance from a lake to the closest fluvial source (Eq. 4.2) determines both the *khasyrei* radii distribution because it could stop the lake growth (see the first term under the integral), and the probability that the lake overgrows this distance (see the second term under the integral):

$$F_h\left(x,t\right)=\frac{\int_0^x 2\pi\gamma u e^{-\pi\gamma u^2}\left[1-F_0\left(u,t\right)\right]du}{\int_0^{+\infty}2\pi\gamma u e^{-\pi\gamma u^2}\left[1-F_0\left(u,t\right)\right]du},$$

where $F_0(x,\ t)$ is the radii distribution of a free growing thermokarst lake at the moment, correspondently the distribution density is

$$f_h\left(x,t\right)=\frac{2\pi\gamma x e^{-\pi\gamma x^2}\left[1-F_0\left(x,t\right)\right]}{\int_0^{+\infty}2\pi\gamma u e^{-\pi\gamma u^2}\left[1-F_0\left(u,t\right)\right]du}.$$

With time ($t\to\infty$) the *khasyrei* radii distribution tends to some limiting distribution. It is easy to see that since the distribution of lake radii with unlimited growth tends to zero for any x, then after a long time the limiting distribution density of the *khasyrei* radius is determined by the expression:

$$f_h\left(x,\infty\right)=2\pi\gamma x e^{-\pi\gamma x^2},$$

and the distribution itself results from the following expression:

$$F_h\left(x,\infty\right)=1-e^{-\pi\gamma x^2}.$$

In other words, the long time after the process has started the *khasyrei* radii distribution should obey the Rayleigh distribution. Hence, we can get the *khasyrei* area distribution (*sh*). Taking into account that:

$$F_{sh}\left(x,\infty\right)=F_h\left(\sqrt{\frac{x}{\pi}},\infty\right),$$

it is easy to conclude that the *khasyrei* area distribution obeys the exponential distribution.

$$F_{sh}\left(x,\infty\right)=1-e^{-\frac{x}{\overline{q}}},$$

where $\overline{q}$ is an average area of khasyreis.

The time distribution for thermokarst lake existence before it turns into *khasyrei* is another important dynamic characteristic of the morphological pattern. The probability that the lifetime of a thermokarst lake μ is less than given time (t) results from the probability that during this time a lake overgrows the distance to the closest fluvial source. Thus,

$$F_\mu\left(t\right)=\int_0^{+\infty}2\pi\gamma x e^{-\pi\gamma x^2}\left[1-F_0\left(x,t\right)\right]dx.$$

Radii distribution for thermokarst lakes is one more parameter of interest. At any time, the radii distribution of thermokarst lakes depends on the distribution of the corresponding radius in case of unlimited growth but under a condition that it would not turn into *khasyrei*, that is, the distance to the fluvial source is more than the radius of the lake. Thus, the radii distribution for thermokarst lakes is:

$$f_l(x,t) = \frac{f_0(x,t) e^{-\pi \gamma x^2}}{\int_0^{+\infty} f_0(x,t) e^{-\pi \gamma x^2} dx}.$$

Using the equation of the unlimited growth (4.1) and simplifying it due to the same terms in the numerator and denominator we get

$$f_l(x,t) = \frac{x^{\frac{a}{\sigma^2}-1} e^{-\pi \gamma x^2} e^{-\frac{\ln^2 x}{2\sigma^2 t}}}{\int_0^{+\infty} x^{\frac{a}{\sigma^2}-1} e^{-\pi \gamma x^2} e^{-\frac{\ln^2 x}{2\sigma^2 t}} dx},$$

where a, σ are distribution parameters.

After the long time ($t \to +\infty$) of growth this equation tends to its limit distribution:

$$f_l(x,\infty) = \frac{x^{\frac{a}{\sigma^2}-1} e^{-\pi \gamma x^2}}{\int_0^{+\infty} x^{\frac{a}{\sigma^2}-1} e^{-\pi \gamma x^2} dx},$$

which is a well-known chi-distribution. Thus, taking into account of roundish shape of lakes, we get the expression for the distribution density of lake area (*sl*):

$$f_{sl}(x,\infty) = \frac{1}{2\sqrt{\pi x}} f_l\left(\sqrt{\frac{x}{\pi}},\infty\right),$$

from which follows that the limiting distribution for the lake area is the gamma distribution:

$$f_{sl}(x,\infty) = \frac{x^{\frac{a}{2\sigma^2}-1} e^{-\gamma x}}{\int_0^{+\infty} x^{\frac{a}{2\sigma^2}-1} e^{-\gamma x} dx}.$$

The spatial distribution of lakes obeys the Poisson law (just like in the model of lacustrine thermokarst plains (Victorov A.S. 2006) during the whole time of its development. It comes from the first assumption of the model. However, the average density distribution of lakes is decreasing though all the time since lakes evolve to *khasyreis*. It depends on distribution of the lakes lifetime, which we have got above.

In fact, this distribution is a probability that a lake evolves to *khasyrei*, and the average density distribution is:

$$\tau_l(t) = \tau_l^0 \int_0^{+\infty} 2\pi\gamma \, xe^{-\pi\gamma x^2} F_0(x,t) \, dx.$$

Correspondently, this is the *density distribution of khasyreis:*

$$\tau_h(t) = \tau_l^0 \left[1 - \int_0^{+\infty} 2\pi\gamma \, xe^{-\pi\gamma x^2} F_0(x,t) \, dx \right].$$

We conducted empirical testing of some obtained results.

According to the suggested theoretical model radii and, hence, diameters of *khasyreis* should obey the Rayleigh distribution, while their areas should obey the exponential distribution. We checked these conclusions at several key sites using Pearson's chi-squared test. The first site with predominantly sea sediments located in Taz peninsula; the second and third sites belong to Yamal peninsula. The second site stretches northeastward in the center of the peninsula between the Yasaveyakha and Nyadeyseyakha rivers. Marine sediments (mainly sands) form its surface. The third site occupies the eastern seaside of the northern part of the peninsula. Alluvial and marine sands of the second terrace form its surface. Figure 4.2 represents an overview scheme with the key areas arrangement.

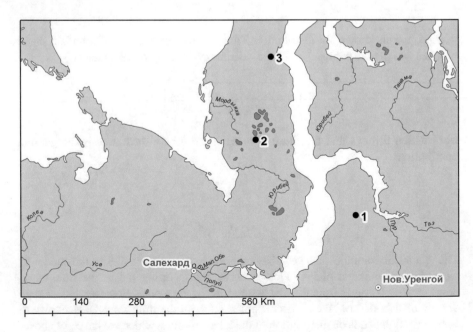

Fig. 4.2 An overview scheme of the key areas for thermokarst plains with fluvial erosion

Verification of the *khasyrei* diameter distribution similarity with the theoretical Rayleigh distribution by Pearson's test for a key site at Taz peninsula gives 8.769 at the critical value 19.675. By this, our hypothesis for the type of its distribution is confirmed.

We tested all the key sites (Fig. 4.3) for compliance between the *khasyrei* area distribution and theoretical exponential distribution with the number of samples ranging from 73 to 122. The analysis showed the similarity between theoretical and empirical distributions at p value 0.95. There is no similarity with some other common distribution at the same p value.

At the same key sites we collected empirical data for area distribution of thermokarst lakes with number of samples ranging from 53 to 59. We studied similarity between the empirical data and the gamma distribution, as is true for the synchronous start. The obtained data show that empirical data do not contradict the established model at p value 0.99 (Table 4.1). At the same time, we have found a certain closeness to the lognormal distribution.

Thus, based on the theoretical analysis and obtained empirical data we managed to get general mathematical expressions describing the structure and dynamics of

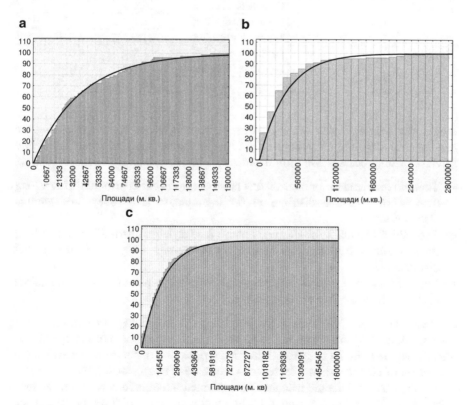

Fig. 4.3 An example of the similarity between empirical and theoretical exponential distributions for the area of the khasyreis. (**a**—key site 1, **b**—key site 2, **c**—key site 3)

Table 4.1 The similarity between theoretical and empirical distributions

Area of *khasyreis*

Key site	Sample volume	Type of the distribution	χ^2	$\chi^2_{0.95}\left(\chi^2_{0.99}\right)$
Key site 1	73	Exponential distribution	8.697	12.592
		Lognormal distribution	11.372	11.070
		Normal distribution	20.319	11.070
Key site 2	122	Exponential distribution	7.315	14.067
		Lognormal distribution	6.661	12.592
		Normal distribution	149.362	11.070
Key site 3	76	Exponential distribution	1.324	7.815
		Lognormal distribution	8.750	5.991
		Normal distribution	58.936	9.488

Area of lakes

Key site	Sample volume	Type of the distribution	χ^2	$\chi^2_{0.99}$
Key site 1	95	Gamma distribution	5.799	6.635
		Lognormal distribution	2.848	9.210
		Normal distribution	27.423	9.210
Key site 2	53	Gamma distribution	6.317	9.210
		Lognormal distribution	6.464	6.635
		Normal distribution	149.362	15.086
Key site 3	93	Gamma distribution	3.802	6.635
		Lognormal distribution	10.807	11.341
		Normal distribution	27.350	18.475

the morphological patterns of thermokarst plains with fluvial erosion in case of the synchronous start and correspondently interrelation between the thermokarst and fluvial erosion processes. We formulated laws of the morphological pattern, such as:

- The radii distribution for thermokarst lakes obeys the chi-distribution after a long time, while the area distribution for thermokarst lakes obeys the gamma distribution;
- The radii distribution for *khasyreis* obeys the Rayleigh distribution after a long time, while the area distribution for *khasyreis* obeys the exponential distribution;
- The distribution of a number of lakes and *khasyreis* (separately) within a random trial plot obeys the Poisson distribution.

Hence, using the approach of the mathematical morphology of landscape, we revealed mutual interrelations of the processes and laws of the morphological pattern development in the case of unidirectional processes (a constant increase in the proportion of khasyreis and a decrease in the share of thermokarst lakes).

For the case of dynamic balance, we examined interrelations between the processes and laws of the corresponding morphological pattern development on the example landscapes with the broad development of landslide process, which is a cyclic one.

We used the term "cyclic processes" for those exogenous geomorphological processes which are characterized with the repeating sequence of changes. This process is not strictly periodical, but with some features of periodicity. These processes include landslides, snow slips, and some others where activation of the process is followed by the stage of its recovery, such as microrelief recovery, vegetation, and soil cover recovery. These processes are characterized with many foci like landslide bodies, for instance, each of them undergoes independent quasiperiodic changes with alternation of activations and recoveries.

Let us consider an area with homogeneous environment and development of the cyclic geomorphological process like landslides. Let the activation of the process occur under the influence of two groups of factors: local factors independently activating separate sites (we call it conditionally "the local activation") and a factor acting within the whole area (we call it conditionally "the heavy activation").

The first group of factors might include local rains and local ice melting inside permafrost rocks. The second group of factors might include earthquakes, an abnormal increase in air temperature as well as abnormal liquid or solid precipitation throughout the study area. Under the "heavy activation" factor impact actual activation of the process at a certain site can be either take place or not. This process is referred to as complex cyclic process.

Within the areas under the process influence (i.e., foci), a certain change in the microrelief, soil, and vegetation cover takes place due to their restoration between activation periods. Accordingly, we can observe changes of nature units (modifications of natural units) corresponding to the process stages.

In this case, the time of the last activation determines the stage of soil and vegetation cover transformation as well as the nature unit as a whole, that is, it determines the characteristics of the nature unit. Suppose that the researcher has got the information about the date of the last activation according to the appearance of nature units in different stages of restoration, data on the landscape morphological pattern, dendrochronology, geology, geobotany indicators, soil. The task is to get dynamic parameters of the process, to make a probability forecast of the next activation for a selected site.

Let us take the interval between neighboring local activations of the focus of the process as a sequence of random variables (ξ_k, $k = 1, 2...$), characterized with a certain distribution ($q(i)$, $i = 1, 2...$, $q(i) > 0$) which is stable in time. For different activation cycles, these variables are independent due to influence of the local factors. Similarly, we take the interval between adjacent heavy activations of the process as a sequence of random variables (μ_k^0, $k = 1, 2...$), characterized with a certain distribution ($r^0(s)$, $s = 1, 2...$, $r^0(s) > 0$), which is stable in time and differs from the first one. At the same time, for different activation cycles these random variables are also independent.

The model is based on the following assumptions (Fig. 4.4):

1. Local activation of different sites does not depend on each other, and its distribution is stable in time ($q(i)$, $i = 1, 2...$, $q(i) > 0$).

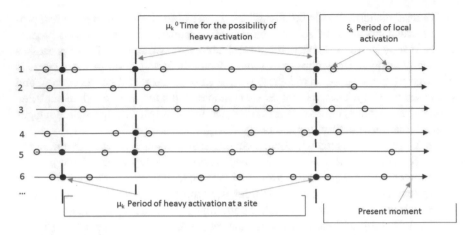

Fig. 4.4 A scheme of cyclic process activations
○—local activation,
●— heavy activation,
|— time for the possibilty of a heavy activation,
2—the number of a site (focus) with the development of the process (explanation in the text).

2. Distribution of the period of possibility for a heavy activation is stable in time
   ($r^0(s)$, $s = 1, 2...$, $r^0(s) > 0$), while actual activation at a certain site does not
   depend on the other sites and has a constant probability (α).

Let us determine the distribution of a period of actual activation at a site due to
the possibility of heavy activation. It does not coincide with the distribution of possibility of heavy activation, because the actual activation is not oblique in this case.

The equation provides the opportunity to establish the time of the actual activation due to the possibility of heavy activation:

$$\mu = \mu_1^0 + \mu_2^0 + \cdots + \mu_v^0, \quad v = 1, 2...,$$

where v is a number of periods between possible heavy activation, within a single
period between actual activations due to the possibility of heavy activation at a
given site.

In other words, the time of an actual activation due to the possibility of heavy
activation is a sum of a random number of random variables μ_k^0, $k = 1, 2...$.

It is easy to get from the second assumption that the value v of the number of
periods between the possible heavy activation occurring per period between actual
activations obeys the geometric distribution:

$$p_v(k) = (1 - \alpha)^{k-1} \alpha,$$

where α is the probability of actual activation in case of the possibility of heavy
activation.

Distribution of a period of an actual activation ($r(k)$, $k = 1, 2...$) at a site, due to the possibility of heavy activation (μ_k, $k = 1, 2...$), can be obtained using generating functions (e.g., Korolyuk et al. 1985).

The analysis shows that the general model of the process described above can be regarded as the Markov chain, and it can be the base for the solution of a general problem.

We call a pair of numbers(i, s), i, $s = 0, 1, 2...$, $i \leq s$, a possible activation of a cyclic process, where s is time since the last heavy activation at a given site, i is time since the last local activation. Taking into account the model assumptions we get three variants of transition:

$(i - 1, s - 1) \Rightarrow (i, s)$, $i, s = 0, 1, 2...$, $i \leq s$—duration of the recovery process (probability p_1),

$(i - 1, s - 1) \Rightarrow (0, s)$, $i, s = 0, 1, 2...$, $i \leq s$—local activation (probability p_2),

$(i - 1, s - 1) \Rightarrow (0, 0)$, $i, s = 0, 1, 2...$, $i \leq s$—heavy activation (probability p_3).

The probability of the first transition is, in fact, the probability that there would not be any activation at the next step (i,s) in case it did not occur before. The probability of the second transition can be determined if we take into account the necessary absence of the heavy activation.

We have estimated analytically the probabilities for every transition. From the equations it is obvious that the transition probabilities do not depend on time. Hence, the chain under investigation is the Markov chain with a countable number of states.

Using the above transition probabilities, we can write the expressions for changing states.

So, the probability $p_0(i, s, t)$ at the moment after the state $(i - 1, s - 1)$ is estimated by the expressions:

$$p_0\left(i,s,t\right) = p_0\left(i-1,s-1,t-1\right)\frac{G(s+1)}{G(s)}\frac{D(i+1)}{D(i)},\qquad(4.3)$$

Expression gives the probability of local activation at moment t:

$$p_0\left(0,s,t\right) = \frac{G(s+1)}{G(s)}\sum_{i=0}^{s-1}p_0\left(i,s-1,t-1\right)\left[1-\frac{D(i+2)}{D(i+1)}\right]\qquad(4.4)$$

and the probability of heavy activation at the moment is estimated by the expression:

$$p_0\left(0,0,t\right) = \sum_{s=1}^{+\infty}\left[\left[1-\frac{G(s+1)}{G(s)}\right]\sum_{i=0}^{s-1}p_0\left(i,s-1,t-1\right)\right],\qquad(4.5)$$

where $G(s) = \sum_{k=s}^{+\infty}q(k)$, $D(i) = \sum_{k=i}^{+\infty}r(k)$, $p_0(i, s, t)$ is a probability of event at a certain state at the moment.

Let us find the characteristics of the chain under consideration. We can show that the transition between any two states is possible in it with positive probability, that is, any states are communicating, and accordingly, the chain is irreducible. We also can show that return to any state is possible in one step, and the whole chain is aperiodic. At least, it is easy to show that the average activation period is finite. Therefore, the chain is recurrent. Irreducible, aperiodic, positively recurrent Markov chain is an ergodic one due to the corresponding theorem (Korolyuk et al. 1985). Hence, there are the probability limits in any state after a long time of the process development (final probabilities).

Final probabilities are in fact a share of areas comprised by the process under consideration (foci) with different external appearance (microrelief, soil, vegetation cover, etc.) because they have different times of the last local and heavy activation. In other words, this is a share of the plots at different stages of recovery. The final probabilities show that despite permanent changes the landscape under consideration preserves the quantitative relationships between the various developmental areas of the process, which are at different stages of recovery, that is, in a state of dynamic balance.

Indication opportunities provided by the suggested model were studied as well. To indicate dynamics of current processes one needs to know the distribution of the activation periods as the necessary parameter, since it provides all the information about the dynamics of the process under consideration, including the information which is needed for forecasting on a probabilistic basis. From Eqs. (4.3)–(4.5) passing to the limit at $t \to +\infty$ and taking the ergodicity into account, we get:

$$p(i,s) = p(i-1,s-1)\frac{G(s+1)}{G(s)}\frac{D(i+1)}{D(i)},$$

$$p(0,s) = \frac{G(s+1)}{G(s)}\sum_{i=0}^{s-1}p(i,s-1)\left[1 - \frac{D(i+2)}{D(i+1)}\right],$$

$$p(0,0) = \sum_{s=1}^{+\infty}\left[\left[1 - \frac{G(s+1)}{G(s)}\right]\sum_{i=0}^{s-1}p(i,s-1)\right],$$

where $p(j, r), j, r = 0, 1, 2..., j \le r$ are final probabilities.

Hence, after transformations, we can obtain the expression that describes the distribution of the heavy activations period:

$$r(s) = G(s) - G(s+1) = \frac{\sum_{k=0}^{s-1}p(k,s-1) - \sum_{k=0}^{s}p(k,s)}{p(0,0)},$$

and the distribution of the local activations period is described by the expression:

$$q(s) = D(s) - D(s+1) = \frac{p(s-1,s-1)}{\sum\limits_{k=0}^{s-1} p(k,s-1)} - \frac{p(s,s)}{\sum\limits_{k=0}^{s} p(k,s)}.$$

Thus, the quantitative analysis of the morphological pattern of a landscape and its components enables us to estimate these probabilities; they are physiognomic elements, and, using resulting equalities one can determine distribution of the activation periods.

Now we have already made the first steps in empirical testing of this model. The landscape in the vicinity of Seattle (USA) was taken as the example of an area under landslide process.

The formation and activation of landslides are mainly confined to the unconsolidated sediments of Pleistocene–Holocene. Lacustrine-glacial deposits are dense silty clays with inclusions of small lenses of fine-grained sands, often covered with fluvioglacial fine- and medium-grained sands with local gravel inclusions. Quaternary deposits include alluvial gravelly sands, deluvial loam and sandy loam, beach deposits, colluvial deposits, as well as anthropogenic deposits. Landslide activations are determined, in particular, by precipitation regime, groundwater, and infiltration. The Seattle landslides can be divide into four groups: the group of shallow colluvial landslides (slip landslides) is the most numerous and interesting for our study.

An extensive database was created (Shannon & Wilson Inc. 2000), containing information on several hundred landslides, as a result of monitoring observations lasting for about 100 years. 1050 landslides are recorded in the database, 1326 cases of all landslide activation for all periods of observation, and 263 cases of repeated landslide activations. Landslide activations within the territory occur both under the local factors and under the influence of storm phenomena typical for the area. The possibility of synchronous large-scale landslide activations appears during storms.

After the data processing, we obtained the distribution of landslide activation time (Fig. 4.5). The graph shows vertical segments on the curve that corresponds to the heavy activation of landslides because of storms (the exponential distribution curve is also shown for comparison).

Figure 4.6 shows the distribution of the activation period without heavy activation.

Further empirical verification is going on. Thus, as a result of the predominantly theoretical investigation, we came to the following conclusions:

- The mechanism of the relationships between dynamic parameters of geomorphological processes and the territory morphological pattern at any time ("a single time slice") was theoretically substantiated for a wide class of geomorphological processes which can be named the complex cyclic processes.
- The mathematical model of geomorphological processes, called the complex cyclic processes, was developed on a base of Markov chain technique.

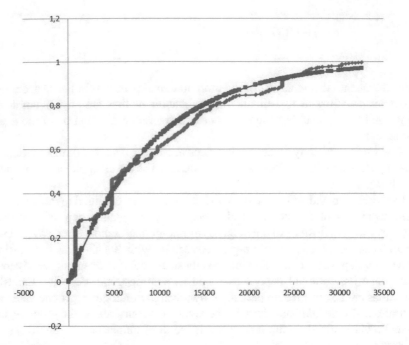

Fig. 4.5 Distribution of landslide activation time (blue line) and a graph of exponential distribution (red line) (sample size 897); the ledges respond to heavy activation under the influence of strong storms (on the abscissa scale, the time in days from the beginning of observation)

• The expressions were obtained allowing us to estimate dynamic parameters of the current process using quantitative characteristics of the morphological pattern from a single time slice.

Thus, using the approach of the mathematical morphology of landscape we revealed interrelations of the processes and laws of the morphological pattern development in case of dynamic balance of the morphological pattern development, taking as an example landscapes with the broad occurrence of landslides. This consideration still theoretical to a large extent; we plan to perform empirical verification in the course of our future research.

The obtained regularities (laws) have both scientific and practical use. One can use them for thermokarst impact risk assessment for a linear object such as gas pipelines, oil pipelines, and. We suggested such assessment for lacustrine thermokarst plains (Victorov 2007).

Within a lacustrine thermokarst plain, we tested the site with the homogeneous physiographic environment, especially its topography and permafrost features. In this case, the impact probability of a linear structure of the length L with at least one thermokarst focus (lake) is equal to:

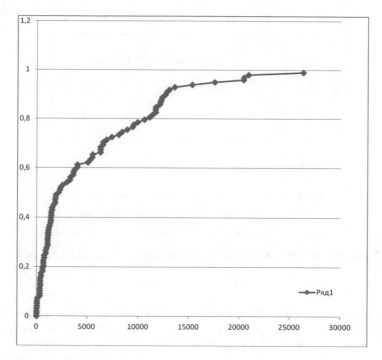

Fig. 4.6 Distribution of landslide activation period without heavy activation (sample size 118, on the abscissa the time in days)

$$P_{dl}(L,t) = 1 - e^{-\mu(t)\overline{pr}(t)L}.$$

where $\overline{pr}(t)$ is the mathematical expectation of the projection of the thermokarst focus (lake) at the time t on the direction perpendicular to the linear structure; $\mu(t)$ is an average number of foci per unit area at the time t. If the lakes have a roundish shape the formula is following:

$$P_{dl}(L,t) = 1 - e^{-\mu(t)\overline{d}(t)L},$$

where $\overline{d}(t)$ is an average diameter of a thermokarst lake at time t. Experimental data prove the suggested solution (Kapralova 2014; Victorov A.S. et al. 2016) (Fig. 4.7).

4.4 Conclusions

Thus, using the mathematical morphology of landscape, we can make a new step in solving the problem of reflection of laws of process development in morphological pattern laws as well as in quantifying regularities (laws) by strict mathematical analysis of morphological pattern models.

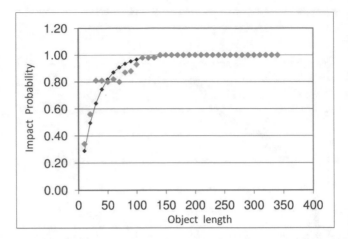

Fig. 4.7 Comparison of theoretical and empirical thermokarst impact probabilities for a linear structure

We obtained new quantitative regularities both for process development and the structure of morphological patterns, as well as the interrelations between quantitative dynamic parameters of processes without long-term stationary observations, but by processing of the corresponding landscape metrics from the one-time section.

Acknowledgments The research was financially supported of Russian Scientific Foundation (project 18-17-00226).

References

Abaturov, B. D. (2010). Microrelief of depressions of the Caspian lowland and the mechanisms of its formation. *Arid Ecosystems, 16*(5), 31–45. (in Russian).

Kapralova, V. N. (2014). *Regularities of the development of thermokarst processes within the lacustrine-thermokarst plains based on approaches of the mathematical morphology of the landscape*. Ph.D. thesis. Moscow: IEG RAS. (in Russian). http://nauchkor.ru/pubs/zakono-mernosti-razvitiya-termokarstovyh-protsessov-v-predelah-ozerno-termokarstovyh-ravnin-na-osnove-podhodov-matematicheskoy-morfologii-landshafta-59be91ef5f1be704abd32346. Accessed 4 Aug 2018.

Korolyuk, V. S., Portenko, N. I., Skorokhod, A. V., et al. (1985). *A handbook on probability theory and mathematical statistics*. Moscow: Nauka. (in Russian).

Nikolaev, V. A. (1978). *Classification and small-scale mapping of landscapes*. Moscow: MSU Publishing House. (in Russian).

Seattle Landslide Study. (2000). Seattle: Shannon & Wilson Inc. http://www.seattle.gov/dpd/cs/groups/pan/@pan/documents/web_informational/dpdp025740.pdf. Accessed 4 Aug 2018.

Slagoda, E. A., & Ermak, A. A. (2014). Interpretation of exogenous processes in typical tundra of the Yamal Peninsula (case study of the district in the middle Yuribey river). *Proceedings of Tyumen State University, Series Earth Science, 4*, 28–38. (in Russian).

Victorov, A. S. (1998). *Mathematical morphology of a landscape*. Moscow: Tratek. (in Russian).

Victorov, A. S. (2006). *Basic problems of mathematical landscape morphology*. Moscow: Nauka. (in Russian).

Victorov, A. S. (2007). Risk assessment based on the mathematical model of diffuse exogenous geological processes. *Mathematical Geology, 39*(8), 735–748.

Victorov, A. S. (2011). Landscape metrics as a reflection of landscape dynamics: Mathematical models. In J. Lechnio (Ed.), *The problems of landscape ecology. Vol. XXX. Four dimensions of landscape* (pp. 15–22). Warsaw: Warsaw University.

Victorov, S. V., & Chikishev, A. G. (1985). *Landscape indication*. Moscow: Nauka. (in Russian).

Victorov, A. S., & Trapeznikova, O. N. (2000). The erosion plain mathematical model as a base for space images interpretation methods in geoenvironmental research. In *Proceedings of the 4th international symposium on environmental geotechnology and global sustainable development* (pp. 603–612). Boston: Lowell.

Victorov, A. S., Kapralova, V. N., Orlov, T. V., et al. (2015). An analysis of the morphological structure development of the thermokarst-lake plains on the base of the mathematical model. *Geomorphology, 3*, 3–13. (in Russian).

Victorov, A. S., Kapralova, V. N., Orlov, T. V., et al. (2016). *Mathematical morphology of cryolithozone landscapes*. Moscow: RUDN Publishing House. (in Russian).

Vinogradov, B. V. (1998). *Fundamentals of landscape ecology*. Moscow: GEOS. (in Russian).

Chapter 5
Transformation of the Chernobyl ^{137}Cs Contamination Patterns at the Microlandscape Level as an Indicator of Stochastic Landscape Organization

Vitaly G. Linnik, Anatoly A. Saveliev, and Alexander V. Sokolov

Abstract The issues of assessing heterogeneous structure of the microlandscape are considered using the example of two sites—in a semihydromorphic and automorphic condition. The research is based on the estimation of the ^{137}Cs pattern transformation by the geostatistical analysis, as well as by the simulation of the relationships between radionuclide contamination and the microrelief parameters. We found evidence that the intensity of the ^{137}Cs patterns transformation increases in semihydromorphic conditions as soil hydromorphism increases. Because of the initial heterogeneity of the microlandscape, a variogram analysis was applied to examine the ^{137}Cs distribution patterns. We identified two scale levels of ^{137}Cs contamination patches: 20–30 m and 1.5–2.0 m. In automorphic environments of the interfluve area, a weak transformation of the ^{137}Cs patterns was found, which, in contrast to that of the semihydromorphic site, remains spatially uncorrelated. We analyzed the validity of the linear (LM) and nonlinear general additive models (GAM), which were built to establish the relationships between ^{137}Cs patterns and microrelief parameters measured in two grid systems with resolution 0.1 and 0.25 m. Transformation of ^{137}Cs patterns at different scale levels serves as a relevant tool for analyzing the stochastic self-organization of landscape structures, where the component relations inside are nonlinear.

V. G. Linnik (✉)
Vernadsky Institute of Geochemistry and Analytical Chemistry, Moscow, Russia

Geographical Department, Lomonosov Moscow State University, Moscow, Russia
e-mail: linnik@geokhi.ru

A. A. Saveliev
Institute of Environmental Sciences, Kazan Federal University, Kazan, Russia

A. V. Sokolov
Kharkevich Institute for Information Transmission Problems, Russian Academy of Sciences, Moscow, Russia

© Springer Nature Switzerland AG 2020
A. V. Khoroshev, K. N. Dyakonov (eds.), *Landscape Patterns in a Range of Spatio-Temporal Scales*, Landscape Series 26,
https://doi.org/10.1007/978-3-030-31185-8_5

Keywords Microlandscape · Chernobyl · Patterns · Self-organization ·
Geostatistics · Semivariogram

5.1   Introduction

A high degree of variability of the soil cover structure over short distances, even in
small test areas, is typical for forest soils (Fridland 1972; Karpachevsky 1977). One
of the reasons for such variability in a number of cases can be attributed to spatial
heterogeneity of environmental factors controlling the process of soil formation
(Phillips and Marion, 2005). Since soils reflect the interaction of various natural
factors, such as geology, climate, hydrology, topography, and vegetation, the consideration of different scale heterogeneity of natural components can be the key to
understanding and interpreting the formation and functioning of landscape systems
at various hierarchical levels of organization.

A fine-scale approach to the study of intralandscape geosystems (*urochishche*
and *podurochishche*[1] levels) allows disclosing features and details of internal
landscape evolution as a result of in situ self-organization processes, along with
climatic fluctuations and pH conditions that are modified at the local level by micro-
and mesorelief forms (Dyakonov et al. 2008). A special function of microrelief in
the formation of the structure of the soil cover of forest landscapes was emphasized
in earlier publications (Karpachevsky and Striganova 1981).

The problem of heterogeneity (variability) of an elementary unit in landscape
science was outlined in (Dyakonov and Linnik 2017). In the cited paper, the authors
suggested that the paradigm of heterogeneity should be developed as a criterion to
discover a hierarchy of landscape patterns at various scale levels of organization.
This paradigm is to be applied along with the well-known paradigm of *homogeneity*
of elementary landscapes adopted in classical landscape studies.

The main goal of our study is to test a working hypothesis (model) on mechanisms of self-organization of landscape heterogeneous systems, based on the theory
of random fields. To determine the functional structure of the landscape facies at the
microlandscape level, the effect of the relief parameters on the transformation of
the ^{137}Cs contamination field was investigated. The microlandscape functioning was
examined with focus on lateral migration of ^{137}Cs, which is extremely heterogeneous in forest landscapes.

---

[1] *Urochishche, podurochishche* and *facies* are the Russian terms for hierarchical levels of a landscape morphological units. See Glossary and Chap. 1, this volume, for details.

5.2 Self-Organization of Nonlinear Hierarchical Landscape Structures

The notion of the hierarchical structure of a landscape assumes that at a certain scale level the landscape system encompasses interacting components (objects of lower hierarchical levels), being itself at the same time a component of a larger system (i.e., a higher level of the hierarchy). The concept of hierarchy as applied to landscape studies implies that the behavior of the landscape system actually has predefined limitations concerning (1) the potential interaction of its components that form the landscape structure and (2) the constraints imposed by higher-level systems (O'Neill et al. 1989).

The issue of self-organization in natural systems is highly debated in various disciplines related to landscape studies, for example, in pedology (Targulian and Krasilnikov 2007). Soil, as an independent natural body in the landscape, from the standpoint of the dynamical systems theory, falls in the category of nonlinear systems in which the output parameters of the system (energy and matter flows) are not proportional to the input parameters. Soil plays the critical role in the formation of nonlinear stochastic structures in a landscape. Soil, as a powerful biogeochemical reactor (Richter 1987; Targulian and Sokolova 1996), is structured in the process of evolution resulting in the formation of genetic horizons, and radically changes the intensity of biogeochemical flows in a landscape.

Examples of apparent nonlinearity of flows in the landscape can be observed while examining surface runoff or soil infiltration, which is not proportional to the amount of precipitation (Phillips 1998). An analogous nonlinearity with respect to input parameters occurs in the formation of the humus layer or the productivity of phytocoenosis. Natural landscapes are by their nature nonlinear systems, which allows for the complexity of their behavior, which is impossible in linear systems (Phillips 2003).

The process of landscape development (i.e., formation of the structure of a natural complex and intracomponent relationships) can be viewed as a process of self-organization of a stochastic uncorrelated field, which, because of random local fluctuations, tends to form correlated stable structures having different characteristic times (ranging from years to thousands of years).

5.3 Landscape as a Stochastic System

Spatial variability of landscape variables is associated with the interaction of a complex of factors (causes) that operate for a long time. The fundamental peculiarity of the radioactive contamination (as a factor) lies in its performance as a single case. However, after ^{137}Cs entered the landscape, the processes of its "embedding" into the existing landscape structure began.

According to Burrough and McDonnell (1998), landscape variables considered from the standpoint of the theory of random processes can be represented as the sum of three components at the point s in the following form:

$$Z(s) = Z*(s) + \varepsilon'(s) + \varepsilon'' \tag{5.1}$$

where $Z*(s)$ is a deterministic part of the random variable $Z(s)$, $\varepsilon'(s)$ is a spatially correlated component of the random variable, and ε'' is a purely random variable, uncorrelated, called "white noise." The parameter s characterizes location, in other words, spatial coordinates (x, y) of the modeled object. $Z*(s)$ usually characterizes the trend and is, in fact, a parameter that identifies a higher scale level of organization of landscape structures.

The following part of the random variable $Z(s) - \varepsilon'(s)$ is more interesting, since analysis of it alone can provide the maximum possible information about the internal system integration of natural parameters under analysis. The variable ε'', according to the accepted definition, does not affect the internal organization of the landscape structure and thus can be ignored.

In practice, the spatial distribution of the random variable $Z(s)$ is estimated using a semivariogram calculated according to formula (5.2):

$$\gamma(h) = 0.5E\left[Z(x+h) - Z(x)\right]^2 \tag{5.2}$$

In the two-dimensional case, the random variable Z denotes the point with the coordinates (x_1, x_2) and h is the lag between the points between which the proximity is estimated according to Eq. 5.2. The measure of proximity (similarity) between points can be computed in all the indicated directions. Hence, the variogram $\gamma(h)$ is a function of the modulus of the vector h and its orientation. For a fixed direction, the variogram $\gamma(h)$ shows how the correlation of the random variable Z varies with an increasing distance between points. If we investigate the behavior of the variogram $\gamma(h)$ in different directions, then it can be used to determine anisotropy of spatial dependencies of the object under investigation.

The degree of an increase in the value of the variogram $\gamma(h)$ with the distance shows how quickly the positive connection between the individual points of the microlandscape decreases. If the variogram $\gamma(h)$ reaches the limiting value sill, the correlation between the points becomes zero; if the sill is exceeded, the correlation may decrease and even take negative values (Gringarten and Deutsch 2001). The distance from which the sign of correlation changes from positive to negative is called a range.

The presence of a sill assumes that covariance is stationary, when the covariance exists and depends only on the vector h (the second-order stationarity hypothesis). The variogram models $\gamma(h)$ without a sill correspond to the case when the size of a modeling domain is smaller than that of the range. A detailed description of geostatistics with reference to ecological and geographic data is presented in publications (Bellehumeur and Legendre 1998; Hengl 2007).

We propose an algorithm for analyzing the estimation of the rate of self-organization at a microlandscape level, using the transformation of ^{137}Cs patterns of "Chernobyl" deposition. The theoretical assumption discussed in this paper is as follows: the spatial distribution of ^{137}Cs at the initial time, which was observed in the Bryansk region after the disaster in late April–early May 1986, can be represented as a random, uncorrelated field.

There are two other basic assumptions: (1) the primary deposition of ^{137}Cs on the soil surface does not correlate with the parameters of the microrelief—the correlation coefficient equals zero; (2) there is no spatial correlation of the primary fallout field of ^{137}Cs, that is, the variogram represents a typical "white" noise (because the size of the test site is small enough).

It is hypothesized that because of integral interaction of various landscape factors, a transformation of the ^{137}Cs pattern will be observed, which occurs as a joint action of lateral and radial migration of radionuclides. These processes can lead to structuring, that is, to self-organization of the ^{137}Cs contamination field, which will be displayed by emergence of spatial correlation between the terrain parts (semivariogram analysis), as well as by dependence of ^{137}Cs contamination levels on the DEM (digital elevation model) parameters (generalized additive model [GAM] and linear model [LM]). The higher the correlation of ^{137}Cs with the parameters of the microrelief is, the more intense self-organization of the microlandscape is.

According to the accepted assumption, the initial ^{137}Cs patterns did not depend on microlandscape factors, that is, in modeling, the ratio of the explained variance could be zero. Since measurements of ^{137}Cs were conducted in 1993; that is, 7 years after the Chernobyl accident, any increase in the explained variance (growth of the determination coefficient) can be treated as an effect of the rearrangement of the ^{137}Cs pattern attributed to a set of landscape factors.

In traditional methods of pedometrics (McBratney et al. 2018) applied in mathematical modeling of the soil cover pattern, various methods of statistical analysis are used for systems in a state of equilibrium or dynamic equilibrium. The validity of a mathematical model is determined by the value of the determination coefficient, the value of which is to be maximized.

In this study, we observed a stage of emergence of relationships between landscape components. We suggest that the role of landscape factors, as displayed by transformation of the ^{137}Cs contamination patterns, can be assessed, using the following two indices: (1) the correlation of the ^{137}Cs patterns and (2) a percentage of the explained variance (the coefficient of determination) in ^{137}Cs relations with a set of landscape factors. In this case, an increase in the percentage of the explained variance indicates an increase in linkage density. By analyzing various test sites within the microlandscape, a degree of heterogeneity of ^{137}Cs "incorporation" into the landscape structure can be estimated.

5.4 Study Area

The study area is located in the western part of the Bryansk Region, Russia, southwest of the town of Novozybkov, near the village of Barky. There are *Polessie* landscapes with sandy and loamy sandy Umbric Podzols, as well as minor areas of Gleyic soils (Vorobyov 1993). The area is about 170 km away from the Chernobyl Nuclear Power Plant. The ^{137}Cs radioactivity represented by a condensation type of aerosols was removed by rainout and deposited within the forest landscape.

Two sites referred to as B1 and B2 below were examined. A detailed description of the landscape properties and statistical parameters of ^{137}Cs distribution at these sites is presented in (Linnik et al. 2007a), while for spatial modeling of ^{137}Cs distribution, see in Linnik et al. (2007b).

Site B1 is located in the lower part of a gentle south-facing slope. This area has slightly undulating relief with an inclination southward and local depressions up to several square meters wide and 20–40 cm deep, where in springtime the ground waters come up to the soil surface. The surface runoff of ^{137}Cs from such depressions could be carried out in springtime along rills, covered by a peat-like litter. The soil cover is rather complex, and soils are hydromorphic in depressions and semi-hydromorphic on the background surface.

In the arboreal layer of site B1, spruce is common (about 70–80 years old), 28–30 m high, with rare occurrence of oak (*Quercus robur*) and birch (*Betula pendula*). Low shrubs layer with *Vaccinium myrtillus* occurs sporadically as well as mosses (20–30% of the area) that do not form a continuous cover. In the lowest central part of the test site, an aspen (*Populus tremula*) forest with *Sphagnum* spots can be found in depressions.

The second test site B2 is located on a flat, slightly inclined interfluve surface. The microrelief is represented by numerous rills up to 20 cm deep. Umbric Podzols with no evidence of gleyization dominate in soil cover. Vegetation is represented by *Pinetum myrtilloso-hylocomiosum* forest.

5.5 Simulation of the Spatial Structure of ^{137}Cs and Transformation at the Microlandscape Level

Multilevel splines were used to build DEM (Saveliev et al. 2005). For spline interpolation, two grids differing in step of calculation (0.1 m and 0.25 m) were chosen. This provides sufficient accuracy of the DEM construction (mean error 0.018 m). With a step of calculation of 0.1 m, the accuracy of restoring DEM details accounts for approximately 0.3–0.5 m. The generalized relief (with a step of 0.25 m) identifies the morphometric relief elements with a characteristic size of several meters.

The following microlandscape parameters controlling the distribution of ^{137}Cs were identified: (1) the value of the Laplace operator; (2) height (H); and (3) X- and Y-coordinate. The value of the Laplace operator was calculated by the formula:

$$\nabla(\nabla f) = \Delta f = \left(\frac{\partial^2}{\partial x^2} + \frac{\partial^2}{\partial y^2}\right) f(x,y) \tag{5.3}$$

Laplacian is a close approximation of the mean curvature; therefore, it may be used to describe the rate of convexity/concavity of the surface (Jost 2011).

It was assumed that the distribution of ^{137}Cs can be affected by microrelief forms. The landscape henceforth transforms the original contamination patterns in agreement with physical laws of matter migration: the negative values of the Laplace operator characterize the outwash zones (convex forms of the microrelief), the positive ones refer to accumulation zones (concave forms).

Two types of models were used to model the relationships between radionuclide contamination and landscape parameters.

A generalized additive model (GAM) is a generalized linear model (GLM) in which the linear predictor is given by a user-specified sum of smooth functions of the covariates plus a conventional parametric component of the linear predictor (Wood 2006). A simple example is:

$$\begin{aligned} G(x_1,x_2) &\sim \mathrm{Norm}\left(m(x_1,x_2),\sigma^2_{err}\right) \\ m(x_1,x_2) &= a_0 + s_1(x_1,\Theta_1) + s_2(x_2,\Theta_2) \end{aligned} \tag{5.4}$$

where $G(x_1, x_2)$ is the response variable, a_0 is a constant (intercept), and s_1 and s_2 are smooth functions of covariates x_1 and x_2; and Θ_1, Θ_2 are the parameters that define the smoothness (the effective degrees of freedom) of the functions.

$$G = a_0 + a_1 \cdot \mathrm{Laplas}_1 + a_2 \cdot H + a_3 \cdot X + a_4 \cdot Y \tag{5.5}$$

$$G = a_0 + s(\mathrm{Laplas}_2) \tag{5.6}$$

$$G = a_0 + s(\mathrm{Laplas}_1) + s(H_{mba1}) \tag{5.7}$$

The use of the linear model of the form (5.5) for the analysis of data on the ^{137}Cs distribution (10 × 10 m radiometric survey grid) throughout site B1 (Fig. 5.1) showed that all parameters of the model are not statistically significant, with the residuals being approximately normally distributed. Therefore, the use of the linear model (5.5) to describe the distribution of ^{137}Cs (over a 10 × 10 m grid) for the entire site B1 as depending on the terrain parameters is not applicable.

Similar modeling results throughout B1 test site were obtained when nonlinear models (5.6) and (5.7) were applied: all variables for nonlinear models (5.6 and 5.7) for the two simulation grids were also found to be statistically insignificant.

However, the geostatistical analysis carried out across site B1 (Fig. 5.1) revealed the spatial correlation associated with the impact of landscape factors. The variogram with small lags (at a distance of 20–30 m) significantly exceeds the dispersion of ^{137}Cs ($\sigma^2 = 39.7$), identifying the presence of negative correlation, more precisely— the presence of the "chessboard" effect when high values alternate with small ones.

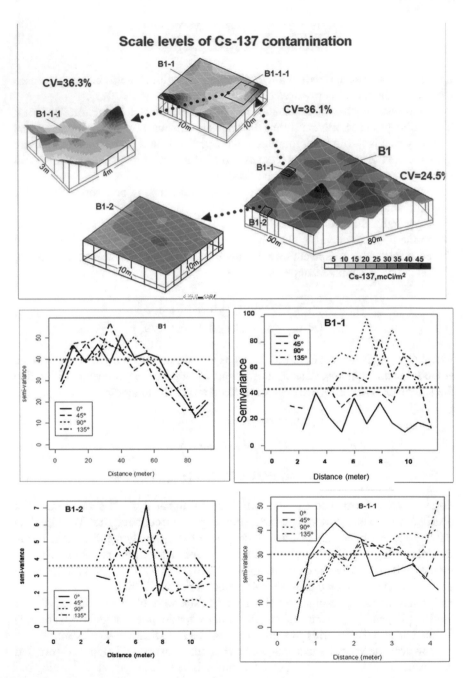

Fig. 5.1 Scale levels of ^{137}Cs contamination and coefficient of variation (CV) at the micro-landscape level. A variable step of radiometric survey of ^{137}Cs with the "KORAD" device: plot B1–10 m; microplots B1–1, B1–2 – 2 m; and microplots B1–1-1 – 0.5 m

Hence, depending on the alternation of sites with varying degrees of hydromorphism, the initial ^{137}Cs patterns underwent uneven transformation, which indicates
a significant heterogeneity of the lateral processes within *facies*.

To test the hypothesis about various intensities of lateral processes, it is necessary
to switch to a more detailed level of research. To do this, we analyzed the results of
modeling on two contrasting plots B1–1 and B1–2 with a size of 10 × 10 m, where
radiometric measurements of ^{137}Cs were carried out with a step of 2 m. The area
B1–1 was chosen in the hydromorphic part of site B1, complicated by hummocks
and hollows, while B1–2 was on a hill with a moss cover.

The analysis of variograms for contrasting microplots (B1–1 and B1–2) showed
the following regularities. For B1–1 in each direction the variogram of ^{137}Cs demonstrates wavy pattern, which indicates the presence of some spatial structure with
a characteristic size of about 5 m with a random character of the ^{137}Cs distribution
as a whole. Both the mean and maximum values of the variogram (dispersion)
depend on the direction and reach a minimum for the 0° direction. This type of
semivariogram is a clear indication that the ^{137}Cs variability in this direction is lower
than in the direction of 90° and 135°.

The use of LM of the form (5.5) to analyze the ^{137}Cs relation with the relief
parameters for B1–1 site demonstrated that all variables are insignificant for the
detailed and generalized relief models. Nevertheless, two nonlinear models (5.6 and
5.7) proved to be valid (Fig. 5.2).

For model (5.6), a nonlinear (quadratic) dependence of the ^{137}Cs distribution on
the Laplacian parameter, significant at the 5% level, was obtained, with explained
variance equal to 30%. Interestingly, the functional relationship between ^{137}Cs and
Laplace parameters takes an unusual form: radioactive contamination grows both in
accumulation zones and in washout zones, approaching zero in flat areas. It should
be noted here that in the graphical representation of the results of the GAM simulation in Fig. 5.2 the zero value on the Y axis corresponds to the average value of the
modeled parameter.

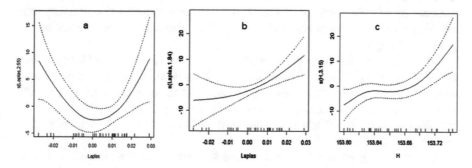

Fig. 5.2 Microplot B1–1. Visualization of the nonlinear smooth function (5.6) s (*Laplase*): (**a**) for
the generalized relief model. Nonlinear smooth function (5.7) for dependence of ^{137}Cs on two
parameters: s (*Laplase*) (**b**) and s(H) (**c**). The pointwise 95% confidence intervals are shown by
dashed lines

Then, the dependence obtained characterizes the deviation from the mean value for the selected predictor. Thus, there is no one-to-one correspondence between the ^{137}Cs deposit and the Laplacian parameter, which characterizes the average curvature of the microrelief forms. This is convincing evidence to support the idea that already at the microlandscape level there are complicated nonlinear dependencies on rates of lateral processes.

When model (5.7), with the parameter "height" being included as a predictor, is applied, some changes in simulation dependences obtained can be registered (Fig. 5.2b, c). Both Laplacian and height (H) are statistically significant at the 5% level, the model becomes quite representative. The more linearized form of the dependence between ^{137}Cs and Laplacian can be due to the fact that model (5.6) for Laplacian incorporated variability depending on height. This graph (Fig. 5.2b) shows that on the convex forms of the microrelief, which could be at different altitudinal levels, ^{137}Cs was removed, whereas on concave forms (depressions), there was accumulation of ^{137}Cs.

The analysis of the graph in Fig. 5.2c (the dependence between ^{137}Cs and the altitude position) produced interesting results. Test site B1–1 is located in the altitude range of 153.6–153.76 m. The ^{137}Cs activity growth was observed in the altitude range of 153.7–153.76 m, which is the local removal zone, the transit zone is found in the altitude range of 153.64–153.68 m. In the depressions (H < 153.84 m), there is an inverse effect of ^{137}Cs concentration (removal of radioactivity because of the intralitter flow) reported earlier (Linnik et al. 2007b).

Local "patchiness" of ^{137}Cs distribution due to the formation of local patterns up to 2 m resulting from lateral processes is also confirmed by the variogram analysis on B1–1-1 microplot (part of B1–1 site) 3 × 4 m in size (Fig. 5.1), where the measurements of ^{137}Cs were carried out with a step of 0.5 m.

As to site B1–2 (Fig. 5.1), located on the less hydromorphic part of the moss-covered area, the variogram analysis does not show any spatial correlation in all four directions. Hence, the nature of ^{137}Cs distribution can be considered absolutely random. The situation can be accounted for by the presence of moss, which prevents lateral transfer, thus allowing ^{137}Cs patterns to remain practically unchanged since the moment of radioactive aerosol deposition. It can be concluded that the hypothesis that there is no spatial correlation at the time of radioactive contamination origin is confirmed.

Weak processes of ^{137}Cs patterns transformation are also confirmed by the results of GAM modeling. Thus, the use of model (5.7) showed that the model parameters were statistically insignificant (the dependence was absent even at the 10% significance level), but the percentage of the explained variance reaches 30%. Hence, the transformation of the ^{137}Cs patterns due to the impact of the Laplacian parameters and height (H) does occur.

Thus, the methods of mathematical modeling and geostatistical analysis for the semihydromorphic site B1 revealed significant heterogeneity in terms of intensity of lateral migration at short distances (on the order of several tens of meters), which is controlled by the conditions of the microrelief, as well as by the nature of the land cover.

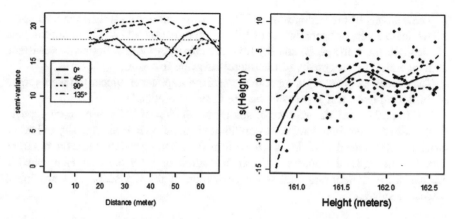

Fig. 5.3 Site B2. The variogram of ^{137}Cs distribution in four directions (on the left) and nonlinear dependence between ^{137}Cs and height (model 5.7)

For the interfluve site B2 (Fig. 5.3), located in automorphic environments, a radiometric survey was performed on a grid of 10 × 10 m. It made it possible to estimate only those changes in the structure of the ^{137}Cs patterns, which were tens of meters in scale. Evaluation of the ^{137}Cs functional relation with the terrain parameters was performed according to models (5.5, 5.6, and 5.7). Within the linear model (5.5), all parameters were not statistically significant, and the residuals were distributed approximately normally. The variogram in four directions demonstrates (Fig. 5.3) that there is no dependence on direction or distance, and, hence, the ^{137}Cs distribution process can be considered completely random.

Consequently, it can be concluded that the prediction about lack of spatial correlation of ^{137}Cs in origin depositions is valid under automorphic landscape conditions. The nonlinear model (5.7) was used to analyze the ^{137}Cs relation with the terrain parameters (Fig. 5.3, right) and it showed that the only statistically significant (5%) variable is height, with the explained variance of 14%. However, the too low value of explained variance is not enough to record a noticeable transformation of the ^{137}Cs patterns as a function of the altitude level.

5.6 Conclusions

The assumption of the random nature of ^{137}Cs distribution with respect to the relief at the time of the origin of radionuclide contamination is supported by evidence presented in the study. Depending on the landscape conditions (semihydromorphic and automorphic), the lateral migration processes vary significantly in intensity. Semihydromorphic landscapes are extremely heterogeneous in terms of intensity of lateral migration and, accordingly, in the formation of correlated contaminated patterns. The intensity of the transformation of ^{137}Cs patterns increases in semihydromorphic conditions as soil moisture content grows.

The analysis of the variograms of ^{137}Cs distribution demonstrated the formation of two scale levels of "patchiness" of ^{137}Cs contamination: (1) 20–30 m (alternation of semihydromorphic plots) and (2) 1.5–2.0 m (in microhollows), that is consistent with the initial heterogeneity of the microlandscape structure.

The processes of functioning at the microlandscape level appear to take a rather sophisticated nonlinear form as it was predicted theoretically.

The hypothesis of the absence of spatial correlation of ^{137}Cs in primary deposition under automorphic landscape conditions, indicating a weak intensity of lateral processes at the local level, has been confirmed by both qualitative and quantitative data. To conclude, the analysis of transformation of ^{137}Cs patterns established a mechanism of functioning and self-organization of the landscape as a nonlinear stochastic system.

Acknowledgments The study was performed with partial financial support of the Russian Foundation for Basic Research, grant No. 16-05-00915.

References

Bellehumeur, C., & Legendre, P. (1998). Multiscale sources of variation in ecological variables: Modeling spatial dispersion, elaborating sampling designs. *Landscape Ecology, 13*(1), 15–25.

Burrough, P. A., & McDonnell, R. A. (1998). *Principles of geographical information systems.* New York: Oxford University Press.

Dyakonov, K. N, Abramova, T. A, Seregina, I. P., & Bezdelova, A. P. (2008). *Swamping middle taiga moraine and fluvioglacial landscape in the Holocene.* In Proceedings of Moscow University, series 5 Geography (Vol. 2, pp. 28–34). (in Russian).

Dyakonov, K. N., & Linnik, V. G. (2017). *Some problems of the science of landscape in the 21st century.* In K. N. Dyakonov (Ed.), Landscape science: Theory, methods, landscape and environmental support for nature management and sustainable development. Proceedings of the XII international landscape conference (Vol. 1, pp. 19–24). Tyumen: Tyumen State University Publishing House.

Fridland, V. M. (1972). *Structure of the soil cover.* Moscow: Mysl. (in Russian).

Gringarten, E., & Deutsch, C. V. (2001). Teacher's aide variogram interpretation and modeling. *Mathematical Geology, 33*(4), 507–534.

Hengl, T. A. (2007). *Practical guide to geostatistical mapping of environmental variables.* Luxemburg: Office for Official Publications of the European Communities.

Jost, J. (2011). *Riemannian geometry and geometric analysis* (6th ed.). Berlin/Heidelberg: Springer.

Karpachevsky, L. O, & Stroganova, M. N. (1981). Microrelief – function of forest biogeocenosis. *Pochvovedeniye (Soil Science), 5*, 83–93. (in Russian).

Karpachevsky, L. O. (1977). *Diversity of the soil cover in forest biogeocenosis.* Moscow: Publishing House of Moscow State University. (in Russian).

Linnik, V. G., Saveliev, A. A., Govorun, A. P., & Sokolov, A. V. (2007a). Spatial analysis and modeling of Cs-137 distribution at the microlandscape level (the Bryansk region). In K. N. Dyakonov, N. S. Kasimov, A. V. Khoroshev, & A. V. Kushlin (Eds.), *Landscape analysis for sustainable development. Theory and applications of landscape science in Russia* (pp. 190–199). Moscow: Alex Publishers.

Linnik, V. G., Saveliev, A. A., Govorun, A. P., Ivanitsky, O. M., & Sokolov, A. V. (2007b). Spatial variability and topographic factors of ^{137}Cs soil contamination at a field scale. *International Journal of Ecology & Development, 8*(7), 8–25.

O'Neill, R. V., Johnson, A. R., & King, A. W. (1989). A hierarchical framework for the analysis of scale. *Landscape Ecology, 3*(3/4), 193–205.

McBratney, A. B., Minasny, B., & Stockmann, U. (Eds.). (2018). *Pedometrics*. Cham: Springer.

Phillips, J. D. (2003). Sources of nonlinearity and complexity in geomorphic systems. *Progress in Physical Geography, 27*(1), 1–23.

Phillips, J. D. (1998). On the relations between complex systems and the factorial model of soil formation (with discussion). *Geoderma, 86*, 1–21.

Phillips, J. D., & Marion, D. A. (2005). Biomechanical effects, lithological variations, and local pedodiversity in some forest soils of Arkansas. *Geoderma, 124*, 73–89.

Richter, J. (1987). *The soil as a reactor: Modelling processes in the soil*. Cremlingen: Catena Verlag.

Saveliev, A. A., Romanov, A. V., & Mukharamova, S. S. (2005). Automated mapping using multi-level B-splines. *Applied GIS, 1*(2), 17/1–17/19.

Targulian, V. O., & Krasilnikov, P. V. (2007). Soil system and pedogenic processes: Self-organization, time scales, and environmental significance. *Catena, 71*(3), 373–381.

Targulian, V. O., & Sokolova, T. A. (1996). Soil as a biotic/abiotic natural system: A reactor, memory, and regulator of biospheric interactions. *Eurasian Soil Science, 29*(1), 30–41.

Vorobyov, G. T. (1993). *Soils of the Bryansk region: (genesis, properties, distribution)*. Bryansk: Bryansk Survey Center for Chemicalization and Radiology of Agriculture "Agrochemicalradiology". (in Russian).

Wood, S. N. (2006). *Generalized additive models: An introduction with R*. Boca Raton, FL: Chapman and Hall/CRC Press.

Chapter 6
Determination of the Order Parameters of the Landscape at the Regional Level

Mikhail Yu. Puzachenko

Abstract The chapter focuses on methodological basis and methods of multidimensional statistical analysis in landscape research. We demonstrate the opportunities to use information on variance of the relief properties and multispectral reflection for evaluating fractions of the spatial variance of landscape attributes measured in the field. On this basis, we established a limited number of linear combinations of measured properties that are mutually independent and physically explicit variables that allow explaining spatial differentiation of most properties to a large extent.

Keywords Hierarchy · Relief · Dimensionality · Spatial differentiation · Order parameter · Discriminant analysis

6.1 Introduction

To describe and understand the rules that determine the relationships between the phenomena of nature in space and time is the traditional goal of geography. According to L.S. Berg (1915, p. 469), "the purpose of geographic research is to reveal links and regularities that exist between separate objects of interest to the geographer, the main question being, 'How do particular sets of objects and phenomena affect each other, and what are the results of this spatially?'" Methods for solving the problem of searching for "relations and laws" have evolved significantly from qualitative (Berg 1915; Dokuchaev 1948; Solnetsev 2001) to quantitative ones (Sochava 2005; Armand 1975; Puzachenko 1971; Puzachenko and Skulkin 1981;

M. Y. Puzachenko (✉)
Institute of Geography, Russian Academy of Sciences, Moscow, Russia
e-mail: m.yu.puzachenko@igras.ru

© Springer Nature Switzerland AG 2020
A. V. Khoroshev, K. N. Dyakonov (eds.), *Landscape Patterns in a Range of Spatio-Temporal Scales*, Landscape Series 26,
https://doi.org/10.1007/978-3-030-31185-8_6

Puzachenko et al. 2012) based on the theory of systems using information analysis, ordination, and statistical analysis methods. Nowadays, modern sources of information on the properties of geosystems became available, such as multispectral remote sensed data (MRSD) and digital elevation models (DEM), as well as powerful personal computers. This advancement provided a lot of new opportunities to determine the spatial relationships between landscape properties via combining modern and traditional field sources of geographical information, using a set of statistical analysis methods. This approach can be defined as multifunctional (multidimensional), since the set of spatial linkages, determined in the form of functional dependencies between the geosystems attributes, is studied through their relationship with relief and reflected radiation. The implementation of this approach is based on the spatially referenced analysis of the links between a large number of properties. The procedure involves step-by-step reduction in the number of properties by means of linear statistical transformations and the determination of the most informative of them. The identification of integrating properties that reflect most connections in a system follows the concept of invariant proposed by V. B. Sochava as well as with the concept of elementary soil and landscape processes by F. I. Kozlovsky (2003) based on interconnected properties of soils and landscape. The aforementioned concepts are closely related to the general idea of synergetics that a complex dynamic system can be displayed through a limited number of independent order parameters (Haken 1980, 2005).

Thus, the development of scientific methodology, measurement systems, and technical facilities for the analysis provides the basis for the transition of landscape research to quantitative analysis, which is believed to give new meaningful objective results about the causes and mechanisms of spatial–temporal differentiation of geosystems.

This chapter deals with the problem of analyzing the rules of organization of complex spatiotemporal geographic dynamic systems. We focus on the selection of stable system-forming relations between the observed properties starting with the hypothesis of their random variation. To achieve the purpose of the study, we concentrate on the following tasks: (1) to establish general laws of spatial differentiation of geosystems for planning field surveys; (2) to conduct field measurements of geosystem properties and components, taking into account the scale of research; (3) to identify the hierarchical structure (organization) of relief to reveal its properties at various scales; (4) to determine and interpolate in space the functional relations between the field data, relief, and reflected radiation; (5) to reduce dimensionality, that is to reduce sequentially the number of functional relations between field characteristics, aimed at establishment of generalized elementary processes (order parameters) at several levels of integration (intracomponent, component, and landscape levels); and (6) to obtain physical interpretation of the order parameters at the component and landscape levels of integration and to establish corresponding independent processes responsible for the spatial variation of the basic properties of geosystems relying on their connection with the DEM characteristics, MRSD, and field data.

6.2 General Methodology of Analysis: Definition of the System as an Object of Research

Historically, two main methodological approaches to research have been developed in geography and other Earth sciences such as botany and soil science. The first approach—"top-down," or holism—postulates rigid interrelations between natural phenomena similar to organism. Accordingly, the study is aimed at identification of integral structures using interdependencies on the whole set of phenomena distributed in space. The second, "bottom-up," approach, or reductionism, is based on the study of phenomena localized in the same space and possible relationships between them followed by identification of relatively separate spatial formations. The first approach can be historically associated with small amounts of available information about the natural phenomena and the impossibility of implementing quantitative methods for analysis of relations between phenomena. The lack of adequate mathematical methods and technical tools resulted in the need for empirical generalizations based on the principles of uniformism and actualism.

The development of the second approach, aimed ultimately at quantitative studies of the relations between observed phenomena in nature, is associated with the elaboration of the probability theory as the foundation for the development of mathematical statistics. However, their application in research of relations between natural phenomena in space became possible only with the advent of personal computers with high computing capabilities. This is due to the fact that the study of real natural phenomena in space and/or in time requires a large number of recorded attributes of these phenomena aimed at the most possible complete description within the existing measurement systems and general hypotheses about their organization. The study of a large set of properties of natural phenomena is necessary, at least, to highlight those properties or their groups that allow the most adequate description of the observed phenomena. This, in turn, allows us to assess their spatial–temporal relationships and identify driving mechanisms. Identification of mechanisms that determine the observed phenomena of nature in space allows, on the one hand, establishing the possible changes generated by them in time, and, on the other hand, using natural resources with greater efficiency.

Thus, the holistic approach has arisen on the basis of empirical generalizations and postulate of deterministic relationships, while the reductionist approach is based on quantitative methods of analysis of the measured properties with strong focus on probabilistic relationships that can generate relatively isolated spatiotemporal systems. It follows that for research (rather than inventory) problems, reductionist methodological approach is preferable.

To formalize the research object, its parts and their space, and the relationships between them, it is useful to apply the approach and terminology of the system theory. Within the framework of this approach, general unambiguous rules of actions, definitions of variables, and their properties and relations between them were introduced (Armand 1988; Klir 1985; Puzachenko 1998; Puzachenko and Skulkin 1981).

The development of system theory encouraged the formation of a new science of synergetics (Haken 1980, 2005) as a methodological approach to the study of complex natural systems. In synergetics, the system is considered through the interaction of its parts, which, in common case, are also systems. Thus, the concept of complex systems with nonequilibrium wave dynamics, self-organization, and hierarchy emerged. The state of such a system, close to the stationary, can be described by a limited set of independent variables derived by methods of mathematical statistics. Hermann Haken called them *order parameters*.

The study presented in this chapter is based on reductionism tenet, using the methodology, methods and terminology of the theory of systems, synergetics, and mathematical statistics.

6.3 Methodology and Methods for Determination of the Hierarchy and the Order Parameters of the Relief

The emergence of hierarchy in nature, including relief, is associated with several possible mechanisms (Puzachenko 1986) that permit quantification of relief hierarchy. It was shown (Turcotte 1997; Puzachenko et al. 2002) that the relief formed as a result of interaction of tectonic movements, accumulation of glacial deposits, and denudation obeys the general rule: the larger is the spatial interval of observations, the greater is the amplitude of elevations. This property was referred to as fractality (Mandelbrot 1982) or discrete continuity, fractal dimension being not integer.

Strictly speaking, fractal process does not carry statistically significant regular components and is determined by the stochastic behavior of the system. However, many processes with a regular component generate self-similar hierarchical structures with noninteger dimensionality (Rabinovich and Trubetskov 2000). Spectral analysis (Marple 1987) with a set of additional methods allows obtaining the necessary information about the relief structure, to measure its most important parameters, to estimate the possible number of order parameters generating the relief pattern, and to identify the main hierarchical levels of its spatial organization. Implementation of this method is possible with regular mapping of elevations (transect, grid). Average linear dimensions of hierarchically subordinated relief structures, determined in the course of the analysis, are used to calculate the height levels and attributes of the relief (slopes, curvature, illumination, etc.).

6.4 Methods of Multidimensional Spatial Analysis and Order Parameters Calculation

Quantitative analysis of the relationship between the landscape component properties measured in the field is possible within the framework of mathematical statistics on the basis of two approaches—direct and indirect. Direct approach involves the analysis of the relationships between directly measured field properties of the components. This approach can be implemented by methods of nonparametric multidimensional scaling (Kruskal and Wish 1978) or various ordination methods. Along with the advantages of this approach, which allows reflecting the nonlinearity of relations, it has limitations associated with the ability to interpolate the results and analyze large data sets.

The indirect approach is based on the analysis of the relations between the properties of the components, through their connections with external variables—MRSD and DEM, which themselves are measured properties with a quasi-continuous character. If the chosen property of the component is directly or indirectly controlled by the properties of the relief and determines the transformation of reflected solar energy, then, using the methods of multivariate statistical analysis, it is reproducible through these external variables with a certain accuracy. In this approach, we consider the mutual reflection of two types of properties: measured by a discrete field sample and obtained from independent quasi-continuous measurement sources. The latter provide a basis for interpolating both the properties measured in the field and the independent functional relations describing the variation of properties at sample plots, which can be interpreted as order parameters.

The issue of interpolation can be solved by means of multiple regression, factor, discriminant, neural network, genetic analyses, etc. However, in accordance with the formulated approach to the analysis of the organization of the geodynamic system, in order to describe, it is necessary to determine the order parameters (functional relations). The latter are expected to describe the state of each analyzed property in the best way on the basis of the accepted statistical criteria. The discriminant analysis, which is the method of finding a linear combination of variables that best separates two or more classes of the phenomenon state (Kim et al. 1989), meets these requirements to the greatest extent.

With a large number of variables, the method of canonical discriminant analysis with a stepwise procedure is used to select the most significant and independent external variables describing the field characteristic. Cross check with up to half of the sample is used to independently assess the quality of the statistical models.

Uncertainty in the results of discriminant analysis, in the framework of the demonstrated research, arises from various sources as follows: errors in field descriptions, insufficient accuracy of their positioning, irregularity and incompleteness in coverage of field descriptions for the various grades of the investigated properties, the incompleteness of the information obtained from the remote sensing and topography for the properties in whole or for certain intervals of their values, different scales of measurement at sample plots for vegetation (commonly, circular area) and soil (commonly, pit, and drilling), medium-scale research in which part of the local variation of the properties is beyond the scale of remote information and terrain.

Thus, each property of the geosystem measured in the field is represented by k-coordinates of particular order parameters (discriminant axes), both for sample plots and for the whole territory. Each particular order parameter is hypothesized to be associated in a certain way with the properties of relief and/or distance information and to have its specific pattern in space.

The purpose of multivariate analysis is to identify the most common "landscape-forming" order parameters of the geosystem. The simplest way to obtain them is integration with the method of principal components of factor analysis (Haken 2005). In this case, we use such a function of factor analysis as the reduction of the number of variables with the revealing of the main integrating factors (Aivazyan et al. 1989). The number of significant factors is estimated on the basis of statistical criteria, provided that the factor is related to two or more characteristics (order parameters of components/landscape).

According to the adopted methodology of reductionism, in order to select order parameters that influence the properties to the greatest extent, it makes sense to analyze the relationships between properties sequentially: first, within the framework of their functional and/or structural blocks; second, the results of their integration as part of the landscape component and then of the landscape as a whole. This procedure results in the reduction of the initial particular order parameters of individual characteristics, with the sequential establishment of the order parameters for their structural and functional blocks, components of the landscape, and the landscape as a whole. Based on the connection of order parameters with MRSD, DEM, field data, and other sources of information, they can be put in line with the elementary processes that determine the variation of the state of most properties included in the block, component, or landscape as a whole. Sometimes order parameters can be an indirect reflection of processes, sometimes a direct reflection, or a reflection of two opposing processes. Anyway, a few order parameters describe spatial distribution of the large number of the field attributes with the highest possible proportion of explained variance. Along with the analysis of properties measured in the field, properties obtained from thematic maps can be used in a similar way. In this work, information about the boundaries of the late Pleistocene (Valdai, or Würm) glaciation is analyzed in the interpretations of various authors. In this case, all plots of the territory are considered as a training sample, for which the authors of different interpretations of the Valdai glaciation defined their genesis.

6.5   Study Area and Materials

Research was conducted at the area of about 20,000 km², located in the south–west of the Valdai hills, which is water divide for the Volga and the Zapadnaya Dvina (Daugava) basins. Administratively, the territory belongs to eight districts of Tver and one district of Pskov region.

The study area is located within the western part of the Moscow syneclise and is confined to the Nelidovo-Orsha tectonic megablock of the Central (Moscow) medium-altitude geo-step. In Quaternary period this territory was repeatedly exposed to land ice-sheet (Kaplin et al. 2005). The last two glaciations—Moscow (Riss II) and Valdai (Würm)—contributed much to sedimentation and shaping the modern structure of relief. Moscow glaciation left a typical glacial hilly and undulating relief, later converted to erosion-hilly and rolling, during the subsequent interglacial period and the Valdai glaciation (Sokolov 1946). As a result, the boundary of the maximum ice-sheet of the Valdai glaciation for the study area and the number (Moskvitin 1967) of late Pleistocene glaciations after the Mikulin (Riss-Würm) interglacial remain a matter of debates. Cover deposits play a particular landscape-forming role in the transition zone of two glaciations. The 30–130-cm thick mantle masked contrasts in the texture of the underlying sediments. In general, silty cover deposits is almost a universal feature of the study area. Their thickness varies from the first tens of centimeters to the first meters; a texture varies from fine sand to medium-heavy loamy, with a gradual increase in thickness and loam content eastward.

The study area is located in the Atlantic-continental region of temperate climate on the border of cold and temperate thermal zones (Milkov and Gvozdetsky 1985). The climate is moderately humid with relatively mild winters and cool summers. In general, the annual amount of positive temperatures decreases eastward by 500 °C, and the annual amount of precipitation increases by 70 mm due to the barrier effect of the Valdai hills.

According to the botanical–geographical zoning, the territory belongs to the zone of oak-tree shrub forests of the East European geobotanical province (Shennikov and Vasiliev 1947) or is included in the central part of the belt of coniferous-broad-leaved forests of the East European plain (Semenova-Tian-Shanskaya and Sochava, 1956). Forest area accounts for 70% (Atlas of forests of the USSR 1973). Mires occupy about 15% of the territory and are mainly represented by mesotrophic and oligotrophic types. The vegetation cover of the territory has been significantly transformed as a result of forestry and, to a lesser extent, that of agricultural activities. In poorly drained flat areas, gentle slopes, and crests of moraine ridges, windfalls occur periodically each 20–30 years (Skvortsova et al. 1983; Vasenev and Targulyan 1995; Puzachenko 2007).

The territory belongs to the Baltic province (Valdai district) of Umbric Albeluvisols (sod-podzolic soils) and Stagnic Albeluvisols, locally with remnants of carbonates; textures are various with predominance of loamy on moraine deposits (Classification and diagnosis of soils of the USSR 1977, 2004). The spatial distribution of soils is closely related to landforms and reflects, first of all, the redistribution of moisture between its elements and the peculiarities of water regime in an ecotope (soil-geological conditions of the non-Chernozem region 1984).

We collected three sets of data: (1) quasi-continuous information—MRSD and DEM—used both to characterize the own properties of geosystems and as a basis to interpolate data from sample plots; (2) discrete information—field data—used for the study of the relationships between geosystem properties; and (3) thematic maps depicting boundary of late Pleistocene glaciations for the study area in the interpretation of various authors as an independent source of information about the genesis of the surficial deposits.

Multichannel satellite scanner images from Landsat satellites were used as a source of multispectral remote information. The choice of this series of satellites was determined by their spectral, spatial, and temporal coverage and resolution (30–60 m) as well as by free access. Based on the individual Landsat scenes, three mosaics were created for the research area for May, September, and October. To reflect the nonlinear relations between spectral bands and increase the information content of images, we calculated indices, which are, in most cases, normalized values of differences between neighboring or close bands, or their relations.

Information about the relief of the territory was obtained from topographic maps at scale 1:100,000. Digital elevation model (DEM) was created by nonlinear interpolation from vectorized isohypses, the edges of the lakes, and the elevation points. The size of the DEM grid (raster) cell was chosen according to the scale of the study and multiple of Landsat images resolution. Thus, the size of the grid cell (pixel), or the size of the system element, was established as 114 m. Derived characteristics of the relief (slope gradients, exposure, different types of curvatures, Laplacian) were calculated from DEM considering the hierarchical organization of the relief.

Measurement of landscape components attributes was carried out in the forms of integrated and track route descriptions and integrated descriptions on transects with a regular step (see Fig. 6.1). The descriptions were performed by graduate students from Faculty of Geography of Lomonosov Moscow State University and researchers from Laboratory of Biogeocoenology of Institute of Ecology and Evolution,

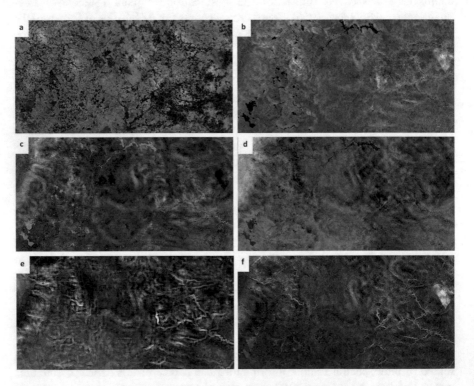

Fig. 6.1 Vegetation order parameters (VOPs): **a** – first, **b** – second, **c** – third, **d** – fourth, **e** – fifth, and **f** – sixth

Russian Academy of Sciences. All descriptions (1459 sample plots) are GPS-referenced with positioning accuracy of 5–15 m. Data were arranged in a database in the Microsoft Access environment with a system of queries for the organization and transformation of data for further analysis. The study looked at 123 attributes of field descriptions, which united in nine structural–functional blocks.

For the purposes of integrated analysis, field data were combined with MRSD and DEM. Thematic cartographic materials were used to analyze the genesis of the relief and sediments, namely the boundaries of the maximum advance of the Valdai glaciation. Interpretations of the latter differ in the works by different authors. The most well-known state geomorphological maps from 1959 to1960 as well as maps in publications (Chebotareva et al. 1961; Moskvitin 1967; Faustova 1972) meeting the requirement of scale not coarser than 1:2,000,000 were vectorized and combined with other data for further analysis similar to data from sample plots. Five interpretations of boundaries were considered both separately and jointly.

6.6 Analysis of the Hierarchical Organization, Order Parameters of Relief, and the Allocation of Its Characteristics

Analysis of the relief spectra for the western, central, and eastern sectors of the study area showed statistically nonsignificant differences in fractal dimension, which ranged from 1.97 to 2.05. These values correspond to the dimension of black noise. After removing the logarithmic trend and the high-frequency component, the presence of nonrandom components in the spectrum residuals was analyzed. As a result, the following linear dimensions of hierarchical levels of relief were established, conventionally classified into four groups: (1) trend that is more than 18 km and not less than 55 km; (2) macrorelief, 12.5 km (10–18 km), 8 km (6–10 km); (3) mesorelief, 3900 m (2700–6000 m), 2400 m (2250–2700 m), 1900 m (1750–2250 m), 1140 m (1050–1500 m); and (4) microrelief, 680 m (525–1050 m) and 340 m (228–525 m).

The analysis of the own spectrum of smoothed residuals from the relief spectrum allowed us to determine the basic frequencies or the number of order parameters that determine the formation of relief structures. We distinguished one main basic frequency and one weakly significant, generating relief structures with the sizes less than 3 km and less than 540 m, respectively. The main order parameter of relief was explained as the result of the interactions between endogenous mechanisms affecting the relief structure and exogenous mechanisms of glacial accumulation. The weakly significant order parameter indicates sedimentation and shaping of slopes during Moscow and Valdai glaciations.

6.7 Analysis of the Properties of Vegetation

The analysis of 69 vegetation attributes obtained from MRSD and DEM was performed separately for six blocks of properties. The main findings are shown in Table 6.1.

Thus, at the level of integrating vegetation *blocks of properties* into the vegetation *component*, 26 integrated order parameters were derived. They were, in turn, integrated into six vegetation order parameters (VOPs) describing 67% of the variation of integrated order parameters. Let us now turn to interpretation of their physical meaning.

The first VOP (26% of variation) is most positively correlated with the slope gradient at mesorelief level, spring biomass of vegetation, heights of tree stand, thickness of peaty horizons, moss PC, and thickness of eluvial horizons, and negatively correlated with the moisture content in vegetation and substrate in autumn, the thickness of the soil. Based on this and taking into account the spatial localization (Fig. 6.1a) of this VOP, it can be interpreted as a manifestation of successions in boreal (southern taiga) forests and anthropogenic transformations (agricultural, forestry, settlement) of land cover.

The second VOP (15% of variation) is positively correlated with the slope of mesorelief, height of relief, biomass of vegetation, basal area (BSA) for *Alnus incana* and *Betula pendula*, and thickness of mor humus horizon, and negatively correlated with the curvature of the thalwegs of the microrelief, moisture content in vegetation and substrate, moss PC, BSA of *Picea excelsa*, and PC of *Pinus sylvestris*. Hence, this order parameter reflects the succession of small-leaved forests (Fig. 6.1b).

The third VOP (9% of variation) is positively correlated with the slope of the terrain, the curvature of mesorelief, PC of *Alnus incana*, BSA of *Corylus avellana*, PC of herbs, and PC of undergrowth of *Fraxinus excelsior*, and negatively correlated with the moisture content in vegetation and soils in autumn, the curvature of the relief, the autumn biomass, PC of *Sphagnum*, and thickness of the soil. This VOP was interpreted as the succession of forests with broad-leaved species and bogging (Fig. 6.1c).

The fourth VOP (6% of variation) is positively correlated with the slope gradient, the curvature of the microrelief, moisture content of vegetation and substrate, PC of *Pinus sylvestris*, and PC of first and second layers of stand, and negatively correlated with the height of the terrain, the slope of the microrelief, the height of first layer of stand, PC of *Acer platanoides* and its undergrowth, thickness of mor humus horizon, and thickness of cover loam. As a result, this order parameter was interpreted as a succession of mesophilic (pine) forests, on the one hand, and the succession of the southern taiga forests, on the other hand (Fig. 6.1d).

The fifth VOP parameter (6% of variation) is positively correlated with the presence of stony-gruss sediments in the lower part of the soil profile and negatively correlated with the curvature of the thalwegs, the slope gradient of mesorelief, the presence of medium loam in the middle part of the profile, and development of

Table 6.1 Results of deriving individual and integrated order parameters for vegetation properties

Blocks of properties	Number of raw variables	Groups of raw variables	Percentage of correct discrimination	Number of individual order parameter	Number of integrated order parameters	Proportion of variance of individual order parameters explained by integrated order parameters
Integrated projective cover (IPC) of vegetation	8	Basal area (BSA) of the stand	59	18	3	66
		Canopy cover for stand	51			
		Canopy cover for layers of the stand	61–63			
		Canopy cover for undergrowth	43			
		Coverage for herbs	44			
		Coverage for mosses	44			
Projective cover (PC) of forest stand species	29	BSA, species	50–92	42	6	62
		PC, species	50–91			
Height and age of the stand	5	Average height of the stand	52	11	3	79
		Average height of the layers in the stand	50–60			
		Average age of the stand	82			
Projective cover of species in undergrowth	22	PC of species in undergrowth	49–89	29	6	66
Groups of herb species and projective moss cover	4	Groups of herbs by species abundance	49	14	4	57
		PC of moss species groups	41–68			
Types of land cover	1	Nine types (forest, windfalls, forest swamps, open bogs, glades, overgrown meadows, meadows, arable land, quarries)	60	–	4	93

floodplain forest–shrub–wet meadow vegetation. This VOP reflects the development of vegetation on a coarse substrate, typical for floodplain vegetation complexes (Fig. 6.1e).

The sixth VOP (5% of variation) is positively correlated with the height and curvature of the relief, with the moisture content in vegetation and soil, autumn biomass, CPC of herbs, PC of undergrowth of *Sorbus aucuparia*, the thickness of organic content horizons, and PC of *Alnus incana*, and negatively correlated with height, spring biomass of vegetation, BSA of *Acer platanoides*, and PC of *Pinus sylvestris* undergrowth. Thus, it reflects the development of shrub–meadow vegetation of river valley slopes (Fig. 6.1f).

The VOPs are the best predictors for the BSA and the PC of the stand, the types of land cover, the height of the stand and its layers, 6 of the 16 groups of herbs, PC of mosses and sphagnum mosses, BSA and PC of birch, PC of undergrowth, PC of spruce, and sorb undergrowth. Thirty-nine of the 69 considered vegetation characteristics are almost not described by VOPs.

6.8 Analysis of Soil Properties, Soil-Forming Depositions, and Their Genesis

Analysis of the soil characteristics and soil-forming depositions (54 variables) for the three blocks of properties showed the following (Table 6.2).

A total of 21 integrated order parameters for blocks of soil properties were used to derive seven soil order parameters (SOPs) describing 72% of the variation of the integrated order parameters.

The first SOP (24% of variation) is positively correlated with the moisture content in vegetation and substrate in autumn and the thickness of oligotrophic-peat horizons and negatively correlated with biomass of vegetation, slopes of mesorelief, and average height of forest stand. As a result, this order parameter can be interpreted as the process of oligotrophic bogging and gleyization (Fig. 6.2a).

The second SOP (14% of variation) is positively correlated with the slope gradients of the macrorelief, spring biomass of vegetation, the Munsell value of the humus horizon, the presence of medium loam in the lowest soil horizons, PC of *Alnus incana*, and PC of first layer of the stand and negatively correlated with the relief elevations, soil thickness, BSA of spruce, height of the first and third layers of the stand, and the presence of sandy sediments in the lowest soil horizons. Hence, the second-order parameter is related to a process of humus accumulation and podzolization (Fig. 6.2b).

The third SOP (9% of variation) is positively correlated with the curvature of the macrorelief talvegs, moisture content invegetation and substrate in spring, PC of *Sphagnum*, the thickness of peat horizons, and the height of the first layer of the stand and negatively correlated with the moisture content in vegetation and substrate in autumn, the slopes of the macrorelief, and the Munsell value of the humus

Table 6.2 Results of deriving individual and integrated order parameters for soil properties

Blocks of properties	Number of raw variables	Groups of raw variables	Percentage of correct discrimination	Number of individual order parameter	Number of integrated order parameters	Proportion of variance of individual order parameters explained by integrated order parameters
Thickness of soil horizons and soil-forming depositions	18	Thickness of different combinations of organic and eluvial horizons	45–81	53	8	67
		Soil thickness	59			
		Number of soil horizons and soil-forming depositions per 1 m	43			
		Depth of reaction with hydrogen chloride (HCL)	55			
		Thickness of cover deposits	43			
Colors of soils and soil-forming rocks	21	Mansell min/max hue, chroma, value for humus horizon	55–63	41	8	82
		Mansell min/max hue, chroma, value for eluvial horizon	47–58			
		Mansell min/max hue, chroma, value for illuvial horizon	47–58			
		Mansell min/max hue, chroma, value for the lowest described horizon	33–45			
Texture of soil horizons and soil-forming deposits	15	Texture of cover deposits	77–83	19	5	79
		Texture of illuvial horizons	66–79			
		Texture of the lowest described horizon	73			
		Average and maximum degree of reaction with HCl	48–49			

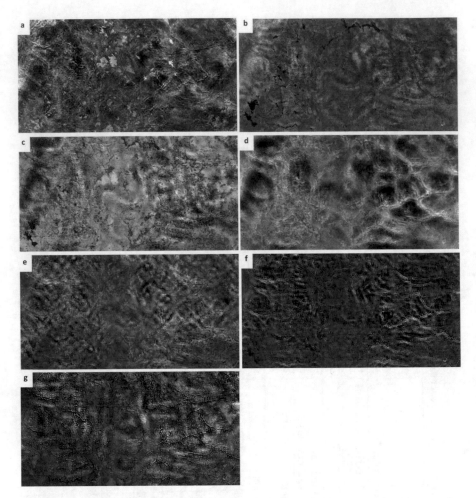

Fig. 6.2 Soil order parameters (SOPs): **a** – first, **b** – second, **c** – third, **d** – fourth, **e** – fifth, **f** – sixth, and **g** – seventh

horizon. Thus, the order parameter reflects mesotrophic swamping and development of eluvial (paleo podzolic) horizons with gleyization (Fig. 6.2c).

The fourth SOP (7% of variation) is positively correlated with the moisture content in vegetation and substrate in autumn, the curvature of the microrelief interfluves, the light textures of the lowest soil horizon and negatively correlated with the moisture content in vegetation and substrate in spring, elevation, slope gradient and curvature of the macrorelief talwegs, height of the first layer of the stand, PC of moss, Munsell value of the lowest horizon, and the thickness of the mor humus horizons. The order parameter reflects the process of lateral eluviation (Fig. 6.2d).

The fifth SOP (7% of variation) is positively correlated with the height of the mesorelief, slopes and curvature of the microrelief interfluves, Munsell chroma of humus horizons, SA of *Acer platanoides*, and the height of the first layer of the stand and negatively correlated with the moisture content in vegetation and substrate in autumn, the curvature of the macrorelief thalwegs, and BSA of *Picea excelsa*. Thus, the order parameter was interpreted as anmoor humus accumulation (Fig. 6.2e).

Sixth SOP (6% of variation) is positively correlated with the biomass of vegetation in autumn, the presence of soil carbonates, the presence of gruss sediments at the lowest soil horizon, and Munsell chroma of humus horizons and negatively correlated with the height of the microrelief, the curvature of the thalwegs of microrelief, and Munsell value of humus horizons. As a result, this order parameter was interpreted as alluvial soil formation (Fig. 6.2f).

The seventh SOP (5% of variation) is positively correlated with the curvature of the macrorelief interfluves, vegetation biomass in autumn, the Munsell chroma of humus horizons, and the thickness of mor humus horizons and negatively correlated with the curvature of the microrelief interfluves, the presence of sand deposits in the lowest horizons, and the Munsell value of humus horizons. The responsible process is the accumulation of mor humus horizons (Fig. 6.2g).

The order parameters of soils and soil-forming depositions to the greatest extent determine the characteristics of the Munsell color of soil horizons, the thickness of organic horizons, the thickness of humus–accumulative and eluvial–humus–accumulative horizons, the thickness of oligotrophic–peat horizons, soil thickness, the number of horizons per meter, the texture of cover deposits, the presence/absence of medium loamy deposits in the lowest part of the soil profile, and the texture of the lowest horizon. Only 16 out of 54 attributes under consideration are practically not described by the order parameters of soils and soil-forming depositions.

Discriminant analysis of the different interpretations of maximum spreading of the last glaciation showed that discrimination increases as the hypothetic boundary is moved eastward. Quality of discrimination accounts for 83% in case we adopt the maximum according to the geomorphological map, 80% according to Moskvitin's version of the middle-Valdai glaciation, 78% according to Chebotareva's version of the maximum Valdai glaciation, 84% according to Faustova's version of the maximum Valdai glaciation, 84% according to Moskvitin's version of the maximum Valdai glaciation, and 59% according to the sequential encoding of the boundaries from west to east.

The integration of 10 individual order parameters derived from discriminant analysis resulted in three-order parameter for the genesis of soil-forming deposits. The first-order parameter (50% of variation) reflects the formation of the glacial deposits in the west of the territory (the presence of Valdai glaciation) (Fig. 6.3a), the second (17%) the formation of ridges of the Moscow glaciation (Fig. 6.3b), and the third (12%) the formation of a complex of glacial deposits in the southern and central sectors (Fig. 6.3c). The areas of glaciation in the interpretation of Faustova and geomorphological maps are described in the best way from the order parameters.

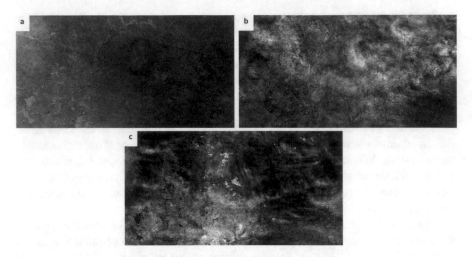

Fig. 6.3 Order parameter for the genesis of soil-forming deposits: **a** – first, **b** – second, and **c** – third

6.9   Integration of Component Order Parameters

Thus, as a result of the analysis of the components, 16 order parameters were identified. Their integration resulted in six order parameters corresponding to a landscape level. Landscape order parameters describe 73% of the variation of the order parameters of components. Interpretation of landscape order parameters is carried out on the basis of their relations to RSD, DEM, and field data, and with the previously determined order parameters of the components.

The first landscape order parameter (21% of variation) reflects the process of succession of boreal forests on Stagnic or Gleyic Albeluvisols, mainly for moraine ridges of the Moscow and Valdai glaciations (Fig. 6.4a), the second one (16%) succession nemoral–boreal forests on Umbric Albeluvisols on a light texture substrate, mainly for the area of the Valdai glaciation (Fig. 6.4b), the third one (13%) succession of boreal–nemoral forests on mor and anmoor Umbrisols on heavy texture substrate, mainly outside the Valdai glaciation (Fig. 6.4c), the fourth one (10%) oligotrophic bogging and gleyization, primarily for the moraine ridges of the Moscow glaciation (Fig. 6.4d), the fifth one (8%) succession of nemoral forests on Umbric Albeluvisols, mostly for moraine ridges of the Moscow glaciation (Fig. 6.4e), and the sixth landscape order parameter (7%) development of floodplain shrub–meadow vegetation on Fluvisols and anmoor Umbrisols (Fig. 6.4f).

The list of attributes that are relatively fully described by landscape order parameters includes BSA of the trees, BSA of *Betula pendula* and *Picea excelsa*, the land cover types, PC of *Alnus incana* and *Picea excelsa*, PC and height of layers of the stand, PC of undergrowth, PC of *Picea excelsa* and *Sorbus aucuparia* undergrowth, group of herbs, PC of moss, and PC of *Sphagnum* mosses, Munsell chroma of

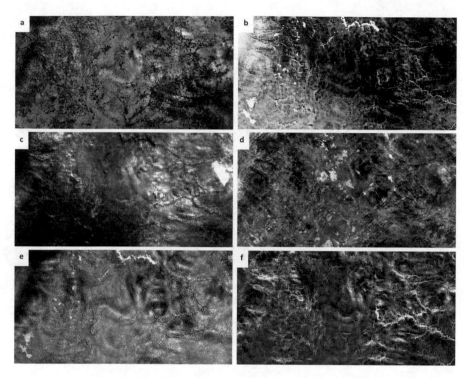

Fig. 6.4 Landscape order parameters (LOPs): **a** – first, **b** – second, **c** – third, **d** – fourth, **e** – fifth, and **f** – sixth

humus accumulation horizons, Munsell value of illuvial horizons, thickness of humus, peat and eluvial horizons, the number of horizons per meter, the presence of cover deposits of medium-loamy and sandy loam textures, and the Valdai glaciation edge in the interpretation of Faustova. In general, only 32 of the 129 considered characteristics have no statistically significant relationships with the landscape order parameters.

6.10   Conclusion

The multidimensional integrated analysis of the field data characteristics, thematic maps, reflection in different spectral bands, and characteristics of the relief of the study area, showed the following.

1. The variation in the state of the characteristics of landscape components measured in the field can be largely (on average 65%) described by RSD and DEM. Most properties depend on the relief-driven redistribution of solar energy and moisture and, to a lesser extent, on the transformation of solar energy by the

landscape cover. In total, 230 individual order parameters describing spatial state of field characteristics and boundaries of glaciation were identified.

2. At the level of integration of individual order parameters in the frame of structural–functional blocks of the properties, we derived 46 integrating order parameters. At the level of integration of blocks into components, 16 order parameters were derived, and at the landscape level, 6 order parameters were derived.

3. The order parameters of components and landscape mainly reflect the processes of forest succession of different types (including anthropogenic), bogging of different types, development of shrub–meadow and meadow–field vegetation, organic matter accumulation, eluviation, etc. The leading process at the landscape level is the succession of boreal forests on Histic Albeluvisols, mainly for moraine ridges of Moscow and Valdai glaciations.

4. More than half of the characteristics of the field data included in the analysis are determined to a great extent by the landscape order parameters.

Thus, quantitative statistical modeling of field data at the medium spatial scale level allows us to distinguish integrating order parameters that reflect independent processes and determine the spatial variation of the properties of components and the landscape as a whole. Knowledge of the localization of the processes that determine the basic properties of the landscape in a particular area allows not only organizing new research aimed at studying the processes and mechanisms of formation of these properties, but also to take them into account when planning economic activities.

Acknowledgments The research was performed within the framework of RSF № 18-17-00129 by metodology development and data analysis by State Task of Institute of Geography RAS № 0148-2019-0007.

References

Aivazyan, S. A., Bushtaber, V. M., Enukov, I. S., & Meshalkin, L. D. (1989). *Applied statistics. Classification and reduction of dimensionality*. Moscow: Finance and Statistics. (in Russian).

Armand, A. D. (1975). *Information models of natural complexes*. Moscow: Nauka. (in Russian).

Armand, A. D. (1988). *Self-organization and self-management of geographical systems*. Moscow: Science. (in Russian).

Atlas of the forests of the USSR. (1973). Moscow: Main Directorate of Geodesy and Cartography under the Council of Ministers of the USSR. (in Russian).

Berg, L. S. (1915). The objectives and tasks of geography. *Proceedings of the Imperial Russian Geographical Society, 51*(9), 463–475. (in Russian). See also in J. A. Wiens, M. R. Moss, M. G. Turner, & D. J. Mladenoff (Eds.) (2006). *Foundation papers in landscape ecology* (pp. 11–18). New York: Columbia University Press.

Chebotareva, N. S., Summet, E. Y., Znamenskaya, O. M., & Rukhina, E. V. (1961). Stratigraphy of the Pleistocene. In *Relief and stratigraphy of Quaternary deposits of the northwest of the Russian Plain* (pp. 101–138). Moscow: Academy of Sciences of the USSR Publishing House. (in Russian).

Classification and diagnostics of soils of the Russia. (2004) Moscow: V.V. Dokuchaeva Soil Institute Publishing House. (in Russian).

Classification and diagnostics of soils of the USSR. (1977) Moscow: V.V. Dokuchaeva Soil Institute Publishing House. (in Russian).

Dokuchaev, V. V. (1948). *Teaching about the zones of nature*. Moscow: Geografgiz. (in Russian).

Faustova, M.A. (1972). *The relief and deposits of the Lovatian lobe of the last glacial cover*. Diss. kand geogr. sciences. Moscow. (in Russian).

Haken, G. (2005). *Information and self-organization. Macroscopic approach to complex systems*. Moscow: KomKniga. (in Russian).

Haken, G. (1980). *Synergetics*. Moscow: Mir. (in Russian).

Kaplin, P. A., Svitoch, A. A., & Sudakova, N. G. (2005). Mainland glaciations and marginal marine basins of Russia in the Pleistocene. *Proceedings of Moscow University, Series 5 Geography, 1*, 55–65. (in Russian).

Kim, J.-O., Mueller, C. W., & Klecka, W. R. (1989). *Factor, discriminant, and cluster analysis*. Beverly Hills, CA: Sage.

Klir, G. J. (1985). *Architecture of systems problem solving*. New York: Plenum Press.

Kozlovsky, F. I. (2003). *Theory and methods of studying the soil cover*. Moscow: GEOS. (in Russian).

Kruskal, J. B., & Wish, M. (1978). Multidimensional scaling. In *Sage University Paper series on Quantitative Application in the Social Sciences*, 07–011. Beverly Hills/London: Sage.

Mandelbrot, B. B. (1982). *The fractal geometry of nature*. New York/Oxford, UK: W.H. Freeman.

Marple, S. L., Jr. (1987). *Digital spectral analysis*. Englewood Cliffs: Prentice-Hall.

Milkov, F. N., & Gvozdetsky, N. A. (1985). *Physical geography of the USSR*. Moscow: Vysshaya shkola. (in Russian).

Moskvitin, A. I. (1967). *Stratigraphy of the Pleistocene of the European part of the USSR*. In Proceedings of GIN AN SSSR, 56. Moscow: Nauka. (in Russian).

Puzachenko, M.Yu. (2007). Landscape accessory of windfalls in the Central Forest Reserve. *Proceedings of Central Forest Reserve, 4*, 304–324. Tula: Grief and K. (in Russian).

Puzachenko, M. Y., Chernenkova, T. V., & Basova, E. V. (2012). Analysis of the natural anthropogenic heterogeneity of the vegetation cover in the central part of the Murmansk region. In A. S. Isayev (Ed.), *Diversity and dynamics of forest ecosystems in Russia* (Vol. 1, pp. 371–383). Moscow: Tov-in scientific publications KMK. in Russian.

Puzachenko, Y. G. (1998). *Methodological basis of geographical forecast and environmental protection*. Moscow: Publishing House URAO. (in Russian).

Puzachenko, Y. G. (1986). Spatial-temporal hierarchy of geosystems from the position of the theory of oscillations. *Issues in Geography, 127*, 96–111. Moscow: Nauka. (in Russian).

Puzachenko, Y. G. (1971). *The study of the organization of biogeocenotic systems*. Doctoral dissertation. Moscow: MSU Publishing House. (in Russian).

Puzachenko, Y. G., Onufrenia, I. A., Aleshchenko, G. M (2002). Analysis of the hierarchical organization of the relief. *Proceedings of Russian Academy of Sciences, Geographical Series, 4*, 29–38. (in Russian).

Puzachenko, Y. G., & Skulkin, V. S. (1981). *Structure of vegetation in the forest zone of the USSR. System analysis*. Moscow: Nauka. (in Russian).

Rabinovich, M. I., & Trubetskov, D. I. (2000). *Introduction to the theory of oscillations and waves*. Moscow-Izhevsk: SRC "Regular and chaotic dynamics, R & C Dynamics". (in Russian).

Semenova-Tyan-Shanskaya, A. M., & Sochava, V. B. (1956). *Plant cover of the USSR. Coniferous-broad-leaved and deciduous forests*. Moscow: Publishing House of Academy of sciences of the USSR. (in Russian).

Shennikov, A. P., & Vasilyev, Y. Y. (1947). Eurasian coniferous Taiga region. In *Geobotanical zoning of the USSR* (pp. 25–60). Moscow: Publishing House of Academy of sciences of the USSR. (in Russian).

Skvortsova, E. B., Ulanova, N. G., & Basevich, V. F. (1983). *The ecological role of windfalls*. Moscow: Forest Industry. (in Russian).

Sochava, V. B. (2005). *Theoretical and applied geography. Selected works*. Novosibirsk: Science. (in Russian).

Sergeev, E. M. (Ed.). (1984). *Soil-geological conditions of non-black earth region*. Moscow: MSU Publishing House. (in Russian).

Sokolov, N. N. (1946). Relief and quaternary deposits of the central Forest reserve. *Proceedings of Leningrad State University, Series of Geographical Sciences, 6*, 52–155. (in Russian).

Solnetsev, N. A. (2001). *The doctrine of the landscape. Selected works*. Moscow: MSU Publishing House. (in Russian).

Turcotte, D. L. (1997). *Fractals and Chaos in geology and geophysics* (2nd ed.). Cambridge: Cambridge University Press.

Vasenyov, I. I., & Targul'yan, V. O. (1995). *The windfalls and the taiga soil formation*. Moscow: Nauka. (in Russian).

Chapter 7
Land Cover Thermodynamic Characteristics Defined by Remote Multispectral Data Based on Nonextensive Statistical Mechanics

Robert B. Sandlersky, Yury G. Puzachenko, Alexander N. Krenke, and Ivan I. Shironya

Abstract This chapter reviews the values of landscape cover thermodynamic variables based on multispectral measurements of reflected solar radiation by Landsat 5 and 7 satellite systems at the local level (Central Forest State Nature Reserve, Tver Region, 1 pixel = 30 × 30m) and MODIS Terra at the regional level (East-European plain, MOD09A1 product, 1 pixel = 1 × 1 km). The calculation of the thermodynamic variables of the landscape cover is based on the nonextensive statistical mechanics model, which is a more general case of the classical mechanics and allows estimating the degree of nonlinearity of relations in a system and the degree of its organization.

Keywords Exergy · Useful work · Q-Parameter · Organization · Nonequilibrium

7.1 Introduction

The calculation of thermodynamic variables of the landscape cover based on multi-spectral remote sensing data allows us to estimate the main components of the energy balance in the spectral range as follows: energy consumption, exergy of solar radiation (i.e., amount used for evapotranspiration), energy dissipation with heat flux and entropy (bound energy), and internal energy increment (i.e., internal energy storage in organic matter). Besides, it evaluates the nonequilibrium transformation of solar energy and entropy flux from the system, that is, how far is the system from equilibrium. This approach was applied in several studies (Puzachenko et al. 2013,

R. B. Sandlersky (✉) · Y. G. Puzachenko · I. I. Shironya
Severtsov Institute of Ecology and Evolution, Russian Academy of Sciences, Moscow, Russia

A. N. Krenke
Institute of Geography, Russian Academy of Sciences, Moscow, Russia

© Springer Nature Switzerland AG 2020
A. V. Khoroshev, K. N. Dyakonov (eds.), *Landscape Patterns in a Range of Spatio-Temporal Scales*, Landscape Series 26,
https://doi.org/10.1007/978-3-030-31185-8_7

111

2016) and was built on the vision of geosystems functioning as open dissipative structures far from equilibrium. It was first elaborated by Jorgensen and Svirezhev (2004), based on the classical Boltzmann-Gibbs-Shannon thermostatics. The approach describes those systems, whose extensive properties are equal to the sum of the system components (fazes), and, hence, are additive. Properties of these systems increase with the growth of a system due to entropy and internal energy.

Additivity is the direct result of entropy, defined by Boltzmann as a casual interaction between molecules or particles, which, given that the thermodynamic equilibrium is reached, may equiprobably appear in any point of a system, and the system itself may appear in any point of the phase space. The particles of the additive system interact only with the neighbor elements and its present state is defined exclusively by the ratio of extensive properties at the particular moment of time. Therefore, thermodynamic equilibrium for such systems is a state with maximum entropy, that is, with a maximum uncertainty of the system microstates. As a result, the phase space of such a system does not contain forbidden states and has the regular properties of continuity, smoothness, and Euclideanism (Kolesnichenko 2015). Complex self-organized systems in nature and society are characterized by hierarchical structure, fractal properties, memory, and emergent properties. They cannot be fully described based on the assumption of the casual character of interactions between elements.

Approaches of nonadditive (nonextensive) thermostatic (Tsallis 2009) developed over the last decades allow moving away from the view of casualty (linearity) of interactions between elementary particles in a system by introducing a parameter (q), which describes the degree of correlation (nonlinearity) of interactions. When $q = 1$, the system under consideration is extensive and thus is described as a particular case of a nonextensive model. The parameter q is a real number, which lays between zero and infinity. Thermostatic studies probabilistic regularities in a system with big numbers of particles, which arise in equilibrium and nonequilibrium states. These patterns are manifested in the distribution of particles between different discrete states i (energy levels), which are characterized by the distribution f_i. Tsallis (2009) generalized the Boltzmann-Gibbs-Shannon formula of entropy (S_b) into the entropy of Tsallis (S_q):

$$S_b = -K_b \sum_{i=1}^{n} f_i \ln f_i =>$$

$$S_q = \frac{K_b}{q-1} \sum_{i=1}^{n} f_i \left(1 - f_i^{q-1}\right),$$

where K_b is the Boltzmann constant; n is a number of possible discrete states i with a sum of probabilities $\sum_{i=1}^{n} = 1$, where q is a parameter of the phase space deformation.

"Deformation" of the logarithmic function of the Boltzmann entropy explains the behavior of complex systems ("thermodynamically abnormal" systems with long-range connections) when the probability of the realization of the large energy values (high energy levels) decreases (when $q > 1$) not exponentially fast, but gradually. Therefore, the Tsallis statistics can describe events, which are impossible in simple systems. Thus, the parameter q is a measure of the system's nonadditivity, which characterizes additional degrees of freedom typical for complex systems and has to be determined in each particular case. Cases with $q < 1$, $q = 1$, and $q > 1$ correspond, respectively, to super-additivity, additivity, and subadditivity of a system. The internal correlations are inherent to systems with $q > 1$, and the higher is q, the bigger is their contribution to the system's organization (Puzachenko 2016). As a result, other things being equal, the higher is q, the lower is the value S_q.

In the Tsallis system the information, as well as entropy, is nonadditive, but the scheme of the energy balance stays the same as in the Boltzmann thermostatic:

$$R = Ex_q + TWS_q + DU_q,$$

where R is consumed radiation, Ex_q is exergy (free energy), TWS_q is the product of heat flow and entropy (bound energy), DU_q is the internal energy. As with additive system, exergy is evaluated by the Kullback information increment (Puzachenko 2016).

7.2 Materials and Methods

In this chapter, we review the assessments of the thermodynamic variables of the southern taiga landscape cover on the example of the Central Forest State Nature Reserve (CFSNR), Tver Region, Russia. We used Landsat 5 and 7 images (resolution 30 m) and MODIS Terra (MOD09A1 product, resolution 1 km) images obtained for the East-European Plain. Both satellite systems have seven shooting channels: for Landsat six channels in short-wave subspectrum and one band in long-wave thermal field and seven short-wave bands from MODIS. According to calibration constants of the sensors, the values in the channels were converted into radiation, reflected by the active surface and detected by survey system sensors (W/m²). For Landsat data brightness, values of the thermal channels were converted into heat flux from the active surface and its temperature. Solar radiation income in the channels was calculated according to the solar constant for each channel adjusted to the solar altitude and the distance between the Sun and the Earth at the time of the survey. Accordingly, the energy absorbed in each channel was calculated as the difference between the solar radiation income and outcome.

The thermodynamic characteristics for land cover were calculated by the procedure suggested by Jorgensen and Svirezhev (2004) and adjusted for the calculation of bound energy, which were initially absent in the model (Puzachenko et al. 2011, 2013). Nonequilibrium in land cover to incoming solar radiation spectrum was measured in accordance with Kullback information increment, measuring the distance between the distribution of energy in the spectrum of the incoming and outcoming solar radiation. The more similar these spectra are, the more equilibrium is the receiver of incoming energy flow (i.e., land cover), and accordingly, the smaller the information increment is. Kullback information increment (K, nit) is:

$$K = \sum_{v=1}^{6} p_v^{\text{out}} \ln \frac{p_v^{\text{out}}}{p_v^{\text{in}}},$$

where $p_v^{\text{in}} = \dfrac{e_v^{\text{in}}}{E^{\text{in}}} / p_v^{\text{out}} = \dfrac{e_v^{\text{out}}}{E^{\text{out}}}$ is fraction of incoming/outcoming energy in spectral range v of total incoming/outcoming energy, $e_v^{\text{in}} / e_v^{\text{out}}$ is incoming/outcoming energy in spectral range v. Exergy of solar radiation (Ex) was calculated as:

$$\text{Ex} = E^{\text{out}} \left(K + \ln \frac{E^{\text{out}}}{E^{\text{in}}} \right) + R$$

where E^{in} is incoming solar energy, W/m², E^{out} is outcoming solar energy, W/m², $R = E^{\text{in}} - E^{\text{out}}$ is absorbed energy, and $E^{\text{out}}/E^{\text{in}}$ is albedo.

To evaluate the bound energy—the energy dissipation with heat flux and entropy, it is necessary to evaluate the entropy of the outcoming radiation. The greater the entropy of outcoming radiation flux is, the more equilibrium is the system, which converts solar energy (S_{out}, nit):

$$S_{\text{out}} = -\sum_{v=1}^{6} p_v^{\text{out}} \ln p_v^{\text{out}}.$$

Bound energy (STW, W/m² nit):

$$\text{STW} = \text{TW} * S_{\text{out}},$$

where TW is heat flux of active surface, recorded by the thermal channel.

The internal energy increment of the system (DU), which is the transition of the absorbed solar energy into the internal energy of the system, is estimated as the residual of the absorbed energy balance equation (R):

$$\text{DU} = R - \text{Ex} - \text{STW}.$$

The measure of organization (Rq) by Von Foerster (1964) is:

$$R_q = 1 - \frac{S_q}{S_q^{\text{max}}},$$

where S_q is Tsallis entropy and S_q^{max} is maximal measured Tsallis entropy.

Thus, the following thermodynamic characteristics were calculated: those forming the balance of the absorbed solar energy (W/m²), which are exergy (W/m²), bound energy (W/m² nit) and internal energy increment (W/m²); structural system characteristics describing its nonequilibrium that are information increment (nit) and entropy of outcoming radiation (nit); system heat flux (temperature), and vegetation index (W/m²), q-parameter—a measure of the system's nonadditivity, organization.

For large scale, we used 22 Landsat scenes, performed in the morning hours with the cloudless sky in different seasons from 1986 to 2011 for Central Forest Reserve (56°30′ N 32°53′ E), located in the southwestern part of the Valdai Hills, within the main water divide of the East-European plain between the Caspian, the Baltic, and the Black Seas basins. Natural complex of the reserve is typical for the southern taiga subzone and is a model of the primary land cover for a vast area of the morainic relief inherent for the central part of the East-European plain. The study area is almost entirely comprised of the reserve core and a part of the protective zone. The total area of the territory under study is 1392 km². For small scale, we used one MOD09A1 product scene for the East-European Plain (July 28, 2002) with the corners 63.56°N 30.64°E, 63.56°N 50.20°E, 50.20°N 30.64°E, 50.20°N 63.56°E (1430 × 1180 km).

7.3 Results

Analysis of the seasonal variation of energy variables for CFSNR as a whole showed that the energy conversion is determined by income solar radiation and, therefore, is fundamentally different for the snow cover period and vegetation period. In winter, the system converting the energy is at most close to equilibrium. In the snowless period with an increase in income of solar radiation, expenditure of energy on exergy is maximum, entropy decreases, and the increment of information and products as well as the biological production related to the increase in system nonequilibrium increases. In general, for the landscape the solar energy absorption, exergy, and nonequilibrium are maximal in summer (June). In winter, from December to early April, the mean for the landscape is $q < 1$. It follows that the system is in a state of disintegration. From the beginning of the vegetation period $q > 1$, with a maximum in June. This indicates active self-organization processes. The change in organization time corresponds to Klimontovich theorem: self-organization is the maximum of entropy and the minimum of entropy production. The organization has two maxima related to the two states of the system: in winter and summer with two transition periods in October–December and in April.

We evaluated spatial variation in thermodynamic variables in the southern taiga landscapes. In winter, meadows and bogs, unlike forests, are in a state close to thermodynamic equilibrium. Only forest communities maximize the absorption of solar radiation and exergy and minimize heat flux and the energy dissipation. Meadows maximize biological production, internal energy, and the information increment at the same time, but for the entire snowless period. Bogs maximize heat flux and energy dissipation and minimize the information increment while supporting the biological production at average level. The average temperature over the bog during

Fig. 7.1 Parameter q, calculated from Landsat 5 TM (30 × 30 km) for 21.06.2002. Black—lowest value, white—highest value

the entire snowless period is 3.4° as high as in the forest and 1.8° as high as in the meadows. In summer, the bog is warmer than the forest by the average of 4.5° and the meadows by 2.2°. The minimum value of the parameter q (Fig. 7.1) with the low level of organization corresponds to the old spruce forests. In the overgrown wind-fall forests, the value slightly increases and reaches its maximum in the young forests. As a forest gets older, the value of q and the organization level gradually decline. On meadows, it reaches its absolute maximum. The organization level and the q parameter decrease from recent logging areas to arable lands and settlements. Bogs represent unique formations with the minimal organization level and low level of q. On the one hand, these relations reflect the difference in self-organization of bogs, forest, and meadow communities. On the other hand, forest communities show that the ability to organize is decreasing as they grow old.

The overall image for the regional level calculated from MODIS data is demonstrated on Fig. 7.2. The q parameter decreases it towards the north and the south. It reaches maximum in the forests of the central European Russia. One more typical feature is that cities and open water both have minimal q. In full accordance with the theory, large values of the index indicate forest landscapes with maximum productivity, transport of moisture and, accordingly, self-organization. On this basis, we can evaluate the whole set of the thermodynamic parameters: exergy, free energy, entropy, Kullback information, the associated energy, and the contribution of self-organization to the functioning, which completely determine the operation of elementary geosystems.

Fig. 7.2 Parameter q, calculated from MOD09A1 (1 × 1 km) for 27.07.2002. Black—lowest value, white—highest value

7.4   Discussion

The applicability of the nonextensive statistical mechanics for the mapping of the solar energy transformation by the landscape cover is first of all confirmed by the correspondence between the spatiotemporal dynamics of the parameter q and the conception of self-organization of various forms of ecosystems. The obtained results are fully valid for both global and regional levels: q is minimal during winter (less than one), when the vegetation cover is less active (system farthest from equilibrium) and maximal during the most active vegetation period (more than 1). The open ground surface has a minimum value of q. Analysis of the seasonal dynamics of the q-parameter as a function of meteorological data showed that at the local level q is positively related to the height of the sun and the sum of temperatures for 24 days. This reflects the fact of the system's memory in relation to the previous conditions. At the regional level, the maximum organization (q) is characteristic of mixed forests in the period of active vegetation.

In the framework of nonextensive statistical mechanics, it is possible to calculate the energy expenditure for evaporation for the equilibrium condition in alignment with the linear Boltzmann-Shannon model and the energy expenditure for the nonequilibrium processes of the structure transformation determined by the Kullback q-information. It was established that with the value of the parameter q greater than 1, the energy expenditure on evapotranspiration is smaller than in the linear model, and the transformation of the structure is somewhat larger. The actual advantages of using nonadditive statistical mechanics in comparison with the classical linear model are determined by the deriving several mutually complementary criteria for self-organization and its contribution to the transformation of solar energy. Accordingly, we can evaluate the most important functional characteristics of ecosystems. Obviously, they should be useful first of all for assessing the efficiency of energy use in agriculture and forestry, assessing the level of degradation in the transformation of the energy of urban and man-made systems and finding ways to optimize them. It is possible that they can be useful for searching for nonequilibrium, self-organizing systems in astrophysical studies, for example (Kolesnichenko 2015).

Acknowledgments The work was supported by the grant of the Russian Science Foundation № 18-17-00129 (Landsat data processing), Russian Science Foundation № 17-77-10135 (MODIS data processing).

References

Jorgensen, S. E., & Svirezhev, Y. M. (2004). *Towards a thermodynamic theory for ecological system*. Oxford: Elsevier.
Kolesnichenko, A. V. (2015). Modification in the framework of Tsallis statistics of the criteria of gravitational instability of astrophysical disks with a fractal structure of the phase space. *Mathematica Montisnigri, 32*, 93–118. (in Russian).
Puzachenko, Y. G. (2016). Thermostatical foundations of geography. *Proceedings of Russian Academy of Sciences, Geographical Series, 5*, 21–37. (in Russian)
Puzachenko, Y., Sandlerskiy, R., & Svirejeva-Hopkins, A. (2011). Estimation of thermodynamic parameters of the biosphere, based on remote sensing. *Ecological Modelling, 222*, 2913–2923.
Puzachenko, Y., Sandlersky, R., & Sankovski, A. (2013). Methods of evaluating thermodynamic properties of landscape cover using multispectral reflected radiation measurements by the Landsat satellite. *Entropy, 15*(9), 3970–3982.
Puzachenko, Y., Sandlersky, R., & Sankovski, A. (2016). Analysis of spatial and temporal organization of biosphere using solar reflectance data from MODIS satellite. *Ecological Modelling, 341*, 27–36.
Tsallis, C. (2009). *Introduction to nonextensive statistical mechanics*. New York: Springer.
Von Foerster, H. (1964). *About the self-organizing systems and their environment. Self-organizing systems* (pp. 113–139). Moscow: Mir.

Part III
How Patterns Control Actual Processes

Chapter 8
Structure and Phytomass Production of Coastal Geosystems Near Lake Baikal

Yulia V. Vanteeva and Svetlana V. Solodyankina

Abstract Landscape structure has an important control over its functions. Landscape functions are treated as the resulting manifestations of the functioning processes in a landscape system of interrelated components: rocks, soil, water, air, and biota. Geosystem concept formulated by Viktor Sochava and factoral-dynamic approach to facies classification elaborated by Adolph Krauklis were applied as a theoretical and methodological framework. Natural landscapes surrounding Lake Baikal provide functions involved in the protection of water quality, water storage, phytomass production, and so forth. The phytomass production function was estimated in four study areas located in different parts of Lake Baikal shore (the east-northern—Barguzin mountain range; the western—the Priol'khon plateau and Olkhinskoe plateau; the southern—part of Khamar-Daban mountain range) and exposed to different degrees of impact. Fieldworks were conducted in summer seasons from 2010 to 2017. For these areas, the landscape maps at local scale were composed based on the landscape approach, fieldwork data, digital elevation model, and remote sensing. Landscape maps were interpreted to evaluate phytomass stock. Interpretation was based on tree phytomass and aboveground herbaceous phytomass measurements for various landscape types. Correlations between landscape attributes and this function were found.

Keywords Landscape approach · Landscape map · Factoral-dynamic series of facies · Geosystem classification · Function of phytomass production

Y. V. Vanteeva (✉) · S. V. Solodyankina
V.B. Sochava Institute of Geography, Siberian Branch of Russian Academy of Sciences, Irkutsk, Russia

© Springer Nature Switzerland AG 2020
A. V. Khoroshev, K. N. Dyakonov (eds.), *Landscape Patterns in a Range of Spatio-Temporal Scales*, Landscape Series 26,
https://doi.org/10.1007/978-3-030-31185-8_8

121

8.1 Introduction

The demand for the multifunctional use of the landscapes is constantly increasing in modern society. The optimal and sustainable land-use is achieved on the basis of understanding the processes occurring in the natural system and quantitative assessment of its potential to provide natural functions. This approach allows designing measures for restoration and engineering and technical compensation of lost natural potential.

Natural functions and the processes responsible for their emergence are determined by landscape structure. Each function is related to a certain hierarchical level of the geographic system, in other words, a certain level of generalization. For example, climate-regulating functions are manifested mainly in *macrogeochore* (i.e., the series of geochors *sensu*, V.B. Sochava 1978), which is referred to as regional level. The biological productivity of landscapes can be considered at different hierarchical levels (ranging from local to global one), but differentiation at the local level is important for decision-making in the field of land-use planning and economic activity at a certain site.

The study presented here was carried out within the framework of the geosystem approach proposed by Viktor Borisovich Sochava (1905–1978) and on the basis of the model of factoral-dynamic series of facies of Adolph Albertovich Krauklis (1937–2006).

8.2 Theoretical Foundations

The interest of science for the landscape spatial structure per se and as an indicator of landscape-forming processes is increasing due to the availability of remote sensing data and computer processing methods. To describe the relationship between the landscape structure and the natural processes in the geographic system, it is necessary to apply different approaches to the identification of structural units of landscapes. One of such approaches was proposed by the Russian scientist V.B. Sochava. He developed a geosystem concept (Sochava 1972, 1974, 1978), based on a general system theory, and introduced the term "geosystem." It became an essential contribution to the development of physical geography and landscape science. Since then, the term "geosystem" has been widely used in physical geography in Russia (Isachenko 2004; Konovalova et al. 2005; Lastochkin 2011) and abroad (Bastian et al. 2015; Christopherson and Birkeland 2015). The geosystem concept has been used in applied aspects of landscape study, landscape ecology, and landscape planning (Khoroshev and Koshcheeva 2009; Khoroshev 2016; Bastian et al. 2014, 2015; Grunewald et al. 2014; Semenov 2014; Semenov and Lysanova 2016; Istomina et al. 2016, etc.).

A general overview of the theoretical aspects of the geosystem and landscape concepts is given in the paper by Bastian et al. (2015). We will consider the main principles of hierarchical classification of geosystems and the factoral-dynamic approach elaborated within the framework of this concept and give examples of how to apply this approach to mapping heterogeneity of geosystem structure on fine-scale and medium-scale maps.

According to the geosystem concept, geosystem is a system of interrelations of various components, including water, air, rocks, soil, and biota. Geosystems can be represented as a double-row hierarchical classification model, where the ideas of homogeneity and heterogeneity of geographical space are combined at the local, regional, and global levels. In this model, *geomer* is treated as a homogeneous typological unit and *geochore* as a heterogeneous chorological unit.

An *elementary geomer* is a minimal homogeneous space, which includes all components of a given geosystem. It is correlated with the concept of homogeneity (Neef 1963a) that has been used to define ecotopes. Geomers (facies, groups of facies, class of facies, etc.) do not occupy a continuous space but are distributed as a mosaic within certain boundaries limiting their distribution area. *Facies* is the set of similar biogeocoenoses.

Geochore (*podurochishche, urochishche, mestnost*,[1] etc.) is a heterogeneous spatial system formed by geographically adjacent geomers, which together represent a structural-dynamic and functional entirety. The term "geochore" is close to the notion of the chorological dimension of geosystems (nano-, micro-, meso-, macro-ecochore; E. Neef 1963b; G. Haase 1973).

An important aspect of the geosystem approach is the idea of the dynamic properties of geosystems. Geosystems are represented by different kinds of primary structures and variable states obeying a particular invariant, the change in which implies the evolution of a geosystem. Variable states of geosystems are treated as different modifications to the primary structure: serial geosystems, recovery stages, and seasonal phases of primary and serial geosystems (Frolov 2015). The main aims of the dynamic method are to trace the transition of a geosystem from one state to another, to reveal the variety of its inherent state variables, and to determine and measure the factors causing the observed changes as correctly as possible.

The most appropriate model for analyzing the dynamic transformations of geosystems at the local level is the model of the *factoral-dynamic series* of facies developed by A.A. Krauklis (1972, 1979). The main purpose of this model is to make clear distinction between: (a) intact etalon facies, which are typical for given zonal, sectoral, high-altitude, and other features of the landscape; and (b) facies, modified under the dominating strong influence of a single factor or a combination of factors. The "norm" or "standard" for typical facies (according to A.A. Krauklis) is, for example, the *plakor* facies (flat or slightly inclined part of the interfluve or terrace) at a well-drained location with loamy soils. A.A. Krauklis distinguished several basic trends (series) in transformation of the etalon facies under the influence of the environmental factors:

---

[1] *Podurochishche, urochishche, and mestnost* are the Russian terms for the landscape morphological units at various hierarchical levels. See Glossary and Chap. 1 for definitions.

(1) The lithomorphic series is developed due to the critical influence of mineral substrate in the geosystem, which causes an increase in stoniness and a decrease in soil thickness.
(2) The hydromorphic series is formed due to constant moistening.
(3) The cryomorphic series is characterized by a decrease in the amount of heat and the appearance of a permafrost horizon that exerts a permanent influence.
(4) The stagnant series is formed due to stagnation of matter in the geosystem and mire development.
(5) The psammomorphic series is characterized by an increase in physicochemical passivity and the biological and environmental impoverishment of soil (gradual transition toward barren sand).

Overlapping of various factors is observed often in nature. Their interaction results in the development of complex "multifactoral" series, for example, hydrolithomorphic, cryohydromorphic ones, etc.

By the degree of transformation, facies are divided into categories reflecting the dynamic state of the geosystem: primary, pseudo-primary, and serial. The primary facies are stable geosystems most corresponding to the latitudinal zonal, longitudinal sector, and high-altitude "standard." Serial facies are geosystems where the stabilization of the structure and regimes is not achieved under the influence of various disturbance environmental factors. The pseudo-primary facies is an intermediate stage between the primary and serial facies.

In general, the series of transformations described by A.A. Krauklis in their movement toward the primary state is to some extent similar to the succession series of F. Clements (1949) and the geographical cycles of W. Davis (1954). The main difference between the factoral-dynamic series of facies from a succession series and geographical cycles is that they are considered as open systems.

The factoral-dynamic series of facies can be considered as one of the variants of the model reflecting the positional-dynamic type of the landscape structure. The positional component is the ordering of the geosystems along the factor axis, depending on the degree of the transforming influence. The dynamic component is the changes in the variable state of geosystems (primary, pseudo-primary, and serial facies).

The study of factoral influence is widely developed in ecology, where environmental factors are considered from the viewpoint of environmental effects on living organisms. An ecological niche is the totality of all environmental factors within which a species can survive in nature. The ecological niche allows describing position that a particular species occupies in relation to the gradients of external factors: temperature, humidity, soils features, etc. (Ramensky et al. 1956; Landolt 1977; Ellenberg 1996; Heams et al. 2015). This idea is associated with the factoral-dynamic series, where the facies are also located in compliance with gradient of environmental factors. It should be noted that at present, in landscape ecology, the gradient approach is actively developing and involves studying gradual changes in landscape properties along ecological gradients (Evans and Cushman 2009; Cushman and Huettmann 2010).

Next, we will consider the application of the described approach to the study of the geosystem structure and the assessment of the phytomass accumulation function of geosystems at the local level by the example of study areas in the Baikal region.

8.3   Methods

Research of the geosystem structure was carried out at the study areas in several steps:

(1) Collecting information about the territory (thematic maps, space images, etc.)
(2) Conducting field works, desktop data treatment, and building Geographic Information System (GIS) databases
(3) Development of the geosystems classifications; analysis and processing of remote sensing data
(4) Compilation of landscape maps

The initial data were topographic maps (1:50000, 1: 200000), digital elevation models (DEM) based on the shuttle radar topography mission (SRTM—with a resolution of 30 m), geological maps (1: 200000), multispectral space imagery (Landsat 5, 8, ASTER—Advanced Spaceborne Thermal Emission and Reflection Radiometer), the landscape map of the southern East Siberia (1: 1500000) (Mikheev and Ryashin 1977), etc.

Fieldwork involved the conduction of integrated physical-geographical descriptions of the most representative sites (sample plots) characterizing the geosystem diversity on the study areas. Sample plots (10 × 10 m for steppes and meadows, 15 × 15 m for forests) were chosen at the main relief elements at various locations. At each sample plot, we recorded the geographical coordinates, altitude, the features of the relief (landform, microrelief, steepness, and aspect of the slopes) and lithology (the presence of rock outcrops), herbaceous and tree vegetation (species composition, vegetation cover, average diameter, height, and density of tree stand), and soil characteristics (for each horizon: thickness, soil texture, color, structure, density, humidity, inclusions, and neoformations).

Classification of geosystems was carried out for a number of geomers: examples of facies (the lowest hierarchical level), groups of facies, classes of facies, and geoms (highest hierarchical unit at the local level). To determine the factoral-dynamic series of facies and groups of facies, we performed ordination of test plots along the axes of three parameters as follows: the thickness of the humus horizon (field data), the topographic moisture index (calculated from DEM), and the altitude from DEM. The identified groups of sample plots were provided with the characteristics of the microrelief, vegetation, and soils. After that, they were ordered into typological units (groups of facies).

The groups of facies were combined in classes of facies corresponding to a certain factoral-dynamic series, in which we distinguished that the primary geosystem corresponds to a zonal norm, and the serial groups of facies transformed under the

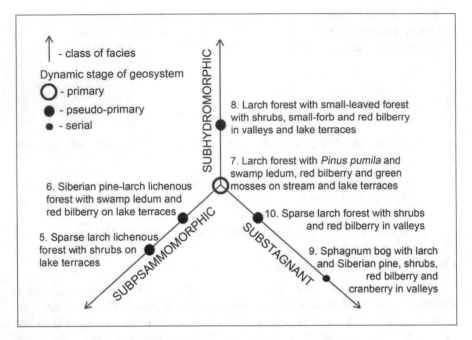

Fig. 8.1 An example of factoral-dynamic series of facies group (for Barguzin mountain range study area, the geom of larch taiga forests of piedmont's and valleys)

strong influence of a certain environmental factor or set of factors (Fig. 8.1). The succession stages of geosystems and their anthropogenic modifications were taken into account in the classification as well. The classes of facies were combined into geoms in accordance with the altitudinal zonation (for mountain areas) with similar structural features of the soil and vegetation cover.

The results of the classification were used to compose the legends of landscape maps. Mapping of the landscape structure was performed based on data obtained from the analysis of the DEM and the automatic supervised classification of space images using field data. Also, additional sources of information were used: photography, geoportals of open spatial data, etc.

The quantitative values of the aboveground phytomass of herbaceous vegetation and tree stand phytomass were estimated on sample plots during fieldwork. The tree stand phytomass was calculated using the values of volume-conversion coefficients for major tree species in Russia, which were published by Zamolodchickov et al. (2003, 2005). The sampling design was chosen to estimate aboveground herbaceous phytomass (Catorci et al. 2009; Deak et al. 2011; Kelemen et al. 2013). The vegetation was clipped at 0.01 m above the ground on the plots 0.25 m^2 and was subsequently field air-dried and weighed to an accuracy of 0.001 kg.

Within the boundaries of the geosystem areas of different hierarchical levels, certain characteristics of the natural functions are homogeneous. This allowed us to apply the geosystem classification and landscape mapping to assess the phytomass accumulation function of geosystems. GIS databases were built on the basis of land-scape maps for each study areas with information on the characteristics of geosystems (typological units, dynamic state, and quantitative indicators—altitude, steepness, and aspects of slopes, etc.). The values of the phytomass were interpolated within the boundaries of the groups of facies taking into account the degree of anthropogenic transformation.

8.4   Study Areas

The geosystem structure of the Baikal region includes combinations of tundra, taiga, and steppe, complicated by high-altitude zones of mountain systems of the region. Also, the intermountain basins play an important role in the differentiation of geosystems. The huge water mass of Lake Baikal has a significant influence on the formation of local climatic features. Mountain larch taiga geoms are a zonal norm in the corresponding range of the energy balance. Mountain taiga covers the main area of the ridges surrounding the basin of Lake Baikal.

To characterize the geosystems of the Baikal region, four study areas have been selected (Fig. 8.2), which allow taking into account the diversity of geosystems formed under the influence of orographic, climatic, and anthropogenic factors (Table 8.1). The first study area (96,600 ha) is located on the southwestern coast of Lake Baikal, in the southern part of the Olkhinskoe plateau, and is distinguished by a combination of cultural (the Circum-Baikal railway—a national-level listed landscape-architectural monument) and natural (at different succession stages) landscapes. The second study area (6910 ha) is located on the northeastern shore of the lake (the piedmont of Barguzin mountain range), where the thermic regime is the most unfavorable for vegetation in comparison with other studied areas. The third study area (1990 ha) is located in the central part of the western coast of the lake, in the northern part of the Priol'khon plateau, and has a minimum annual precipitation. The fourth study area (12,530 ha) is located on the north-facing slope of the Khamar-Daban mountain range with maximum annual precipitation and moderate continentality.

8.5   Results

Landscape maps of study areas (see Fig. 8.2) were compiled based on the results of field and remote sensing data treatment.

The southern part of the Olkhinskoe plateau is presented by light-coniferous mountain taiga and submountain subtaiga. As a result of the construction of the Circum-Baikal Railway (1899–1905), coastal geosystems of the study area were subject to significant transformations (up to blasting operations for terracing slopes and tunneling). The steppe on steep slopes widespread here is dynamic due to

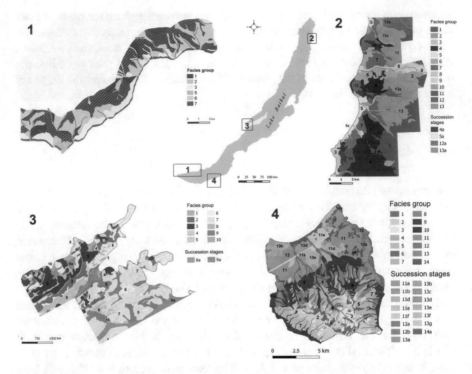

Fig. 8.2 The location of the study areas in the Baikal region and their landscape maps: (1) Olkhinskoe plateau—a fragment (hatching shows geosystems in succession stages), (2) Barguzin mountain range, (3) Priol'khon plateau, (4) north-facing macroslope of the Khamar-Daban mountain range

Fragment of the legend (prevalent types of geosystems)

Olkhinskoe plateau

NORTH ASIA TAIGA

Class of geoms—Mountain-taiga

Geom—Mountain-taiga of coniferous forests on granites and gneiss

Class of facies—subhydromorphic

Group of facies—1. Larch (*Larix sibirica* Ledeb.) with Siberian pine (*Pinus sibirica* Du Tour) forest with small shrub (*Rubus arcticus* L.), small-forb (*Maianthemum bifolium* (L.)) and sedge in herbaceous layer on combinations of Entic Podzols and Haplic Albeluvisols Abruptic on flat surfaces (Primary)

a. Secondary aspen-birch forest with shrubs and herbaceous layer formed by sedge, forb and small reed (60–190 t/ha).

Geom—Subtaiga of light-coniferous forests on granites, gneiss, dikes of dolerites and porphyrites

Class of facies—sublithomorphic

Group of facies—2. Larch-pine (*Pinus sylvestris* L.) forest with herbaceous layer formed by forb, sedge and graminoids on combinations of Entic Podzols and Haplic Albeluvisols Abruptic on slopes (Pseudo-primary) (200 t/ha)

a. Secondary birch forest with herbaceous layer formed by forb, sedge and small reed (90–190 t/ha)

Group of facies—6. Cultural landscapes (rail way and settlements)

Barguzin mountain range

NORTH ASIA TAIGA

Fig. 8.2 (continued)

Class of geoms—Mountain-taiga with a certain Pacific influence
Geom—Mountain-taiga of larch forests on biotite granites and diorites
Class of facies—sublithomorphic
Group of facies—4. Larch-Siberian pine and Siberian pine-larch (*Larix gmelinii* (Rupr.) Rupr.) forest with shrubs (*Rosa acicularis* Lindl., *Salix* spp., *Ledum palustre* L.) with small-forb (*Maianthemum bifolium, Linnaea borealis* L., *Trientalis europaea* L.) and red bilberry (*Vaccinium vitis-idaea* L.) and green mosses on Umbric Albeluvisols Abruptic on slopes or upper slopes (Primary) (145–287 t/ha)
a. Secondary mixed aspen-birch-larch and Siberian pine (*Populus tremula* L., *Betula platyphylla* x *B. pubescens*) forests with red bilberry (*Vaccinium vitis-idaea*) and forb (*Chamaenerion angustifolium* (L.) Scop., *Calamagrostis langsdorffii* (Link) Trin.) in herbaceous layer (212–221 t/ha)
Geom—Piedmont and valleys taiga of larch forests on alluvium
Class of facies—subpsammomorphic
Group of facies—7. Larch forest with *Pinus pumila* (Pall.) Regel and swamp ledum (*Ledum palustre*), red bilberry (*Vaccinium vitis-idaea*) and green mosses on Rustic Podzols on stream and lake terraces (Primary) (90–219 t/ha)
Class of geoms—Mountain-taiga
Geom - Mountain-taiga of pine forests on biotite granites diorites and monzonites
Class of facies—sublithomorphic
Group of facies—13. Pine (*Pinus sylvestris*) forest with red bilberry (*Vaccinium vitis-idaea*) and green mosses or debris layer on Rustic Podzols and Umbric Albeluvisols Abruptic mostly on steep slopes (Pseudo-primary) (204–406 t/ha)
a. Secondary mixed aspen-birch-pine (*Pinus sylvestris, Betula platyphylla* Sukaczev, *Betula pendula* Roth, *Betula divaricata* Ledeb., *Populus tremula*) forest with red bilberry (*Vaccinium vitis-idaea*) and small-forb (146–147 t/ha)
Priol'khon plateau
NORTH ASIA TAIGA
Class of geoms—Mountain-taiga
Geom—Piedmont light-coniferous forests on crystal slate, gneisses and marbles
Class of facies—sublithomorphic
Group of facies—3. Sparse larch (*Larix sibirica*) with cotoneaster (*Cotoneaster melanocarpus* Fischer ex Blytt*)* and forb *(Carex pediformis* C.A. Meyer, *Festuca lenensis* Drobow, *Pulsatilla turczaninovii* Krylov et Serg.) on Leptosols Humic or Dystric Hyperskeletic Leptosols on slopes mostly northern and north-western direction (Pseudo-primary) (57–101 t/ha)
CENTRAL ASIAN STEPPE
Class of geoms—Upland
Geom—Piedmont steppes on crystal slate, gneisses and marbles
Class of facies—subxerolithomorphic
Group of facies—6. Steppe (*Agropyron cristatum* (L.) Beauv, *Festuca lenensis, Artemisia gmelinii* Weber ex Stechm, *Aconogonon angustifolium* (Pallas) Hara, *Phlojodicarpus sibiricus* (Fischer ex Spreng.) Koso-Pol., *Carex argunensis* Turcz. Ex Trev, *Carex korshinkyi* Kom., *Carex pediformis*) with caragana (*Caragana pygmaea* (L.) DC) on Leptosols Humic or Dystric Hyperskeletic Leptosols on slopes (Serial) (0.4–1.5 t/ha)
Group of facies—7. Steppe (*Festuca lenensis, Thalictrum foetidum* L. s. l., *Chamaerhodos altaica* (laxm.) Bunge) on Leptosols Humic or Dystric Hyperskeletic Leptosols on slopes (Pseudo-primary) (0.6–1.3 t/ha)
Group of facies—9. Steppe (*Agropyron cristatum, Stipa baicalensis* Roshev, *Festuca lenensis, Cymbaria dahurica* L.) on Leptosols Humic or Dystric Hyperskeletic Leptosols on gentle slopes and lows (Primary) (0.5–1.6 t/ha)
a. Secondary forb-sagebrush steppe (0.3–1 t/ha)
Northern macro-slope of the Khamar-Daban mountain range
NORTH ASIA TAIGA

Fig. 8.2 (continued)

Class of geoms—Mountain-taiga
Geom—Mountain-taiga of dark-coniferous forests on gneiss and plagiogneiss
Class of facies—sublithomorphic
Group of facies—7. Fir (*Abies sibirica* Ledeb.) with the admixture of spruce (*Picea obovata* Ledeb.) and Siberian pine forests with forb (*Dryopteris expansa* (C. Presl) Fraser-Jenk. & Jermy, *Maianthemum bifolium, Anemone baicalensis* Turcz.) and elephant-eared saxifrage (*Bergenia crassifolia* (L.) Fritsch) in the herbaceous layer on combinations of Humic Leptosols and Dystric Cambisols with rock outcrops on steep slopes (Serial) (198–344 t/ha)
Group of facies—8. Spruce-fir and fir-spruce forests with herbaceous layer formed by forb (*Maianthemum bifolium, Anemone baicalensis, Calamagrostis obtusata* Trin.), fern-forb (*Dryopteris expansa, Phegopteris connectilis* (Michx.) Watt, *Matteuccia struthiopteris* (L.) Tod.) and bilberry (*Vaccinium myrtillus* L.) on steep slopes on combinations of Entic Podzols and Regosols (Pseudo-primary) (600–1450 t/ha)
Geom—Piedmont taiga of dark-coniferous forests on alluvium
Class of facies—subhydromorphic
Group of facies—11. Siberian pine-spruce forest with fir krummholz with small-forb (*Maianthemum bifolium, Linnaea borealis*) and sedge (*Carex appendiculata* (Trautv. & C.A. Mey.) Kuk., *Carex iljinii* V.I. Krecz.) herbaceous layer with abundant peat mosses (*Pleurozium schreberi* (Brid.) Mitt., *Hylocomium splendens* (Hedw.) Bruch et al., *Dicranum scoparium* Hedw.) on combinations of Folic Entic Podzols, Histic Cryosols and Fibric Histisols (Pseudo-primary) (433–463 t/ha)
Notes: in parentheses are the numerical symbols—the values of the tree phytomass and for *Geom - Piedmont steppes on crystal slate, gneisses and marbles* (Priol'khon plateau)—herbaceous phytomass. The names of the soils are given in accordance with The World Reference Base (WRB)

dangerous endogenous and exogenous processes, despite the adoption of thorough and well-considered erosion control measures in the construction of the railway.

Geosystems at the study area in the Barguzin mountain range are presented by bald mountain and taiga types. Most territory is covered with taiga (99.4% of the area).

The study area on the Priol'khon plateau is represented mainly by steppes (75.1% of the area); the other part of the territory is covered by the forest-steppes. As a result of pastoral and recreational degradation, steppe indigenous plant communities are partially replaced by species with a wide ecological amplitude (*Potentilla acaulis, Artemisia monostachya, Artemisia frigida*), and meadow communities replaced by *Elytrigia repens* (Vanteeva and Solodyankina 2015; Znamenskaya et al. 2018).

The geosystems of the study area on the north-facing macroslope of the Khamar-Daban mountain range are represented by bald mountain, mountain meadow, and taiga types. Forests cover 82% of the territory, and subalpine meadows 17.2%.

According to the field data and the method described above, the quantitative values of a tree and herbaceous phytomass were determined for each type of geosystems (groups of facies) for all study areas. To analyze the distribution of quantitative values of the phytomass accumulation function in geosystems, the normal distribution equation was used (Austin 1976; Ter Braak and Looman 1986). The minimum and maximum values, arithmetic mean, standard deviations, skewness, and kurtosis were calculated (Table 8.2; Solodyankina and Vanteeva 2017).

The distribution of the phytomass values of the tree and herbaceous vegetation in the forests of the study area on the north-facing macroslope of the Khamar-Daban

Table 8.1 Environmental conditions and factors of landscape differentiation of study areas

Factors		Study area			
		Olkhinskoe plateau	Barguzin mountain range	Priol'khon plateau	Khamar-Daban mountain range
Total solar radiation per year (mJ/m²)		4200–4600	3800–4200	4400–4600	4200–4400
Temperature, °C	July	14–16	10–12	14–16	14–16
	January (below zero)	16–18	24–22	18–20	18–22
Sum of active temperatures above 10 °C		1200–1600	600–1000	1400–1600	1200–1600
Annual rainfall (mm)		300–500	200–500	100–200	1000–1400
The duration of the frost-free season		90–100	80–100	110	80–120
Snow depth (m)		0.2–0.4	0.3–0.6	0.0–0.1	0.2–2.5
The difference in elevation (m)		460–1260	480–1240	460–760	480–1680
The aspect of macroslopes		Southwestern	Western	Northeastern	Northern
The character of the distribution of permafrost (area, %)		Sporadic (less than 5)	Insular (30–40)	Rare-insular (5–30)	Sporadic (less than 5)

mountain range does not satisfy the three-sigma rule. Therefore, the maximum values of these distributions can be considered statistically improbable. The distribution of phytomass values of herbaceous communities of this study area is smoothed.

The distribution of tree phytomass values in the study areas in the Barguzin mountain range and the southern part of the Olkhinskoe plateau is close to the normal distribution, but the former is relatively peaked while the latter is smoothed. It is explained by the high occurrence of indigenous plant communities on the Barguzin Range as well as their seral stages on the Olkhinskoe plateau (within the boundaries of the study area). The distribution of phytomass values of herbaceous communities in the study area on Priol'khon plateau is close to exponential, which testifies the incompleteness of the sample, that is, high frequency of low values. The maximum values occur rarely, which is probably related to a high degree of degradation of the geosystems in this study area.

The arithmetic means and maxima of all the calculated indicators are the highest for the study area on the Khamar-Daban mountain range, and the lowest for the study area on Priol'khon plateau, which is explained by the significant differences in climatic conditions (see Table 8.1) and the degree of anthropogenic impact. The maximum values of the tree and herbaceous phytomass on the Khamar-Daban mountain range (see Table 8.2) are far from the confidence interval of the distribution of these characteristics and are confined to the valleys with poplar. It should be noted that the foothill plain on this study area is occupied, mainly by secondary forests. If the phytomass value of secondary forests is excluded from the analysis, the average mean of this value increases by 101.9 t/ha.

Table 8.2 Range of phytomass values

Descriptive statistics		Study area			
		Olkhinskoe plateau	Barguzin mountain range	Priol'khon plateau	Khamar-Daban mountain range
Tree phytomass in forests (t/ha absolute dry matter)	Min	44.5	21.8	30.6	44.1
	Max	206.6	554.1	115.4	1728
	Avg.	124.4	196.1	60.3	414.3
	S	57.6	120.7	30.6	384.6
	A	0.0	0.99	0.7	1.6
	E	−1.4	1.03	−0.3	3
Herbaceous phytomass in forests (t/ha air-dry matter)	Min	0.2	0.4	0.4	0.4
	Max	0.6	1.5	1.0	6.8
	Avg.	0.3	0.8	0.7	1.5
	S	0.1	0.4	0.2	1.6
	A	1.0	1.2	0.2	2.8
	E	0.4	1.6	0.9	7.3
Phytomass of meadows, steppes, bogs (t/ha air-dry matter)	Min	0.7	0.4	0.3	1.1
	Max	1.3	0.8	2.9	4.7
	Avg.	0.95	0.6	0.9	2.6
	S	–	0.7	0.6	1.4
	A	–	0.1	2	0.6
	E	–	1.2	5.9	−1.5

Min minimum value, *Max* maximum value, *Avg.* arithmetic mean, *S* standard deviation, *A* skewness, *E* kurtosis

For the study area in the Barguzin mountain range, the phytomass values are relatively high despite unfavorable climatic conditions for vegetation development (see Table 8.1). The maximum values are observed in old-age dark coniferous and pine forests.

8.6   Discussion

To reveal the relationship between the geosystem structure and phytomass accumulation function at the local level, multiple regression analysis was performed based on phytomass data. The phytomass values were assigned to the groups of facies in accordance with the hierarchical classification of geosystems developed for each study area. Sample plots with secondary forests were excluded from the analysis. As independent variables, we selected the attributes characterizing the structure of geosystems as follows: (i) number in the sequence of geoms in the legend indicating the climatic and elevation gradient (1, the highest elevation zone, e.g., *goltsy*[2]); (ii) class

---

[2] Russian-language term – zone of the high mountain tundra above the upper boundary of the forest in Siberia.

of facies related to a certain factoral-dynamic series and with the main driving factor; (iii) the dynamic state of the group of facies, reflecting the degree of influence of this or that factor in order of increasing influence (for the primary state, 0; pseudo-primary, 1,; serial, 2).

Five geoms and five types of factoral-dynamic series (sublithomorphic, subpsammomorphic, subhydromorphic, sublithohydromorphic, substagnant) were identified in the study area in the Barguzin mountain range. Based on the results of the data analysis from 38 sample plots, we found a high correlation ($R = 0.89$, $R^2 = 0.78$, $F = 30.1$, $p = 1.3566E\text{-}10$) between the values of phytomass and the dynamic state of geosystems (Eq. 8.1 in the logarithmic scale, Table 8.3):

$$y = 5.82 - 0.37x_1 - 0.47x_2 - 1.08x_3 - 2.03x_4, \tag{8.1}$$

where y is phytomass (t/ha absolute dry matter), the degree of influence of the factors (from 0 to 2): x_1 is lithomorphic, x_2 is hydromorphic, x_3 is psammomorphic, and x_4 is stagnant.

The function of the accumulation of phytomass in geosystems decreases as degree of factor influence increases. At the same time, the degree of influence of lithomorphic and hydromorphic factors is of low certainty. Hypertrophic influence of psammomorphic and stagnant factors causes development of serial facies with low phytomass stocks.

Geosystems similar to the *goltsy* with low phytomass stocks occur in the foothill part of the Barguzin mountain range. As a result, the sequence of altitudinal zones and the characteristic climatic regimes is disturbed. Therefore, the relationship between the phytomass accumulation function of geosystems and geom structure is poorly expressed.

Five geoms and three types of factoral series (sublithomorphic, subhydromorphic, and substagnant) were identified for the study area on the Khamar-Daban mountain range. As a result of multiple regression analysis (41 sample plots), a high correlation ($R = 0.93$, $R^2 = 0.86$, $F = 74.3$, $p = 1.0137E\text{-}15$) of phytomass (y is

Table 8.3 Parameters of multiple regressions (logarithmic scale)

Variables	Coefficients	Standard deviation	t-test	p (certainty)
Barguzinsky mountain range				
Constant	5.82	0.18	32.06	1.9013E-26
x_1	−0.37	0.27	−1.36	0.18
x_2	−0.47	0.25	−1.84	0.07
x_3	−1.08	0.32	−3.36	0.002
x_4	−2.03	0.19	−10.77	2.4618E-12
Khamar-Daban mountain range				
Constant	2.64	0.39	6.67	7.7171E-8
x_1	0.86	0.1	8.48	3.3469E-10
x_2	−0.83	0.17	−4.96	1.5858E-05
x_3	−5.77	0.41	−14.02	2.2106E-16

phytomass, t/ha absolute dry matter) was revealed with the following structural indicators in a logarithmic scale: x_1 is the number of the geom, x_2 is the degree of influence of the hydromorphic factor, and x_3 is the degree of influence of the stagnant factor (Eq. 8.2, see Table 8.3).

$$y = 2.64 + 0.86x_1 - 0.83x_2 - 5.77x_3. \tag{8.2}$$

In this area, the positive relationship between the phytomass accumulation function and the type of geom is clearly manifested. To some extent, this finding is in agreement with the viewpoint that the identification of this hierarchical unit requires taking into account the biomass productivity of geosystems (Sochava 1980). The hypertrophied influence of hydromorphic and stagnant factors has a negative effect on the phytomass accumulation function of geosystems. As in the previous case, the influence of the lithomorphic factor is poorly correlated with the phytomass stock. Therefore, this attribute was excluded from the analysis.

In the study area on the Priol'khon plateau, the geosystems are represented by two geoms and four types of factoral-dynamic series (subxerolithomorphic, sublithomorphic, subpsammomorphic, and subhydromorphic). As a result of the analysis (55 sample plots), a high pairwise correlation ($R = 0.80$, $R^2 = 0.65$, $t = -9.85$, $p = 1.4489E-13$) of tree phytomass with a geom type was detected. In this case, geoms reflect completely different types of natural environment such as taiga and steppe so the differences in the values of tree phytomass are naturally large.

Regression analysis was not performed for the study area on the Olkhinskoe plateau, as described sample plots are characterized mainly by succession stages of geosystems.

8.7 Conclusion

Analyzing the results of descriptive statistics of variation of phytomass values on four study areas and multiple regression analysis, we conclude that the phytomass accumulation function in geosystems is influenced by various scale factors and features of the geosystem structure.

The influence of climatic features (latitudinal zoning) is expressed in decreasing the maximum values and arithmetic mean of phytomass (see Table 8.2) from the south (the Khamar-Daban mountain range) to the north (the Barguzin mountain range). The topography affects aridity of the climate—from wet northwest-facing macroslopes of the mountain ranges to the central part of the Priol'khon plateau.

In the study areas on the Khamar-Daban mountain range and the Barguzin mountain range, high-altitude zonation (relationship with geom structure) plays an important role in the differentiation of geosystems and, accordingly, the accumulation of phytomass, especially in the Khamar-Daban mountain range (see Eq. 8.2). On the piedmont of the Barguzin mountain range, this relationship is disturbed by the cooling effect of Lake Baikal. The geosystem structure of these areas, determined by high-altitude zonation, is also influenced by local landscape-forming factors associated with the geological structure and the moistening regime. These factors cause a change in the dynamic state of geosystems and the development of factoral-dynamic series of groups of facies. According to the results of regression analysis, the stronger is the influence of psammomorphic, hydromorphic, or stagnant factors on the change in the dynamic state of the geosystem, the lower is the phytomass values in geosystems.

In the study areas on the Olkhinskoe plateau and the Priol'khon plateau, the anthropogenic factor had a great influence on the geosystems structure. Therefore, it is not yet possible to reveal clear dependencies of phytomass accumulation on local landscape-forming factors and structural characteristics with the available data set.

The geosystem structure of the study areas, considered in the framework of factoral-dynamic analysis, reflects both the regional features of the geosystem differentiation of the Baikal region and the local specificity of the study areas formed under the influence of orographic factors and the cooling effect of the water mass of Lake Baikal. In view of the complexity and mosaic of the geosystem structure of the Baikal region, the classification of geosystems by similarity of functional attributes seems to be more appropriate.

Landscape-typological maps, which take into account the dynamic state of geosystems and driving environmental factors, are relevant as the basis for identifying, evaluating, and forecasting the spatial patterns of natural functions at a certain hierarchical level.

Acknowledgments This study was partially supported by the Russian Foundation for Basic Research (project no. 17-05-00588) and the Russian Geographical Society (project no. 17-05-41020).

References

Austin, M. P. (1976). On non-linear species response models in ordination. *Vegetatio, 33*(1), 33–41.

Bastian, O., Grunewald, K., Syrbe, R.-U., et al. (2014). Landscape services: The concept and its practical relevance. *Landscape Ecology, 29*(9), 1463–1479.

Bastian, O., Grunewald, K., & Khoroshev, A. V. (2015). The significance of geosystem and landscape concepts for the assessment of ecosystem services: Exemplified in a case study in Russia. *Landscape Ecology, 30*(7), 1145–1164.

Catorci, A., Cesaretti, S., & Gatti, R. (2009). Biodiversity conservation: Geosynphytosociology as a tool of analysis and modeling of grassland systems. *Hacquetia, 8,* 129–146.

Clements, F. (1949). *Dynamics of vegetation.* New York: Hafner.

Christopherson, R. W., & Birkeland, G. H. (2015). *Geosystems: An introduction to physical geography* (9th ed.). New York: Pearson Education.

Cushman, S. A., & Huettmann, F. (Eds.). (2010). *Spatial complexity, informatics and wildlife conservation.* New York: Springer.

Davis, W. (1954). *Geographical essays.* New York: Dover Publications.

Deak, B., Valko, O., Kelemen, A., et al. (2011). Litter and graminoid biomass accumulation suppresses weedy forbs in grassland restoration. *Plant Biosystems, 145,* 730–737.

Evans, J. S., & Cushman, S. A. (2009). Gradient modeling of conifer species using random forests. *Landscape Ecology, 24*(5), 673–683.

Ellenberg, H. (1996). *Vegetation mitteleuropas mit den Alpen in okologischer, dynamischer und historischer sicht* (5th Aufl). Stuttgart: Ulmer.

Frolov, A. A. (2015). Geoinformtional mapping of landscape variability (exemplified by Southern Cisbaikalia). *Geography and Natural Resources, 36*(1), 99–107. (in Russian).

Haase, G. (1973). Zur ausgleiderung von baumeinheiten der chorischen und regionisen dimensionen – dargestellt an beispielen aus bodengeographie. *Petermanns Geographische Mitteilungen, 117*(2), 81–90.

Heams, T., Huneman, P., Lecointre, G., et al. (Eds.). (2015). *Handbook of evolutionary thinking in the sciences*. Dordrecht: Springer.

Isachenko, A. G. (2004). *Theory and methodology of geographical science*. Moscow: Academia. (in Russian).

Istomina, E. A., Luzhkova, N. M., & Khidekel, V. V. (2016). Birdwatching tourism infrastructure planning in the Ria Formosa Natural Park (Portugal). *Geography and Natural Resources, 37*(4), 371–378. (in Russian).

Grunewald, K., Bastian, O., & Drozdov, A. (Eds.). (2014). *TEEB-prozesse und okosystem-assessement in Deutschland, Russland und weiteren staaten des nordlichen Eurasiens*. Bonn: BfN-Skripten.

Kelemen, A., Torok, P., Valko, O., et al. (2013). Mechanisms shaping plant biomass and species richness: Plant strategies and litter effect in alkali and loess grasslands. *Journal of Vegetation Science, 24*, 1195–1203.

Khoroshev, A. V. (2016). *Polyscale organization of a geographical landscape*. Moscow: KMK. (in Russian).

Khoroshev, A., & Koshcheeva, A. (2009). Landscape ecological approach to hierarchical spatial planning. *Terra Spectra Planning Studies, 2*(1), 3–11.

Konovalova, T. I., Bessolitsyna, E. P., Vladimirov, I. N., et al. (2005). *Landscape-interpretation mapping*. Novosibirsk: Nauka. (in Russian).

Krauklis, A. A. (1972). Factoral-dynamic rows of elementary geosystems as a basis for modeling natural regions. In *International Geography, Vol. 2* (pp. 960–963). Montreal: International Geographic Union.

Krauklis, A. A. (1979). *Problems of experimental landscape science*. Novosibirsk: Nauka. (in Russian).

Landolt, E. (1977). Okologische zeigerwerts zur Sweizer flora. *Veroff. Geobot. Inst. ETH. Zurich, 64*, 1–208.

Lastochkin, A. N. (2011). *General theory of systems*. St. Petersburg: Lemma. (in Russian).

Mikheev, V. S., & Ryashin, V. A. (1977). *Map of landscapes of the south of East Siberia*. Institute of geography of Siberia and Far East SB AS USSR. (in Russian).

Neef, E. (1963a). Dimensionen geographischer betrachtungen. *Forsch Fortschr, 37*, 361–363.

Neef, E. (1963b). Topologische und chorologische arbeitsweisen in der landschaftsforschung. *Petermanns Geographische Mitteilungen, 107*(4), 249–259.

Ramensky, L. G., Tsatsenkin, I. A., & Chizhikov, O. N., et al. (1956). *Ecological assessment of fodder land by vegetation cover*. Moscow: Sel'khozgiz (in Russian).

Semenov, Y. M. (2014). Landscape-geographical support of the ecological policy of nature management in regions of Siberia. *Geography and Natural Resources, 35*(3), 208–212.

Semenov, Y. M., & Lysanova, G. I. (2016). Mapping of geosystems for landscape planning of areas in the Altai Republic. *Geography and Natural Resources, 37*(4), 329–337.

Sochava, V. B. (1972). The study of geosystems: the current stage in complex physical geography. In *International Geography* (Vol. 1, pp. 298–301). Toronto: International Geographic Union.

Sochava, V. B. (1974). Das systemparadigma in der geographie. *Petermanns Geographische Mitteilungen, 118*(3), 161–166.

Sochava, V. B. (1978). *Introduction to the theory of geosystems*. Novosibirsk: Nauka. (in Russian).

Sochava, V. B. (1980). *Geographical aspects of the Siberian taiga.* Novosibirsk: Nauka. (in Russian).

Solodyankina, S. V., & Vanteeva, Y. V. (2017). Variability of the production function of vegetation of coastal geosystems of the Baikal region. *Geography and Natural Resources, 2,* 73–80. (in Russian).

Ter Braak, C. J. E., & Looman, C. W. N. (1986). Weighted averaging, logistic regression and the Gaussian response model. *Vegetatio, 65,* 3–11.

Vanteeva, J. V., & Solodyankina, S. V. (2015). Ecosystem functions of steppe landscapes near Lake Baikal. *Hacquetia, 14*(1), 65–78.

Zamolodchikov, D. G., Utkin, A. M., & Chestnyh, O. V. (2003). Coefficients of conversion of plantation stocks to phytomass for the main forest-forming species of Russia. *Forest Valuation and Forest Inventory, 1*(32), 119–127. (in Russian).

Zamolodchikov, D. G., Utkin, A. M., & Korovin, G. N. (2005). Conversion coefficients of phytomass/reserve in connection with dendrometric indicators and composition of a stand. *Forest Science, 6,* 78–81. (in Russian).

Znamenskaya, T. I., Vanteeva, J. V., & Solodyankina, S. V. (2018). Factors of the development of water erosion in the zone of recreation activity in the Ol'khon region. *Eurasian Soil Science, 2,* 221–228.

Chapter 9
Catena Patterns as a Reflection of Landscape Internal Heterogeneity

Irina A. Avessalomova

Abstract Holistic character of a landscape is ensured by radial migration of chemicals among vertical layers and by unidirectional lateral matter and energy flows in catena. This results in emergence of elementary landscape units differing in degree of geochemical autonomy as well as in complication of catena patterns in a landscape. We applied catena analysis to explain internal heterogeneity and to reveal multiplicity of structures on the examples of two taiga regions in East-European plain. The complexity of catena composition is determined by abiotic template and redistribution of matter with lateral flows. Contrast in migration conditions serves as a criterion of internal landscape heterogeneity. Changes in migration conditions at the boundaries between neighboring units create prerequisites for emergence of barrier zones. In agrolandscapes geochemical barriers were detected in the lowest sections of catenas at the contact of cultivated fields and elements of ecological network. Influence of lateral phytobarriers on ionic discharge depends on their retention capacity and differs in various catena types resulting in variability of hydrochemical properties of river water. We demonstrate landscape-geochemical maps which provide the opportunity to reflect patterns emerging at various hierarchical levels. This enables us to project anthropogenic geosystems with proper consideration for landscape heterogeneity.

Keywords Spatial pattern · Heterogeneity · Catena · Migration · Neighborhood · Geochemical barrier · Retention

9.1 Introduction

Internal heterogeneity emerges as a result of landscape evolution and manifests itself in spatial neighborhoods of diverse catena variants. Catena is a simple (zero-level) cascade landscape-geochemical system which methodologically is treated as a para-

I. A. Avessalomova (✉)
Lomonosov Moscow State University, Moscow, Russia

genetic association of elementary landscape units (ELU)[1] conjugated by migration of chemical elements. It is a three-dimensional body that occupies certain space and is uniform in geochemical links between contrast natural units (Kasimov et al. 2012). Holistic character of elementary landscape units is ensured by radial migration of chemicals among vertical layers, whereas that of catena is by unidirectional lateral matter and energy flows. This results in emergence of ELU differing in degree of geochemical autonomy as well as in complication of catena patterns in a landscape. Studying catena patterns requires application of both structural-genetic and functional approaches to ensure proper explanation of landscape heterogeneity.

Catena analysis involves several principal directions as follows:

- Identification of factors that control structure of geochemical toposequences, diversity, and neighborhoods of ELU. According to the Law of Requisite Variety (Ashby 1956), spatial heterogeneity in catena is a realization of the fundamental principle of landscape organization.
- Consideration for functional contribution of certain ELU and neighborhood effects to formation and transformation of matter flows. Edge effects on the boundaries between units depend on either monotonous or steplike character of geochemical gradients which determine contrasts in catena affecting flow convergence or divergence (Kolomyts 1987; Fortescue 1980).
- Establishment of lateral links peculiarities, elements distribution pattern, and location of transit and accumulation zones for catenas. Based on differences in pathways and forms of elements migration, M.A. Glazovskaya (2012) distinguished 12 catena types classified into two groups depending on dominance of migration in either soluble or particulate forms. This kind of information has a crucial importance since it allows considering migration types and pathways as well as their role in spatial differentiation of neighboring geochemical fields. The approach aiming at revealing differences among catenas in connection with matter input and output in various sections was applied to choose classification criteria (Sommer and Schlichting 1997). As a result, accumulative, leached, and transformational catena types were identified.
- Explanation of reasons for *geochemical barrier* emergence, their spatial location, and role in element accumulation and dispersion. To characterize geochemical barriers, it was proposed to use contrasts in migration situations, functioning regime, retention capacity, and position in catena (Perelman 1972a; Fortescue 1992; Ryszkowski et al. 1999; Glazovskaya 2012).

Basic principles of catena classification involve similarity and dissimilarity of structural and functional organization. It considers not only their formation in certain bioclimatic conditions of various natural zones and physiographic conditions but also ratio between geologically induced and migration-induced differentiation of element content, i.e., either convergent or divergent redistribution (Kasimov and Samonova 2004). The combination of proposed classification criteria provides the

---

[1] Elementary landscape unit (ELU) – the term in landscape geochemistry for the smallest uniform unit with homogeneous bedrocks, soil, moisture. Catena consists of a toposequence of elementary landscape units (autonomous, trans-eluvial, trans-accumulative, super-aqual, sub-aqual) (Eds.)

opportunity to compare different variants of autochthonous and allochthonous catenas by degree of heterogeneity, contrast, ecogeochemical conditions, and stability. The latter is determined by ratio of migration processes supporting balance of input, internal, and output flows.

We applied catena analysis to explain internal heterogeneity on the examples of two taiga regions in East-European plain. The first one is located in the middle taiga subzone in the southern Arkhangelsk region (interfluve of the Kokshenga River and Ustya River). The landscape belongs to the plain within the structural Ustyanskoye plateau composed of Permian marl with mantle of morainic and glacio-lacustrine sediments and shaped by erosion. The landscape has been of great agricultural interest for centuries and now can be referred to as agrolandscape. Spatial organization is characterized by combination of cultivated and abandoned fields with forests and meadows. Units with recovering vegetation form the ecological network. The second study area was chosen in the southern taiga subzone in the Meshchera fluvial and lacustrine plain. To study differences in catenas within the landscapes, we used the results of soil, plant, and water geochemical sampling and landscape-geochemical maps.

9.2   Structural and Functional Organization of Catenas

Catena analysis is performed in several steps differing in objectives and combination of thematic issues. The general logic assumes study of individual catenas followed by description of intra-landscape spatial heterogeneity. The first step is devoted to building information base and involves establishing structural and functional organization in catena with two interrelated issues. The first one deals with peculiarities of structural-genetic frame and the second with specifics of migration processes (Fig. 9.1). The study objects are geosystems of two hierarchical orders, namely, elementary landscape units (*facies*[2] level) and their geochemical toposequences (catena level). The investigation focuses on morphological, migration, and geochemical patterns and their comparison among catenas. The approach follows the concept of structure multiplicity ("polystructuralism"[3]) which enables us to take into consideration simultaneously contributions of factors of differentiation and integration to landscape development and functioning. Their effects are revealed according to the uniform scheme "factor–process–result" required by analysis of intra-component and between-component relations in catena.

When catena structure is being formed, physiographic environment contributes to catena heterogeneity via relief and lithogeochemical specifics of substrate. Relative elevation range, terrain ruggedness, slope gradients, combination of landforms differing in genesis, and their spatial proportions determine a set of autono-

---

[2] See Glossary and Chapter Khoroshev in Part I, this volume, for definition.

[3] The concept in landscape science accepting the possibility of various structural projections of a landscape based on a number of system-forming relations.See Glossary and Chapter 1 for definition (Eds.).

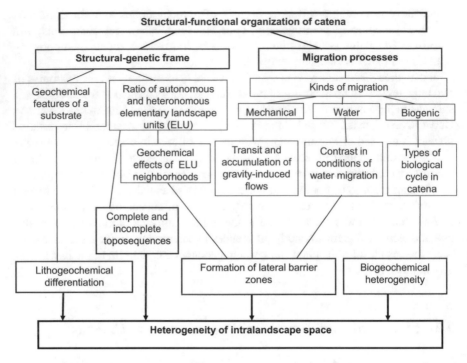

Fig. 9.1 Structural-functional organization of catenas

mous and heteronomous ELU as well as development of full or partial geochemical toposequences. Spatial neighborhood of morphologically different catenas reflects complexity of landscape mosaics and landscape pattern. Increase in structural complexity is favored by substrate heterogeneity resulting in the necessity to distinguish *monolithic* and *heterolithic catenas*. The former were defined as catenas with uniform soil-forming rocks or genetic type of Quaternary sediments along the entire catena and the latter as catenas with soil formation on various substrates due to lithogeochemical contrasts (Kasimov et al. 2012). Intermediate variant (quasi-monolithic catenas) can also distinguished if soil-forming rocks are uniform and contrasts in underlying rocks do not directly affect processes in soils. Parent rocks and Quaternary sediments differ in physical properties, density, resistance to weathering, and matter composition. Their spatial variability in heterolithic catenas encourages heterogeneity of geochemical fields in soils and soil-forming rocks.

Structural-and-genetic frame forms intra-catena space within which migration results in redistribution of chemical elements by biotic, water, and gravity-driven flows. Since biological cycling has a fundamental importance for landscape functioning, study of multidirectional biotic flows in processes of autotrophic and heterotrophic biogenesis is critically needed. F.I. Kozlovsky (1972) treated these flows as the principal migration cycle. To characterize biotic flows, extensive and intensive parameters of biological cycle have been proposed as follows: plant biomass (F), stock of nutrients in plant biomass, bioproduction (P), dead mass (M), and intensity of biological uptake of elements by plants. Changes in ecogeochemical conditions along the lateral gradients are indicated by occurrence of plant communities differing in structural and functional parameters. Phytocoenosis diversity and occurrence of species with various phylogenetic specializations generate biogeochemical heterogeneity in catena.

Biogenesis, as well as lithogeochemical specifics of substrate, exerts influence on conditions of water migration and involvement of elements into lateral flows. Their role is manifold. On the one hand, both processes act as an integrating factor ensuring holistic properties of a catena. On the other hand, they promote increase of spatial differences which entails changes in water supply and growing contrast in conditions of water migration in toposequence. This results in development of transit and accumulation zones. M.A. Glazovskaya (2002) showed than in humid plain landscapes functioning of geochemical toposequences was driven by intimate relations between surface, soils, and groundwater with the resulting high contrast between autonomous and super-aqual landscapes. The reason is high intensity of meltwater lateral surface flows and subsurface flows in other seasons. Impermeable sediments reduce contribution of subsurface flows. In catenas of arid landscapes, water flows during snowmelt do not cause rise of groundwater level. Hence, groundwaters are almost excluded from lateral flows, and links in toposequences are ensured by surface and near-surface flows. In rugged terrains snowmelt provokes both sheet and rill erosion resulting in soil degradation. In this case links in toposequence are related to surface flows and mechanical migration. Significance of the latter greatly increases in areas with high occurrence of gravity-driven falls and taluses, whereas contrast in conditions for water migration is insignificant.

Lithogeochemical differentiation of substrate and matter redistribution generates geochemical structure reflecting heterogeneity of element spatial patterns in catenas. Feedback between geochemical structure and migration structure promotes landscape evolution (Glazovskaya 2007). To characterize geochemical structure *R-L analysis* was proposed. This tool is aimed at revealing principal regularities and reasons for elements spatial redistribution based on calculation of coefficients of radial (R)[4] and lateral (L)[5] migration.

To evaluate radial differentiation of elements, the researcher should consider landscape-geochemical processes localized within the vertical profile of ELU (e.g., detritus production, humus accumulation, gleization, oxides production). Phases of mobilization, transfer, and accumulation are commonly localized in different layers of landscape unit. This is the reason for emergence of leaching and accumulation zones at radial geochemical barriers. Conditions for their development, retention capacity, and layers composition indicate specific character of functioning in individual ELU including ratio between biogenic accumulation and leaching.

Elementary landscape units differ in contrast of element distribution which is expressed by ranked R values and ability of elements to accumulate in soil genetic

---

[4] Ratio between element content in soil horizon and in parent rock.

[5] Ratio between element content in autonomous and subordinate elementary landscape units.

horizons. Several types of radial differentiation were identified. The first type is referred to as surface accumulation which corresponds to accumulation in upper soil horizons with biogeochemical and sorption barriers and gradual decrease of R toward the lower horizons. The second type of differentiation demonstrates the opposite pattern with increased contents in the lower horizons (eluvial-illuvial pattern). Commonly, this pattern can be explained by more active motion of elements under leaching as compared to biogenic accumulation and consequent accumulation in illuvial horizon. The third type is characterized by the more complicated distribution and two accumulation zones (combination of surface accumulation and eluvial-illuvial pattern). Under acid leaching, elements accumulation in illuvial horizons are more well-pronounced at sorption barrier, whereas under neutral one – at alkaline barrier. Contrast of lateral differentiation (L) indicates difference in element dispersion and accumulation among ELUs in toposequence within catena. Similar to radial matter distribution, various types of lateral differentiation were identified depending on occurrence of accumulation zones either in upper or in lower catena sections (Kasimov and Samonova 2004).

Now, we turn to the example of the catena in the agrolandscape within the Ustyanskoye plateau with dominance of cultivated fields and remnant forests and meadows in the lower sections of toposequence (Fig. 9.2). Element distribution pattern is driven by substrate heterogeneity, high intensity of mechanical migration change in pH conditions, and additional input of chemicals with fertilizers. Contributions of these factors to lateral differentiation of various elements are not equal. Although carbonate rocks are poor in mobile phosphorus, high contrast of its radial differentiation in cultivated soils (R=140...230) is explained by fertilizer input and biogenic accumulation in the upper soil horizons. Lateral differentiation is driven by sheet erosion which results in impoverishment of trans-eluvial Ca-class[6] ELU (L=0.03) and enrichment of trans-accumulative ELU on proluvial cones (L up to 0.8). Calcium accumulation in soils is caused by outcrops of Permian marlstones on steep slopes (L=2.4). Its lateral redistribution is controlled by gravity-driven flows as well. Lateral heterogeneity increases much if catena includes sandy alluvial terraces with nutrient-poor soils (low content of P, K, Ca, Mg, etc.) belonging to agricultural and forest H-class geochemical landscapes. In the lower sections of catena, acid leaching in cultivated podzolic soils favors loss of cationogenic elements (Ca, Mg, etc.) which is indicated by decrease of L (0.6–0.8). In general, variability of R and L values testifies heterogeneity of geochemical structure in catena.

Functional roles of particular ELU may be assessed by either individual properties or geochemical effects depending on neighborhoods in catenas. Barrier zones serve as a critical control over lateral differentiation of elements. Combinations and contrasts of ELUs in toposequence determine location of lateral geochemical barriers in autonomous and heteronomous sections. We treat these phenomena as a manifestation of emergent properties in catena. The degree of cascade geosystems "openness" depends on mechanisms of matter concentration and dispersion as well

---

[6] Classes of geochemical landscapes are identified by typical elements and ions in water migration (Ca, H, Fe, Na etc.)

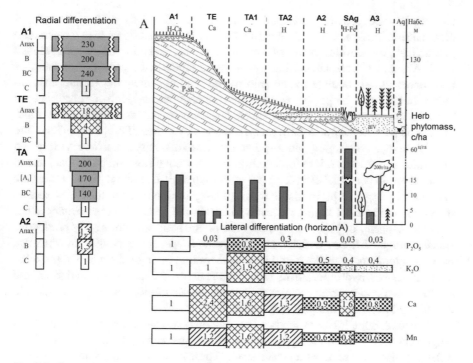

Fig. 9.2 Catena in the agrolandscape within the Ustyanskoye plateau and elements distribution patterns. A1 – the upper (interfluve) autonomous elementary landscape units (ELU), TE – trans-eluvial ELU, TA – trans-accumulative ELU, A2 – the middle (upper terrace) autonomous ELU, A3 – the lower (low terrace) autonomous ELU, SAq – super-aqual ELU. H-Ca, Ca, H, H-Fe – classes of water migration. A, B, BC, C – indices of soil horizons

as on factors which control flows direction. The "openness" is higher in complete geochemical toposequences, where loss of elements is caused by river runoff, and lower in incomplete ones, especially if the final section is super-aqual landscape with inherent lateral barrier. Barrier zones together with lithogeochemical substrate variability and biogeochemical heterogeneity reflect peculiarities of structural and functioning organization of catena and form prerequisites for heterogeneity of intra-landscape space.

9.3 Landscape Neighborhood Effects and Formation of Barrier Zones

The geochemical barrier concept (Perelman 1972a, b) is based on understanding that sharp shift of geochemical conditions in space favors concentration of elements losing their mobility on such a boundary. In catena effects of neighborhood are manifested in emergence of barrier zones. Their peculiar character is influenced by contrast in conditions of water and mechanical migration as well as intensity of plant nutrient uptake. Changes in activity of gravity-driven flows lead to increased probability of development of mechanical barriers in trans-accumulative ELU. Absence of mechanical barriers in these topographic positions encourages involvement of particulate matter into motion and accumulation in bottom sediments. In catenas with perfect water-driven connections and high contrast between autonomous and super-aqual landscapes, boundary effects are generated by functioning of physicochemical barriers which were classified (Perelman 1972b) into alkaline, acid, oxygen, gley, and sorption ones. Such diversity of barriers is explained by instability of water migration conditions (redox conditions, in particular) due to seasonal groundwater level dynamics in super-aqual ELU. Lateral phytobarriers are generated by input of bioavailable forms of biogenic elements to the lower section of catena, and nutrient supply increases in the sites of groundwater emergence. The retention capacity of phytobarriers depends on the abundance of highly productive species, biogeochemical activity, and phylogenetic specialization of dominants that intercept elements from flows and involve them into biological cycle. On the other hand, increased hydromorphism in super-aqual ELU, detritus accumulation, and slowed down decay form prerequisites for emergence of biogeochemical barriers in organic soil horizons like mor and moder.

In general, complexity of landscape spatial pattern can be described in terms of neighborhoods of continual geochemical fields and barrier zones between them reflecting discreteness of intra-catena space. These zones act as mechanisms responsible for transformation of matter flows and affecting landscape geochemical structure.

Differences in localization in catena enable us to distinguish two groups of lateral geochemical barriers as follows: (1) spatial barriers associated with ELUs forming contact buffer strip between autonomous and heteronomous units and (2) linear barriers emerging directly on the boundary between them. Structure of barrier zones is shaped by intensity of autotrophic and heterotrophic biogenesis. Its complexity increases under simultaneous functioning of biogeochemical barriers in vegetation and soil layers. In agrolandscapes diversity of barrier zones increases due to neighborhood of cultivated fields with elements of ecological network.

For example, in middle taiga landscape of the Ustyanskoye plateau, lateral barrier zones develop in the lower sections of forested catenas. At the floodplains, barrier zones are located in swamped oxbow depressions at the toe slopes of terraces where acid low-mineralized (100–135 mg/l) sulfate-hydrocarbonate-calcium groundwaters emerge. Tree layer of spruce-pine forests is the main contributor to the retention of biogenic elements at the phytobarriers. At the same time in extra-humid conditions, slowed down matter decay facilitates nutrient deposition in organic horizons of Histic Gleysols. One more concentration mechanism operates in hollows where dissolved forms of nutrients accumulate. However, retention capacity of such geochemical traps cannot persist to nutrient loss with river runoff (Avessalomova et al. 2016).

The southern taiga landscapes of Meshchera with pine forests are dominated by forest-and-mire catenas with high contrast between the autonomous forest H-class ELUs and the super-aqual H-Fe-class ELUs in fens. In the lower sections of toposequences soil and groundwater acidity decrease, redox conditions change resulting in

Table 9.1 Lateral physicochemical barriers in the Meshchera landscapes

Geochemical indices	Elementary landscape units		
	Autonomous forest H-class	Super-aqual forest H-Fe class	Super-aqual fen H-Fe class
Ph	4.9–5.0	5.3–5.5	6.0–6.9
Water migration class	SO_4–Ca	SO_4–Ca	HCO_3–Ca
pH и Eh (mB) in soils	4.0–4.5	4.5–5.5	5.5–6.5
	400–390	350–250	250–150
Soil concretions	–	Fe in external section, Mn in internal section	–
Manganese content in soils, mg/kg	200–400	600–800	300–400

development of super-aqual forest H-Fe-class ELU in the marginal section of mires (Table 9.1). Seasonal fluctuations of groundwater table cause variations in water flow direction which affect changes in ratio between oxidogenesis and processes (Avessalomova 2002). This determines internal complexity of ringlike barrier zone generated by coincidence of oxygen, gley, and sorption physicochemical barriers. It is detected by soil concretions enriched with Mn, P, Mo, and other microelements. Lateral barriers limit loss of elements with runoff as well. After land reclamations aimed at lowering groundwater table spatial oxygen, geochemical barriers develop on the entire area of the bog expanse.

Man-induced landscape transformation is, commonly, followed by sharp changes in migration structure. Agriculture generated a particular type of biological cycle with fertilizer inputs, growing intensity of gravity-driven flows, and nutrient involvement into flows of particulate matter and dissolved substances.

In the agrolandscapes of the Ustyanskoye plateau, transformation of water flows is detected by increasing mineralization (up to 247–314 mg/l) and contents of K, P, and other elements in rivers. Loss of these elements can be reduced by functioning lateral phytobarriers in the lower sections of catena. To evaluate their retention capacity, we used data on plant biomass, its fractional structure, and ash content. Strip-like barriers are formed by communities of *Filipendula ulmaria* which indicate emergence of groundwaters in the marginal sections of floodplains and superimposed alluvial fans. Such monodominant phytobarriers have large plant biomass (5.04 t/ha) and stock of nutrients (0.28 t/ha) due to high biogeochemical activity of the species and the capacity to interception of nutrients leached from the upper sections of a catena. Strip-like barriers consist of several sections dominated by species with various phylogenetic specializations and element uptake capacities: *Anthriscus sylvestris* (K), *Filipendula ulmaria* (Cu, Zn, Ba), and *Deschampsia cespitosa* (Si).

The lowest sections of catenas have more complex structure if spatial lateral phytobarriers occur at the contact with super-aqual units. Barriers can occur, for example, at the proximal sections of plowed alluvial fans with elevated nutrients and water supply partially superimposed on floodplains. Due to diversity of edaphic conditions in contact zones, phytocoenoses differ in productivity, by this affecting the retention capacity of lateral phytobarriers and their ability to modify discharge

Table 9.2 Influence of lateral phytobarriers on ionic discharge in agrolandscapes

Elementary landscape units			Contents of elements in river water, mg/l		
Trans-accumulative (TA)	Barrier zone (TA/Saq)	Super-aqual (Saq)	K	Mineral P	Si
	Phytomass/stock of ashes elements, t/ha				
Wheat	Forbs-grass meadows, 2.6/0.17	Grass-forbs meadows,1.76/0.22	2.0	0.12	4.2
Barley	Equisetum-forbs meadows, 1.76/0.2	Grass-forbs meadows, 3.12/0.3	1.8	0.015	5.2
Fodder grass mixture	Tall forbs and *Filipendula ulmaria*-dominated meadows, 3.28/0.24	Grass-forbs *Filipendula ulmaria* meadows, 2.68/0.2	1.2	0.013	5.1
Fodder grass mixture	Tall forbs meadows dominated by *Umbelliferae* species, 7.44/0.45	Calamagrostis-tall forbs meadows, 3.37/0.29	0.8	0.010	3.6

(Table 9.2). The lowest retention capacity is characteristic of phytobarriers at south-facing slopes where we detected small plant biomass and ash content in herbal communities with dominance of Gramineae species and *Equisetum* sp. In small catchments dominated by this type of catena, we observed elevated contents of dissolved forms of K, P, and Si in river waters. The most well pronounced barrier effect occurred in contact zone of tall forbs meadows dominated by *Filipendula ulmaria* and various *Umbelliferae* species. It is manifested in decrease of biogenic elements (K, in particular) in river water in summer. Floodplain meadows dominated by tall forbs and *Calamagrostis* sp. contribute to retention of Si at lateral phytobarrier as well due to prevalence of grasses with silica-rich tissues in fractional structure of plant biomass. Thus, role of lateral phytobarriers in formation of ion discharge differs in various types of toposequences. The development of barriers increases complexity of structural and functional organization and internal heterogeneity in catena.

9.4   Catenary Differentiation of Landscapes

Diversity of catenas which are grouped into more complex small catchment geosystems is representative of the landscape territorial heterogeneity. It may be detected on the special fine-scale landscape-geochemical maps. The design of such maps is aimed at depicting different variants of catenas, their internal structure, and spatial neighborhoods. The information derived from the maps can be enriched if we show objects of various hierarchical levels and their combinations. At the first step areas of ELUs are mapped. To this end, we used the classification by A.I. Perelman (1972b) since it takes into account the type of biological cycle, conditions for water migration, degree of autonomy, and lithogeochemical specialization of substrate. The second step is aimed at systematizing and revealing catenas with special focus on internal heterogeneity (Kasimov and Samonova 2004). This procedure is based on criteria as follows: neighborhoods of contrast ELUs, position in a river basin, and geochemical contrast in toposequence. The map analysis enables us to assess landscape neighborhood effects and location of barrier zones.

The fragment of landscape-geochemical map (scale 1:10,000) illustrates heterogeneity of the agrolandscape of interest in the Ustyanskoye plateau (Fig. 9.3, Table 9.3). We established catena subgroups which differ in biogeochemical heterogeneity and neighborhoods between fields and elements of ecological network. For

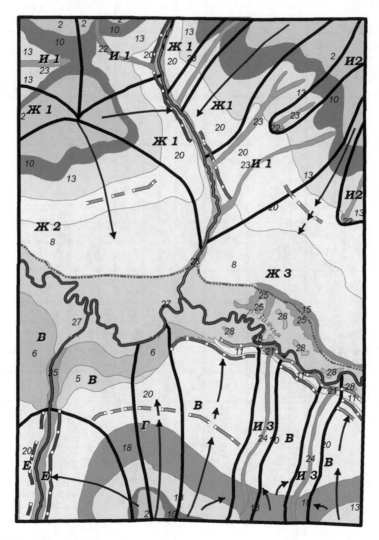

Fig. 9.3 Elementary landscape units and catena types in the Ustyanskoye plateau agrolandscape. Legend see in Table 9.3. Letters and numbers in bold letter– catena types. Numbers in small letters – elementary landscape units. Arrows – direction of flows. Colored strips – barrier zones

Table 9.3 Classification of elementary landscape units in the Ustyanskoye plateau agrolandscape

Genera	Species	Coniferous and small-leaved forests (Middle taiga)			Meadows – Interfluve			Floodplain	Meadow-fen	Fen	Cultivated fields	
		H	H-Ca	H-Fe	H-Ca	Ca	H-Ca-Fe	H-Fe	H-Fe		H – H-Ca	Ca
Autonomous	On morainic loams	1									2	
	On alluvial sands underlain by marl (the second terrace)				3						4	
	On alluvial sands underlain by moraine (the second terrace)										5	
	On alluvial sands (the first terrace)	6			7						8	
Trans-eluvial	On morainic loams	9									10	13
	On marls		11				12					
	On sands underlain by marl		14									
	On alluvial sands	15	16									
Trans-accumulative	On deluvium underlain by marl				17						18	
	On deluvium underlain by sands				19						20	
	On proluvial loams underlain by alluvial loams							21				
Eluvial-accumulative	On alluvium-deluvium		22		23						24	
Super-aqual	On alluvial loams					25	26	27	28	29		

example, we distinguished field-and-forest, field-forest-and-meadow, and field-and-meadow catenas as well as their modifications depending on phytobarrier retention capacity. Within the catchments the map shows allochtonous catenas where super-aqual ELU is subject to impact of lateral flows from the adjacent territories. Catena pattern of the agrolandscape is more complex if it includes both complete and incomplete toposequences and typical catenas neighbor to pseudo-catena (sensu N.S. Kasimov) which emerge due to erosion-induced change in direction of lateral flow. In accordance with matter movement in intra-catena space, we distinguished catenas by contribution of mechanical migration, surface and subsurface flows, and contrast in conditions of water migration. This allowed us to depict location of mechanical and physicochemical barriers.

Structural and functional heterogeneity of agrolandscapes requires applying site-specific approaches to elaboration of optimization measures. The first group of approaches is based on the conception of adaptive land use. For example, interpretation of landscape-geochemical maps showed evidence that slopes with intensive sheet erosion are characterized by low productivity of agrocoenoses and, hence, are ineffective for further plowing. The second approach is constructive since it proposes to consider landscape neighborhood effects in order to regulate flows by intentional construction of lateral barrier zones. Realization of both approaches is based on analysis of catena patterns in agrolandscapes.

9.5   Conclusions

1. Explanation of landscape heterogeneity is based on the concept of hierarchical organization of regional-scale geosystems and considers diversity of catenas. Structural and functional organization of catenas results from combined effects of abiogenic and biogenic factors which determine differentiating role of structural-genetic frame and migration processes.
2. Heterogeneity of intra-catena space depends on combination of elementary landscape units differing in type of biological cycle, degree of autonomy in relation to matter input, lithogeochemical specialization of substrate, and contrast of migration conditions. Radial and lateral differentiation of elements reflects geochemical structure of a catena and location of accumulation and leaching zones. This induces variability in toposequences of elementary landscape units.
3. Changes in migration conditions at the boundaries between neighboring units create prerequisites for emergence of barrier zones which affect lateral flows. In agrolandscapes geochemical barriers were detected in the lowest sections of catenas at the contact of cultivated fields and elements of ecological network. Influence of lateral phytobarriers on ionic discharge depends on their retention capacity and differs in various catena types resulting in variability of hydrochemical properties of river water.

4. High informativity of landscape-geochemical maps is ensured by the opportunity to reflect pattern emerging at various hierarchical levels, namely, catena differentiation of a territory and internal organization of catenas. This enables us to project anthropogenic geosystems with proper consideration for landscape heterogeneity.

Acknowledgments This research was financially supported by Russian Foundation for Basic Research (grant 17-05-00447).

References

Ashby, W. R. (1956). *Introduction to cybernetics*. London: Chapman & Hall.

Avessalomova, I. A. (2002). Geochemical barriers in the marginal zones of the Lake Meshchera mires. In N. S. Kasimov (Ed.), *Geochemical barriers in the zone of hypergenesis* (pp. 162–175). Moscow: Moscow University Publishing House. (in Russian).

Avessalomova, I. A., Khoroshev, A. V., & Savenko, A. V. (2016). Barrier function of floodplain and riparian landscapes in river runoff formation. In O. S. Pokrovsky (Ed.), *Riparian zones. Characteristics, management practices, and ecological impacts* (pp. 181–210). New York: Nova Science Publishers.

Fortescue, J. A. C. (1980). *Environmental geochemistry a holistic approach*. New-York/Heidelberg/Berlin: Springer.

Fortescue, J. A. C. (1992). Landscape geochemistry: Retrospect and prospect – 1990. *Applied Geochemistry, 7*, 1–53.

Glazovskaya, M. A. (2002). *Geochemical foundations for typology and methods of research of natural landscapes*. Smolensk: Oikumena. (in Russian).

Glazovskaya, M. A. (2007). *Geochemistry of natural and technogenic landscapes*. Moscow: Moscow State University Press. (in Russian).

Glazovskaya, M. A. (2012). Geochemical barriers in soils of plains, their typology, functional peculiarities, and ecological significance. *Proceedings of Moscow University, series 5 Geography, 3*, 8–14. (in Russian).

Kasimov, N. S., & Samonova, O. A. (2004). Catena landscape-geochemical differentiation. In K. N. Dyakonov & E. P. Romanova (Eds.), *Geography, society and environment (Functioning and present-day state of landscapes)* (Vol. II, pp. 479–489). Moscow: Gorodets. (in Russian).

Kasimov, N. S., Gerasimova, M. I., Bogdanova, M. D., et al. (2012). Landscape-geochemical catenas: Concept and mapping. In N. S. Kasimov (Ed.), *Landscape geochemistry and soil geography* (pp. 59–80). Moscow: APR. (in Russian).

Kolomyts, E. G. (1987). *Landscape studies in transitional zones*. Moscow: Nauka. (in Russian).

Kozlovsky, F. I. (1972). Structural-functional and mathematical model of migration landscape-geochemical processes. *Pochvovedeniye (Soil Science), 4*, 122–138.

Perelman, A. I. (1972a). *The geochemistry of elements in the zone of supergenesis*. Moscow: Nedra. (Translated form Russian). (Geol. Surv. Canada Trans. No. 1048).

Perelman, A. I. (1972b). *Landscape geochemistry*. Moscow: Vysshaya Shkola. (Translated form Russian). (Geol. Surv. Canada trans. No. 676, Part I and II).

Ryszkowski, L., Bartoszewicz, A., & Kedziora, A. (1999). Management of matter fluxes by bio-geochemical barriers at the agricultural landscape level. *Landscape Ecology, 14*, 479–492.

Sommer, M., & Schlichting, E. A. (1997). Archetypes of catenas in respect to matter – a concept for structuring and grouping catenas. *Geoderma, 76*, 1–33.

Chapter 10
Structure of Topogeochores and Modern Landscape-Geochemical Processes

Yury M. Semenov

Abstract Radial differentiation of matter is a strong control over the differentiation of geosystems. Our research in steppes and dry steppes of South Siberia showed that the structure of geochores (heterogeneous geosystems) at topological level and landscape-geochemical processes are densely interrelated and control each other. Each hierarchical level of geomers (homogeneous geosystems) is characterized by the corresponding range of matter stock increasing at the higher hierarchical levels. Within taxons, the geomers differ in stock of chemical elements or their radial differentiation. Integration of geomers is ensured by lateral matter flows. Each rank order of topogeochores has certain ranks of corresponding geomers with certain limits of matter stock variability. The interfluve catenary and slope microgeochores have negative budget of matter differentiation with dominance of lateral flows over radial ones. Matter output from slope geosystems depends on steepness, properties of soils, and parent rocks. Positional-dynamical automorphic microgeochores on slopes and semi-hydromorphic super-aqual microgeochores often have zero balance. Hydromorphic super-aqual microgeochores are characterized by positive balance with dominance of radial flows. Micro-basin super-aqual and sub-aqual halomorphic microgeochores have positive balance with prevalence of lateral flows. Step-by-step integration of geochores (regionalization) should be based on basin landscape-geochemical synthesis.

Keywords South Siberia · Steppe · Geosystem · Geomer · Geochore · Landscape-geochemical processes · Matter differentiation

Y. M. Semenov (✉)
Sochava Institute of Geography, Russian Academy of Sciences, Siberian Branch, Irkutsk, Russia

© Springer Nature Switzerland AG 2020
A. V. Khoroshev, K. N. Dyakonov (eds.), *Landscape Patterns in a Range of Spatio-Temporal Scales*, Landscape Series 26,
https://doi.org/10.1007/978-3-030-31185-8_10

Spatiotemporal organization of the geosystems is one of the most critical research issues in physical geography. Landscape mapping is intended to display and document results of field studies as well as to generalize physical-geographical information on a variety, differentiation, integration, dynamics, and evolution of natural and anthropogenically changed geosystems. Scale and resolution of cartographic model describing structural and dynamic properties of geosphere are chosen in accordance with purposes, research problems, and hierarchical level of the focus units (Suvorov et al. 2007, 2009).

Hierarchy of geosystems is known to have been generated in process of historical development (Sochava 1978). Therefore, the actual landscape envelope[1] is a polygenetic and heterochronic phenomenon. Hence, one could hardly think of its particular parts as about self-developing landscape units. Geochemical migration is believed to play the major role in its formation for centuries-long period, so that landscape development history is a continuous matter differentiation (Kovda 1973). The notion of "matter differentiation" involves both process and result of migration and is understood as distribution and redistribution of substance.

Matter differentiation along the pathways results in the particular sequence of transformations leading to newly emerged spatial combinations of its indices. Therefore, differentiation of a landscape envelope of Earth, integration of its separate parts, and their interrelated functioning anyway are reflected in matter differentiation. Although application of indices of substance state has not yet gained sufficient dissemination, they are critically needed to make classifications more objective, to enrich knowledge of integration mechanisms, as well as to detect development trends in natural phenomena. Many researchers agree that these indices are highly perspective.

Landscape science and landscape geochemistry cannot yet explain with sufficient accuracy how organization of geosystems is related to substance state indices. To solve this problem, a researcher is expected to get insight into the regularities of matter differentiation that reflect various aspects of the landscape organization. For this purpose, it was necessary to reveal dependencies between substance distribution and landscape spatial patterns (landscape-geochemical analysis), to identify material indicators of geosystem integration (landscape-geochemical synthesis), and to evaluate contribution of migration and matter transformation to dynamics and evolution of geosystems (landscape-geochemical diagnosis).

Long-term topological- and regional-scale studies in the geosystems provided the opportunity to estimate substance differentiation and its contribution to spatial patterns development. The principles of the landscape-geochemical analysis, synthesis, diagnosis, forecast, and assessment were formulated. Relevance of these procedures for implementation was tested on the examples of steppe and forest-steppe geosystems. We evaluated substance stocks and their ranges for each hierarchical

---

[1]Landscape envelope, in Russian geography, is a sphere of strong interactions between atmosphere, hydrosphere, lithosphere and biosphere within the layer 10n meters above and below the Earth surface. Weathering crust, soils, clay minerals, chemical composition of waters are the examples of specific matter generated by this interaction (Eds.).

level and established differences between geomers[2] at topological level. The role of radial substance differentiation for geomer classification and that of lateral differentiation for integrating geochores was shown. Methods for geomers classification, integration of geochores, and diagnostics of a dynamic condition of geosystems, as well as relevant tools for mapping geomers and geochores, were developed.

Geosystems sensu Sochava (1963, 1978), who introduced this concept, were taken as research objects. Classification and integration were performed taking into account differentiation of substance which constitutes these natural units and migrates between them and within their limits.

Geosystem approach to study of substance differentiation assumes interdependency of migration in neighboring topogeosystems. However, in study areas with dominating low- and medium-elevation mountains, plots of the same topogeosystem (even of the lowest rank order) may occur on opposite sides of water divide which is impassable for water as well as for most air and mechanical migrants. This results in a certain geochemical isolation of geochore elements. Therefore, delimitation of the elementary landscape-geochemical basins (EB) was needed to generalize results of substance differentiation studies (Snytko and Semenov 1980, 2006).

Within each EB, the most complex system of substance differentiation was observed in the geosystems located close to junctions of several converging unidirectional substance flows. For example, in trans-eluvial-accumulative microgeochores of the Sharasun River valley in the Onon-Argun steppe (Transbaikalia), geosystems develop with simultaneous contributions of substance input both from slopes and riverbed. In contrast, in geochores of gentle slopes with unilateral flows, landscape pattern is almost monotonous.

The general pattern of substance differentiation in the Sharasun river basin, being the geosystem of the higher rank order compared to the elementary one, involves flows from eluvial EB of the upper course of river through transitional EB toward accumulative EB. Eluvial-accumulative EB devoid of output streams are isolated from the general system of substance migration in low-mountain steppes.

The study of substance behavior in geosystems with focus on mechanisms of differentiation and integration was performed step by step at various scale levels of landscape organization. Regularities of substance differentiation were considered simultaneously both in subjects of landscape-geochemical analysis (geomers) and in objects of landscape-geochemical synthesis (geochores).

Substance stock in geomers and geochores of a topological and regional order turned out to be in certain compliance. In many respects, it is explained by relationships among geosystems of both classification rows. For example, the elementary geochore often is manifested as a mosaic of the elementary geomers relating to the same *facies,*[3] and microgeochore is a mosaic of the elementary geomers belonging

---

[2] *Geomer* is a homogeneous geosystem, and *geochore* is a heterogeneous geosystem. The terms were introduced in Siberian school of landscape science. See Glossary and Chapter by Cherkashin, Part I, this volume for definitions.

[3] *Facies* is the elementary morphological unit of a landscape in Russian terminology. See Glossary and Chapter by Khoroshev in Part I, this volume, for details.

to the same type of facies, etc. The values of substance stock may be related to hierarchical levels of geomers and geochores. Facies approximately corresponds to elementary geochore, the type (subtype) of facies corresponds to microgeochore, the group of facies corresponds to mesogeochore, the class of facies corresponds to topogeochore, and the geom corresponds to macrogeochore.

Of course, the interrelation of two classification rows of geosystems is not absolute. However, it is more convenient to consider the quantitative indices of substance differentiation at the conjugate levels of geosystems: the elementary geomer and the elementary geochore, the facies and the microgeochore, the type (subtype) of facies and the mesogeochore, the group of facies and the topogeochore, and the geom and the landscape province.

Long-term studies showed that differentiation and integration of geosystems are related to various processes of substance distribution and movement.

As it was mentioned above, the huge diversity of landscape patterns emerged as a result of complex history of substance differentiation. At the same time, the present-day differentiation of geosystems is determined by radial distribution of substance. Therefore, hierarchical classification of geomers has to be based on the landscape-geochemical analysis of geosystems through accounting of absolute amounts of substance in their components. Research showed evidence that range of substance stock in soils and phytomass increase from the lowest taxons to the higher ones and are peculiar for each hierarchical level of geomers.

Classification of low-mountain steppe and forest-steppe geosystems of South Siberia for fine-scale mapping was based on the principles proposed by V.B. Sochava (1978) and on concepts of interdependencies between geosystems organization and differentiation of substance of their components (Semenov 1991). The technique of composing landscape maps with related mapping of geomers and geochores (Snytko and Semenov 1981) allows displaying both typological and chorological units of geosystems as well as their dynamic state. On such a map, the mosaic of geochores includes geomer units. If the geochores are shown within natural boundaries (taking into account limitations imposed by scale, of course), geomers are either displayed or omitted from consideration depending on scale and the level of generalization, i.e., depending on geomer rank order.

Differentiation of geosystems is driven by radial differentiation of substance. Each rank order of geomers at topological level has its own characteristic range of substance stock in soils and phytomass increasing with a taxon rank order. This allows us to classify geomers taking into account results of the landscape-geochemical analysis by evaluating absolute amounts of substance in their components since the taxons of geomers differ by absolute contents of elements or by pattern of their radial differentiation (Semenov 1991).

Facies is commonly the main mapping unit at large scale. The *facies* includes the biogeocenoses in different dynamic states with similar plant communities and soils at similar landforms. The facies can be united in subtypes, types, and groups of facies according to the principle of homogeneity. The landscape-geochemical conditions of development and functioning of the geosystems under consideration described by classification of locations serve as a fundamental criterion for uniting

Table 10.1 Maximum ranges of substance stock in top 1-m soil layer in the Onon-Argun steppe for geomer taxons of topological and regional levels, t/ha

Rank of geomer	Organic-mineral mass	Organic carbon	Nitrogen	Carbonates	Water-soluble salts
Elementary geomer	200	5	0.7	200	15
Facies	**200**	5	**0.7**	**200**	15
Subtype of facies	3000	50	9.2	1000	110
Type of facies	8000	100	11.2	1000	**300**
Group of facies	**11,000**	**190**	**15.2**	1000	300
Class of facies	11,000	190	15.2	1000	300
Geom	11,000	190	15.2	1000	300

Note. The limiting minimum and maximal values of range of substance stock that can be used to identify rank orders of geomers are highlighted in bold type

facies into types (e.g., eluvial, trans-eluvial, trans-accumulative, super-aqual)[4] (Glazovskaya 1964; Fortescue 1992). Types of facies are divided into subtypes of facies in which names (para-eluvial, ortho-eluvial, neo-eluvial, trans-eluvial, on steep, very steep, steep and sloping slopes; trans-accumulative, on sloping and gentle slopes; accumulative, on flat surfaces) have reflected character and intensity of matter migration (Semenov 1991; Zagorskaya 2004).

Increase in range of substance stock in a hierarchy of geomers (from elementary geomer up to geom) in the Onon-Argun steppe is shown in Table 10.1.

Identical indicators of substance stock can be used to identify facies and elementary geomers which are their constituents. Stock values increase for each substance in a particular manner for the higher rank orders. Group of facies is the highest rank order which can be identified based on indices of absolute amounts of substance. All geomers in the corresponding taxons differ in stocks of chemical elements or radial distribution of their contents. This fact, in principle, is well known and affords using it in classifications of soils and geochemical landscapes. The similar regularity was detected in geosystems of Nazarovsky intramontane depression.

Spatial combinations of geomers form units of another classification row, namely, geochores. In nature elementary geochores are spatially integrated into the microgeochore, microgeochores into mesogeochore, mesogeochores into topogeochore[5], etc., which together form a landscape cover of the Earth. This integration is caused by lateral differentiation of substance. Each rank of geochores of topological scale level has certain limits of variance of absolute amounts of the substance in components of geomers with areals limited to respective geochores. Every rank of

---

[4] Geochemical classification of landscape units composing catena. See Glossary and Chapter 9 for definitions and details (Eds.).

[5] Macrogeochore, topogeochore, mesogeochore, microgeochore are the terms for various hierarchical orders of geochores adopted in the Siberian and German schools of landscape science. See Glossary for the definition of geochore (Eds.).

Table 10.2 Limits of substance stocks in 1-m soil layer for ranks of geochores in elementary geomers (bold type) and average range of substance stocks in the subordinated geochores (italic type), t/ha

Rank of geochores	Subordinated geosystems	Organic-mineral mass	Organic carbon	Total nitrogen	Carbonates	Water-soluble salts
Microgeochore	Elementary geomers	**11,000**	**50**	**5.2**	**1000**	**250**
	Elementary geochores	*133*	*3*	*1.9*	*350*	*15*
Mesogeochore	Elementary geomers	**11,000**	**50**	**5.2**	**1000**	**300**
	Microgeochores	*4625*	*24*	*3.4*	*500*	*88*
Topogeochore	Elementary geomers	**11,000**	**190**	**15.2**	**1000**	**300**
	Mesogeochores	*6071*	*100*	*8.8*	*719*	*115*
Macrogeochore	Elementary geomers	**11,000**	**190**	**15.2**	**1000**	**300**
	Topogeochores	*3091*	*110*	*9.5*	*705*	*86*

geochores at topological scale level corresponds to rank of geomers with certain stocks of the substance.

Table 10.2 shows the limits of absolute amounts of substance in a 1-m soil layer in the topological-level geochores calculated for elementary geomers inside them. The average range of substance stocks in the subordinated geochores in southeastern Transbaikalia are shown in Table 10.2 as well.

The geochores unlike the geomers are not strictly controlled by absolute amounts of substance, though the increase in a hierarchical rank of geosystems is followed by increase in substance stocks up to topogeochore rank as well. Most likely, topogeochores are the largest taxons of geosystems that can be delimited based on integration of geochores of lower level. Therefore, systematic integration of geochores based on a landscape-geochemical synthesis which takes into account absolute amounts of substance in soils is possible at the topological level only. Delimitation of geosystems at regional level can be performed only by partitioning of the higher order units but also taking into account the differentiation of the substance.

Integration of geosystems is governed by lateral substance fluxes. The unidirectional migration and uniform balance of differentiation of substance are characteristic of the lowest levels of geochores (Semenov 1991).

In redistribution of substance along slopes, the key role belongs to microgeochores. Slope-related differentiation of microgeochores is much more complicated than a traditional schematic picture of the conjugate series of facies (Snytko and Semenov 1979).

The microgeochores of interfluves formed by geomers of the eluvial and adjacent trans-eluvial locations have the negative balance of substance and well-manifested prevalence of lateral flows over radial ones. The balance of substance in most slope microgeochores is also negative, with a dominance of lateral migration, but the mass of the substance migrating downslope varies depending on slope gradient, soil properties, and bedrocks.

Zero balance of substance with a weak dominance of radial flows is often characteristic of the microgeochores of concave parts of the long slopes, the lower parts of gentle slopes, and super-aqual microgeochores of the small rivers valley and streams with sub-hydromorphic facies. At the same time, super-aqual microgeochores with hydromorphic facies have positive balance of substance and dominance of radial flows. The positive balance of substance with the well-manifested domination of lateral streams is typical for super-aqual and halomorphic microgeochores.

Therefore, integration of geochores should be based on landscape and geochemical synthesis of geosystems through evaluating absolute amounts of substance and identification of balances of migration flows in the subordinated geochores (Semenov and Semenova 2006).

Addition of two additional taxons of a topological rank, namely, type of facies and subtype of facies, to the taxonomic system of geomers (Sochava 1978) allowed reaching more perfect compliance between taxons of typological and chorological rows. While traditional generalization of the landscape maps and regionalization schemes can result in crossing geomer units by boundaries of geochores, the integrated map of geomers and geochores gives the chance to generalize geomers step by step, in concordance with generalization of geochores. Therefore, the final map avoids intersections of geomers by chorological boundaries (Snytko and Semenov 1981).

The geosystem dynamics is reflected in legends of landscape maps also by display of dynamic conditions of the mapped elementary geomers. The technique of landscape-geochemical diagnostics of geosystems state involves permanent study of substance behavior. To this end the research can include lysimetric observations, micromorphological diagnostics of the partial soil forming processes and geochemical relicts, the experimental modelling of interaction between natural and technogenic substance flows, and balance sheet calculations on the basis of long-term series of observations of substance migration (Semenov 1991).

Unfortunately, the approaches to landscape synthesis described above are applicable only at the topological level. The single successful example of the medium-scale map in the framework of this concept is the map of geosystems of the Nazarovsky intramontane depression (Semenov 1991). However, the relations between substance differentiation indices and differentiation and integration of landscape structure turned out to be significant for steppe territories only; we found no evidence that this holds true for forest geosystems. First, indices of phytomass differentiation were not valid because of large stocks of substance in trees in contrast to the anthropogenically transformed vegetation due to deforestation, forest fires, and successions. Second, absolute indices of substance differentiation in forest soils do not reflect specifics of geosystems properly since they are insensitive to podsolization, lessivage, gleyization, and other soil processes which are crucial geosystems classification in taiga and sub-taiga.

Besides that, we faced serious difficulties in interpretation of interrelations between substance differentiation and the organization of geosystems after including materials from the vast steppe and forest-steppe regions to the database.

Therefore, we switched the focus of research to determine not simply a role of substance in formation of geosystems pattern but to evaluate contributions of natural and technogenic flows of certain chemical elements and their associations to the formation of a typological diversity as well as to regional and local regularities of integration in geosystems of a mountainous taiga and sub-taiga (Semenov and Semenova 2006).

Development includes evolution, i.e., change of geosystem invariant, and dynamics, which is manifested in change of states within single invariant (Sochava 1974, 1978; Sochava et al. 1974; Isachenko 1991). Each geosystem state can manifest itself only in a particular time interval which is long enough for the full cycle of fluctuations in substance attributes. In this paper we understand geosystem state as not any changes of parameters of their components but only those sets of properties that remain constant for a long time. We agree with N.L. Beruchashvili (1982) who defined a geosystem state as a combination of parameters of structure and functioning in a particular period. At the same time, it is supposed that each change of geosystem state is in compliance with particular change of geochemical parameters. We argue that the critical issues in study of temporary organization of geosystems focusing on substance differentiation are establishment of cycles in changes of landscape-geochemical indexes (or of their absence) and detection of development trends based on diagnostics of geosystem state. All changes in substance attributes leading to structural change of geosystems without change of their invariant (commonly, migration of the mobile compounds) are considered as indicators of geosystems dynamics. Changes of invariant (e.g., transformation of a soil profile or bulk composition) testify evolution of geosystems. The task of crucial importance is to make proper diagnosis of geosystem state.

We conducted the study of radial migration of substance in the Onon-Argun steppe which involved observation of dissolved substance regime, direct evaluation of migrating substance by lysimetric method, and micromorphological investigation of soil processes. This allowed identifying the elementary geomers according to their dynamic state. Diagnostics of dynamic states based on the abovementioned indices allowed us to make distinction between *intact*, *quasi-intact*, and *serial geosystems*.

Intact geomers commonly have the stable regime. Intensity of radial substance migration is rather low. In soil micromorphology there are no clear indicators of the soil forming processes other than the basic one. Although intensity of substance migration in intact geosystems is small in comparison with the other geomers which have similar properties and position in classification, their absolute sizes can be large enough.

Serial geomers are different due to obvious instability and huge intensity of substance migration. The signs of the accompanying or opposite soil forming processes are clearly testified by soil micromorphology.

The processes in *quasi-intact elementary geomers* exhibit features which are intermediate between that in intact and serial geosystems. The substance migration is quite stable in these geomers, but signs of several soil-forming processes are reflected in a microscopic structure of soils quite often.

References

Beruchashvili, N. L. (1982). Issues in classification of states of natural-territorial complexes. In I. I. Mamay & V. A. Nikolaev (Eds.), *Issues in Geography* (Vol. 121, pp. 73–80). Moscow: Mysl. (in Russian).

Fortescue, J. A. C. (1992). Landscape geochemistry: Retrospect and prospect—1990. *Applied Geochemistry, 7*(1), 1–53.

Glazovskaya, M. A. (2002). *Geochemical foundations for typology and methods of research of natural landscapes.* Smolensk: Oikumena. (in Russian). First published: Glazovskaya, M. A. (1964). *Geochemical foundations for typology and methods of research of natural landscapes.* Moscow: MSU Publishing House. (in Russian).

Isachenko, A. G. (1991). *Landscape science and physical-geographical regionalization.* Moscow: Vysshaya shkola. (in Russian).

Kovda, V. A. (1973). *Foundations of soil science* (Vol. 1). Moscow: Nauka. (in Russian)

Semenov, Y. M. (1991). *Landscape-geochemical synthesis and geosystem organization.* Novosibirsk: Nauka. (in Russian).

Semenov, Y. M., & Semenova, L. N. (2006). The role of matter flows in landscape structure development. In K. N. Dyakonov (Ed.), *Landscape science: Theory, methods, regional studies, practice. Proceedings of the XI international landscape conference* (pp. 349–351). Moscow: MSU Publishing House. (in Russian).

Snytko, V. A., & Semenov, Y. M. (1979). Microgeochores as a reflection of matter differentiation in geosystems. *Reports of Academy of Sciences of the USSR, 244*(2), 455–457. (in Russian).

Snytko, V. A., & Semenov, Y. M. (1980). Structure of steppe geosystems and matter differentiation. *Geography and Natural Resources, 2,* 39–50. (in Russian).

Snytko, V. A., & Semenov, Y. M. (1981). Experience of integrated geomers and geochores mapping. *Geography and Natural Resources, 4,* 28–37. (in Russian).

Snytko, V. A., & Semenov, Y. M. (2006). Landscape and landscape-geochemical research. In A. N. Antipov (Ed.), *Geography of Siberia* (pp. 66–73). Delhi: Research India Publications.

Sochava, V. B. (1963). Definition of some notions and terms in physical Geography. *Proceedings of the Institute of geography of Siberia and Far East SB AS USSR, 3,* 50–59. (in Russian).

Sochava, V. B. (1974). Geotopology as a part of geosystem doctrine. In V. B. Sochava (Ed.), *Topological aspects of geosystem doctrine* (pp. 3–86). Novosibirsk: Nauka. (in Russian).

Sochava, V. B. (1978). *Introduction to the theory of geosystems.* Novosibirsk: Nauka. (in Russian).

Sochava, V. B., Krauklis, A. A., & Snytko, V. A. (1974). To unification of the concepts and terms used at complex researches of a landscape. *Proceedings of the Institute of geography of Siberia and Far East SB AS USSR, 42,* 3–9. (in Russian).

Suvorov, E. G., Semenov, Y. M., & Antipov, A. N. (2007). Concept of landscape information renovation for Siberia area. In K. N. Dyakonov, N. S. Kasimov, A. V. Khoroshev, & A. V. Kushlin (Eds.), *Landscape analysis for sustainable development. Theory and applications of landscape science in Russia* (pp. 76–80). Moscow: Alex Publishers.

Suvorov, E. G., Semenov, Y. M., & Novitskaya, N. I. (2009). Landscape-estimated map of the Asian part of Russia: Principles and methodical aspects of cartography. *Geography and Natural Resources, 30*(4), 313–317.

Zagorskaya, M. V. (2004). Landscape structure of the central Olkhon district. *Geography and Natural Resources, 4,* 58–68. (in Russian).

Chapter 11
Modeling of Hydrological and Climatic Resources of the Landscape for Sustainable Land Use at Small Watersheds

Alexander A. Erofeev and Sergey G. Kopysov

Abstract The analysis of conditions for the runoff formation is believed to be the useful example of applying theoretical knowledge to study interrelations between landscape structure and functioning. The practical importance of such knowledge is seen in developing area zoning which takes into account microclimatic conditions and is aimed at sustainable organization of landscape structure and effective land use at the level of the small watershed. To model the water runoff and certain elements of water balance in the study area, we calculated topographically mediated indicative values of water cycle and solar radiation and assessed how they are related to the diversity of landscape conditions affecting runoff formation. The variability in growing conditions at different landforms was studied using the original method of hydroclimatic calculations and Ramenskiy's vegetation scales which provide information on species sensitivity to moisture content. After that, the detected diversity was considered as rationale for modeling optimal distribution of land use types. The application of the method provides new opportunities for effective use of landscape resources as well for the sustainable development of natural and social systems. This method is believed to be of particular relevance for the newly cultivated agricultural lands and buffer zones of nature protected area.

Keywords Land use · Water balance · Landscape conditions · Topographic attributes · Landscape physics

A. A. Erofeev (✉) · S. G. Kopysov
Tomsk State University, Tomsk State University of Architecture and Building, Tomsk, Russia

© Springer Nature Switzerland AG 2020
A. V. Khoroshev, K. N. Dyakonov (eds.), *Landscape Patterns in a Range of Spatio-Temporal Scales*, Landscape Series 26,
https://doi.org/10.1007/978-3-030-31185-8_11

11.1   Introduction

Recent advance in creating integrated maps of land use to a great extent is associated with the rapidly developing approach which involves environmental parameters obtained by remote techniques (Ramamohana and Suneetha 2012; Myshlyakov and Glotov 2015). At the same time, a landscape-geophysical approach applied for small watersheds as a separate economic unit can allow us to solve the problem of sustainable land use organization in a comprehensive way. Dokuchaev (1892) and his followers emphasized the crucial importance of this problem as early as the end of the nineteenth century. He was engaged in environmental management and proposed measures to optimize natural geosystems taking into account the ecological role of various elements of a landscape. Nowadays, it has become possible to look at solution of such problems using the modern methods of GIS-based simulation.

By the early twenty-first century, evidence was shown that particular types of landscapes play a much bigger role in the global emission of evident or latent heat from land to the atmosphere than could be expected (Shmakin et al. 2001). Proceeding from that, we developed a method that is believed to be useful in optimizing land use structure. The method is based on evaluating water balance elements from individual landscapes within river basins by means of *hydroclimatic calculations* (HCC). This method takes into account the landscape conditions of runoff formation which are indicated by water circulation and insolation geophysical indices calculated from digital elevation models (DEM) using GIS tools.

11.2   Study Area

The study object is a small watershed measuring 2.04 km², which is located about 30 km from the city of Tomsk, in the Kyrgyzka river catchment basin (844 km²) on the stream of a second-order tributary (Fig. 11.1).

The registration of runoff was organized in the middle course and in the outlet of the watershed by a triangular spillway with the flooded lower weir. Also, pilot systems for automatic monitoring of water balance elements have been installed in the middle and lower part of the watershed area (Kopysov and Yarlykov 2015). The choice of watershed was determined by the following reasons:

- Diverse relief, which makes it possible to determine the watershed boundaries explicitly and provide a variety of landscape conditions
- Relative proximity to the city, which provides the possibility of regular field observations along with a weak anthropogenic load on the watershed area

The watershed is located within a gently rolling upland and well-drained plain and has an elevation range of 58 m; elevation on the outlet area is 149 m. Within the watershed area, there are two large depressions with the width of 170–250 m, where

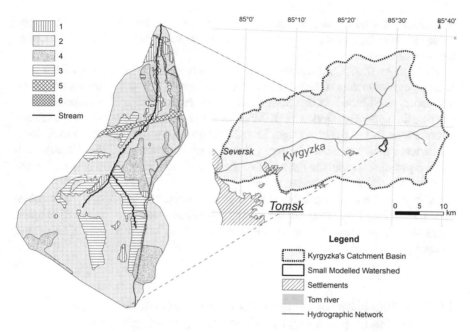

Fig. 11.1 Spatial location of the study object and land use structure: (1) Abietum magnoherbosum forests (4.5%); (2) Betuleto-Populetum magnoherbosum forests; (3) abandoned agricultural lands with dense young birch forests (14.5%); (4) haymaking with high grass meadows (15.5%); (5) soil regenerating roads (1.8%); (6) power transmission line right-of-way with grass meadows (1.7%)

the runoff is concentrated to a large extent. The main types of relief in the study area are as follows: (1) low-lying floodplain, with elevations 150–180 m a.s.l. and slope gradient 0.5–3° in the bottom of depressions; (2) steep slope (5–13°) with elevations 165–195 m a.s.l.; and (3) slightly inclined plains at interfluves with elevations 170–207 m a.s.l. and gentle slopes 0.3–2°.

In the study secondary birch and aspen forests dominate occurring in depressions with an admixture of spruce and fir (Fig. 11.1).

11.3 Methods of GIS-Based Modeling

The important geophysical significance of the earth's relief is known to be associated with the redistribution of matter and energy and, consequently, gravity fields and insolation (Sysuev 2014). While performing GIS-based modeling, we applied the combined method for building digital elevation model (DEM). The SRTM data with a spatial resolution of 30 m and ASTER GDEM with a spatial resolution of 15 m (U.S. Geological Survey ... 2017; Rodriguez et al. 2005; ASTER GDEM 2009) were used as the basic information on the topography. To eliminate errors in determining elevations associated with the presence of vegetation cover, a network of contour lines with a step of 1 meter was generated on the basis of a DEM raster. After that, the flattening of areas with the presence of vegetation was performed in the manual mode.

To increase the spatial resolution of DEM, we included information from fine-scale topographic maps (elevation points and isohypses) as well as the data of leveling and ground-based laser scanning (Erofeev et al. 2017). This allowed us to increase the spatial resolution of DEM up to 10 m and the accuracy of determination of elevations to 1–2 m on average. The resulting DEM was subjected to hydrological correction by Fill Sinks method (Planchon and Darboux 2002). This approach made it possible to analyze the relief of the watershed area with an accuracy sufficient for integral water balance studies.

After that, we calculated the morphometric attributes describing hydrocirculation[1] (the index of potential humidity) and insolation (slope aspect, negative openness) for the purpose of their further transformation into elements of the water balance.

11.4   Integration of Hydrocirculation Parameters in the Calculation of Water Balance Elements

To calculate the water balance for the individual elements of the landscape, we used the HCC method, initially developed by professor Mezentsev (Mezentsev 1982; Karnatsevich et al. 2007; Kopysov 2014). The system of equations for the HCC method, when calculating the hydrological and climatic characteristics, takes into account the redistribution of moisture within a year and partial transfer of moisture from year to year. The HCC method is a mathematical model of the processes of moisture transformation at the level of the active surface of watersheds. It is based on the fundamental laws of energy and matter conservation. In relation to calculating evaporation, HCC is the genetic method (Karnatsevich et al. 2007).

SWAT is one of the most popular scale models for river basin and its watershed (SWAT: Soil and Water ... 2018). It is developed by the Agricultural Research Department of the US Department of Agriculture (USDA) (Gassman et al. 2007). However, the use of this model is limited due to lack of data, which is characteristic of remote and unexplored territories.

In Germany, a similar BAGLUVA method is widely used to calculate water balance characteristics (Glugla et al. 2003). The advantages or disadvantages of the HCC method over the BAGLUVA method can be studied from this paper (Kopysov 2014).

The runoff and evaporation are the main water balance characteristics of the landscape. In the HCC method, evaporation Z is determined by the water equiva-

---

[1] Hydrocirculation is the lateral water flowing downward under force of gravity. Hydrocirculation indicators show the distribution of water under the force of gravity.

lent of the heat and energy resources of evaporation Z_M; the total wetness of the active surface H, consisting of the sum of the corrected atmospheric precipitation and the change in the humidity of the active layer over the settlement period $W_1 - W_2$; and also dimensionless parameters: n which reflects the influence of the landscape conditions of flow formation and r which characterizes the texture-dependent ability of the soil to bring moisture to the surface and to ensure evaporation. Taking into account the abovementioned indices, the formula for calculating the runoff Y is the following (Mezentsev 1982; Karnatsevich et al. 2007; Kopysov 2014):

$$ Y = H - Z = \left(X + W_1 - W_2 \right) - Z_M \left[1 + \left(\frac{X + W_1 - W_2}{Z_M} \right)^{-rn} \right]^{-\frac{1}{n}} \qquad (11.1) $$

The landscape conditions for runoff formation are not only quite diverse, but also closely interrelated. Hence, the best way to evaluate the dynamics of each condition is to use not a separate but integral parameter. In the HCC method, the parameter n is used to take into account the landscape conditions for runoff formation. It reflects mainly the geomorphological conditions and the ability of the active layer of landscape to discharge excess moisture under the action of gravity forces (Mezentsev 1982; Karnatsevich et al. 2007; Kopysov 2014).

In terms of its physical meaning, the composite topographic attribute *Wetness Index* (W_T) (Beven and Kirkby 1979; Moore et al. 1991) is most closely related to the landscape conditions parameter n in the HCC method. It is extracted automatically on the basis of DEM using specialized software, e.g., *SAGA* software (System for Automated ... 2018), *GRASS* (Geographic Resources ... 2018), etc.

The index of potential wetness (Wetness Index) was calculated by the formula (Moore et al. 1991):

$$ W_T = \ln \frac{A_s}{T \cdot \tan \phi}, \qquad (11.2) $$

where T is the permeability of the soil; $\tan \varphi$ is the inclination tangent of slope angle; and As is the specific catchment area.

When applying Eq. 11.2 for large areas with diverse landscape conditions, it becomes difficult to specify the permeability parameters of soil T. In the HCC method, soil permeability r is used as a surrogate for the parameter of soil hydrophysical properties.

If in Eq. 11.2 soil permeability $T = 1$, then it becomes possible to link it to Eq. 11.3 through the following relationship (Kopysov et al. 2015) (Table 11.1):

$$ n = 1{,}1 + \frac{W_T}{6{,}1}, \qquad (11.3) $$

Table 11.1 Physical significance of parameter limits of runoff landscape conditions (n) and the index of potential wetness (Wetness Index - *WT*)

$n \to 0$ $(W_T \to 0)$	$Z/Z_M \to 0$	All the water runs off – there is nothing to evaporate
$n \to \infty$ $(W_T \to \infty)$	$Z = H$	There is no runoff – all the water evaporates
	$Z/Z_M = 1$	Water stagnates in depressions due to the lack of evapotranspiration energy resources

(see Table 11.1 in Kopysov et al. 2015)

The calculation of the maximum possible evaporation in the HCC method is not associated with a specific type of evaporating surface and represents values averaged over large areas (Mezentsev 1982). This problem was partially solved by taking into account the insolation characteristics of the landscape, calculated in SAGA software (System for automated ... 2018). It was assumed that for territories with a pronounced relief, the spatial distribution of heat energy resources of evaporation should be calculated as the ratio of the actual illumination duration j_i at a particular section to the illumination duration at zero slope and open horizon j_0. Based on this, we added the following correction to the sum of active temperatures $\sum t_{10°C}$, considering the slope gradient, slope aspect, and negative openness:

$$\chi = 1 + \log\left(\frac{j_i}{j_0}\right) \tag{11.4}$$

This correction gives results similar to the tabulated data (Kopysov 2014) on the thermal characteristics of various forms of relief. The formula for calculating the water equivalent of the heat energy resources of evaporation (evaporation capacity) (Mezentsev 1982), taking into account this correction, takes the following form:

$$Z_M = 200 \cdot \left(\frac{\chi \cdot \sum t_{10°C}}{1000}\right) + 306 = \frac{\chi \cdot \sum t_{10°C}}{5} + 306 \tag{11.5}$$

The results of calculation of hydrocirculation parameters are shown in Fig. 11.2. For the flat surfaces, the potential evapotranspiration accounts for 678 mm as calculated by the Tomsk weather station for the period from 1881 to 2013 according to the data (Specializirovannye massivy ... 2017).

The calculated spatial distribution of evaporation (Fig. 11.2d) shows the results of the runoff formation similar to the distribution of the landscape condition parameter *n* (Fig. 11.2a). It can be explained by the lack of moisture on the most drained relief elements for areas with optimal humidification (precipitation ≈ evaporation) (evaporation 461 mm). The average long-term value of evaporation in the watershed area accounts for 496 mm, and for the most humidified areas, it rises up to 522 mm.

The value of the local (climatic) model runoff (Fig. 11.2b) has the lowest values where, due to poor conditions of excess moisture discharge (maximum values of

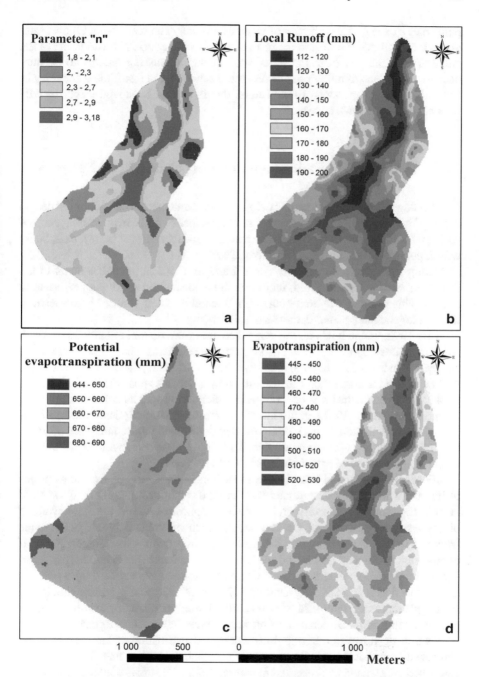

Fig. 11.2 The distribution of the parameter of runoff formation of landscape conditions *n* and the normal annual amounts of water balance at the studied watershed. (**a**) parameter "n"; (**b**) local runoff (mm); (**c**) potential evapotranspiration (mm); (**d**) evapotranspiration

parameter n), the greatest amount of water evaporates. The real channel runoff integrates local (climatic) runoff in the form of a channel network. The variability of the value of precipitation from the relief and vegetation was not taken into account because of the lack of unambiguous recording schemes and insufficient accuracy of precipitation measurements. In the future, this issue will be solved using the data obtained from the snow surveys.

11.5   Hydrological Landscape Planning of Land Use Patterns

In accordance with agroclimatic zoning of the watershed, it is recommended to develop livestock farming and to grow medium-ripened and early-ripening sorts of spring wheat, and winter rye and early- and late-ripening sorts of barley, oats, buckwheat, potatoes, peas, and flax (Shashko 1967).

During the reclamation of the Virgin Land in 1950s, all suitable lands in the catchment area were cultivated, including those located on gentle slopes prone to erosion. This was due to aggregating small-area fields, which resulted in the destruction of biogeocoenoses and decrease in soil fertility.

After 1992, arable lands on the watershed area were no longer cultivated, which led to rapid development of dense birch coppice (Fig. 11.1). This process was much slower on lands used for haymaking. Over the past 25 years, the soils on the studied watershed have restored their fertility. And now it becomes possible to involve them in plowing again on the basis of ecological principles. Proper land use management, according to Odum (1971), is the most important practical application of environmental science. Resuming agriculture on abandoned land will require a lot of financial expenditure for removing coppice. This could have been avoided if herbs on abandoned lands had been cut during the first few years.

In our opinion, fundamental basins of sustainability should be treated as points of departure when distributing land use over the catchment area. One of the main laws in the world is the unity and conflict of opposites within the framework of binary systems (All-Nothing, Order-Chaos), which are stable only within the range from 1/3 to 2/3. This range sets a harmonious proportion – the Golden Ratio. If competing phenomena go beyond these limits, their instability rises sharply, management becomes problematic, and without the forced intervention, the system is doomed to decay. That is why Odum (1971) mentioned that a person should not seek to get more than one-third of gross output if they are not ready to supply energy to ensure long-term maintenance of primary production in the biosphere.

If we want to live in a harmonious world, the distribution of land use over the watershed area should be environmentally reasonable. Not less than 1/3 of the total watershed area should be preserved in natural state, while areas with high anthropogenic loads (arable lands, roads, buildings) should not cover more than 1/3. The rest of the catchment area should be used with moderate loads (hayfields, pastures, gardens, and plantations). Wilson (2016) believes that the same rule holds true for the global scale. However, any ecologist knows well that thinking globally, one needs to act locally. Therefore, we believe that a harmonious world should be built of "bricks" – small watersheds – in state of harmony in relation to each other.

The local ecological conditions of growth are reflected in the best way by the composition of the ground vegetation cover. Vegetation as the component of ecosystem can be characterized quantitatively by the moisture range (MR) according to the Ramensky[2] scale (Ramensky 1971). The other approach is modeling the dynamics of the LAI sheet coating index (Hickler et al. 2012). However, LAI is more suitable for global modeling of woody vegetation, since it does not reflect the local conditions of the ground cover growth.

To simulate the dynamics of natural processes in the watershed areas, of which the ground vegetation cover is an indicator, the moisture range is calculated from DEM and climate humidification ($\beta_H = H/Z_M$) according to the formula proposed by Kopysov (2015):

$$MR = 100 \cdot \beta_H \cdot \left(\frac{r-1}{rn+1} \right)^{1/m}$$

where the hydrophysical properties of soil are taken into account, similar to the HCC method, by the parameter r (Mezentsev 1982; Karnatsevich et al. 2007).

To harmonize land use in the catchment area for agricultural needs, taking into account crop rotations and dividing forest belts, the land units with parameter n values ranging from 2.25 to 2.75 should be used. In fact, 2/3 of the catchment area falls within this range of parameter n. Crop rotation should comply the requirement of limiting cultivation at any moment to no more than 2/9 of the total catchment area. This is due to the fact that the most optimal ratio of arable land to hayfields is 1:3 for the non-destructive management of agriculture. This enables us to preserve natural fertility of lands due to natural geochemical cycles and manure application.

The moisture level in the watershed ranges from 62 to 73 on a 100-point Ramensky scale. The overlay of the maps of humidification steps and zone of economic activity (for the values of parameter n from 2.25 to 2.75) made it possible to compile the map of the recommended land use structure (Fig. 11.3).

The proposed land use structure in the Kyrgyzka catchment area is based on the need to return to a small-field agriculture, i.e., to divide large fields into ecologically and technologically homogeneous areas. Usually, this implies delimiting field plots by forest belts along the contour lines (Prikhod'ko et al. 1994). Our study showed that environmental conditions can be described holistically by parameter of runoff landscape conditions n since it reflects both slope gradient and slope length. At the same time, it is relevant to locate arable lands in highlands with the

---

[2]L.G. Ramensky (1884–1953) is Russian geobotanist and geographer who proposed ecological scales for vegetation ordination along the axes of moisture supply, nutrient supply, salinity, etc. He is also the author of pioneer works concerning theory of landscape and landscape morphological units.

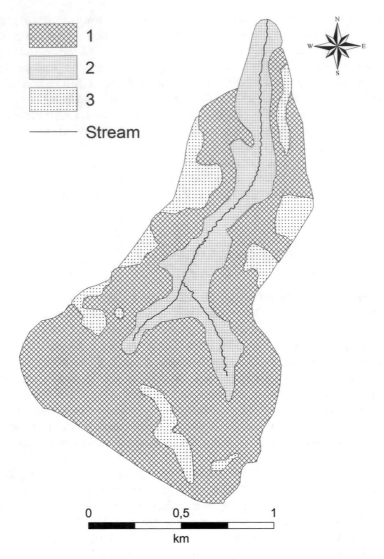

Fig. 11.3 Recommended structure of land use of the modeled watershed in accordance with parameter *n* and moisture range. 1 – the zone of economic activity; 2 – water protection zone of surface runoff; 3 – water protection zone of groundwater supply

parameter *n* values ranging from 2.2 to 2.8. Higher values of n indicate danger of bogging while lower ones danger of erosion. Hence, such lands can be used as either hayfields or pastures. If *n* values exceed 2.9 (water protection zone of surface runoff) or are less than 2.2 (water protection zone of underground runoff), afforestation is needed.

The proposed land use organization is expected to limit surface runoff, to reduce the intensity of soil erosion, to regulate the wind regime and snow accumulation, to reduce the intensity of snowmelt (decrease erosion and maximum river levels) and, as is commonly believed (Prikhod'ko et al. 1994), to increase the yield of agricultural crops and overall biological productivity of the landscape by 10–15%.

At the end of the nineteenth century, the following crop rotation system was practiced in the Kyrgyzka catchment area: Virgin Land was cultivated for 4–5 years, and then the land served for haymaking for 2 years (since in dry years it suffered low productivity of herbs in elevated sites); after that it was commonly abandoned for 10–15 years (Tomilov 2001). Under such farming system, winter crops on thick gray soils accounted for 1.5–2 t/ha, while in the elevated sites in dry years, it did not exceed 0.5 t/ha.

This dependence on the current moistening indicates the need in regulation of the water and heat regime of soils. For such purpose it is reasonable to restore and create new ponds in the ravine and gully network and maintain snow retention on high relief elements by means of forest belts, which, as pointed out by Bosch and Hewlett (1982), activate the supply of groundwater.

The final map shows zones as follows (Fig. 11.3): (1) the zone of economic activity corresponding to the areas of optimal cultivation of field crops which suffer from lack of moisture in dry years in the second half of summer (moisture range 64–68, i.e., predominantly humid soils without signs of gleyization on sufficiently drained territories (Ramensky 1971); (2) water protection zone of surface runoff corresponding to the floodplains interacting with channel runoff by means of matter and energy exchange; (3) water protection zone of groundwater supply, located in local depressions with infiltration of the flow into soil and sediments.

11.6 Conclusion

The solar energy which is our main natural renewable resource is accumulated in the phytomass as well as in the form of potential energy of watercourses, which is unevenly distributed over the landscapes. For its effective use on the basis of the diversity of landscape resources, we proposed zoning of the small catchment area by means of modeling the optimal distribution of land use types. The application of the proposed method allows us to ensure the energy-efficient use of the landscape resources and, hence, sustainable development of the associated natural and social systems. The method is recommended for newly introduced lands and buffer zones of nature protected areas.

Acknowledgments This research was financially supported by The Tomsk State University Competitiveness Improvement Programme (Project No. 8.1.32.2018).

References

ASTER GDEM. (2009). http://gdem.ersdac.jspacesystems.or.jp/index.jsp. Accessed 01 May 2018.

Beven, K. J., & Kirkby, M. J. (1979). A physically based, variable contributing area model of basin hydrology. *Hydrolological Science Bulletin, 24*, 43–69.

Bosch, J. M., & Hewlett, J. D. (1982). A review of catchment experiments to determine the effect of vegetation changes on water yield and evapotranspiration. *Journal of Hydrology, 55*, 3–23.

Dokuchaev, V. V. (1892). *Our steppes in the past and at the present time.* St. Petersburg. Evdokimov's printing house. 128 p. (in Russian).

Erofeev, A. A., Li, V. G., & Kopysov, S. G. (2017). Experience of a terrestrial laser scanning of natural landscapes of the southeast of West Siberia. In *Contemporary problems of geography and geology. Proceedings of the IVth All-Russian scientific and practical conference* (Vol. 1, pp. 57–61). Tomsk: Tomsk State University. (in Russian).

Gassman, P. W., Reyes, M. R., Green, C. H., & Arnold, J. G. (2007). The soil and water assessment tool: Historical development, applications, and future research directions. *Transactions of the ASABE, 50*(4), 1211–1250.

Geographic Resources Analysis Support System (GRASS GIS). (2018). https://grass.osgeo.org. Accessed 01 May 2018.

Glugla, G., Jankiewich, P., Rachinow, C., et al. (2003). *BAGLUVA. Wasserhaushaltsverfahren zur berechnung vieljähriger mittelwerte der tatsächlichen verdunstung und des gesamtabflusses. BfG – 1342.* Koblenz: Bundesanstalt für Gewässerkunde.

Hickler, T., Vohland, K., Feehan, J., Miller, P. A., et al. (2012). Projecting the future distribution of European potential natural vegetation zones with a generalized, tree species-based dynamic vegetation model. *Global Ecology and Biogeography, 21*, 50–63.

Karnatsevich, I. V., Mezentseva, O. V., Tusupbekov, G. A., & Bikbulatova, G. G. (2007). *Study of the dynamics and mapping of elements of energy and water balance and characteristics of energy and water availability.* Omsk: Omsk State Agricultural University Press. (in Russian).

Kopysov, S. G. (2014). Parametrical expression of landscape conditions of runoff in the method of hydroclimatic calculations. *Geography and Natural Resources, 3*, 157–161. (in Russian).

Kopysov, S. G. (2015). Reflection of the ecosystem processes in method of hydro-climatic calculations. In *Reflection of bio-, geo, antropospheric interactions in soils and soll cover* (pp. 417–421). Tomsk: Izdatel'skij dom TGU. (in Russian).

Kopysov, S. G., & Yarlykov, R. V. (2015). Experience in organization of hydrological and climatic observations on small model catchments of West Siberia. *Bulletin of the Tomsk Polytechnic University. GeoAssets Engineering, 326*(12), 115–121. (in Russian).

Kopysov, S. G., Erofeev, A. A., & Zemtsov, V. A. (2015). Estimation of water balance over catchment areas taking into account the heterogeneity of their landscape conditions. *International Journal of Environmental Studies, 72*(3), 380–385. https://doi.org/10.1080/00207233.2015.10 10876.

Mezentsev, V. S. (1982). *Hydrological calculations for melioration.* Omsk: Omsk Agricultural Institute Publishing house. (in Russian).

Moore, I. D., Grayson, R. B., & Ladson, A. R. (1991). Digital terrain modeling: A review of hydrological, geomorphological and biological applications. *Hydrological Processes, 5*, 3–30.

Myshlyakov, S., & Glotov, A. (2015). «Geoanalitika.Argo» – an innovative solution for agricultural monitoring. *Geomatics, 2*(27), 58–62. (in Russian).

Odum, E. P. (1971). *Fundamentals of ecology* (3rd ed.). Philadelphia: W.B. Saunders.

Planchon, O., & Darboux, F. (2002). A fast, simple and versatile algorithm to fill the depressions of digital elevation models. *Catena, 46*(2), 159–176.

Prikhod'ko, N. N., Shaldej, V. V., & Pishchak, D. V. (1994). Issues of rational nature management and ecological optimization of the west of Ukraine landscapes. In *Problems of regional ecology. Vol. 2. Regional nature management* (pp. 26–35). Tomsk: Krasnoe znamya. (in Russian).

Ramamohana, R. P., & Suneetha, P. (2012). Land use modeling for sustainable rural development. *International Journal of Science, Environment and Technology, 1*(5), 519–532.

Ramensky, L. G. (1971). *Problems and methods for studying vegetation cover. Selected works.* Leningrad: Nauka. (in Russian).

Rodriguez, E., Morris, C. S., Belz, J. E., Chapin, E. C. et al. (2005). *An assessment of the SRTM topographic products.* Technical report JPL D-31639. Pasadena, California: Jet Propulsion Laboratory.

Shashko, D. I. (1967). *Agro-climatic zoning of the USSR.* Moscow: Kolos. (in Russian).

Shmakin, A. B., Krenke, A. N., Mihajlov, Y. A., & Turkov, D. V. (2001). Role of the landscape structure of terrestrial surface in climatic system. *Proceedings of Russian Academy of sciences, geographical series, 4,* 38–43. (in Russian).

Soil and Water Assessment Tool (SWAT). (2018). https://swat.tamu.edu. Accessed 01 May 2018.

Specialized data for climatic research. (2017). http://aisori.meteo.ru/ClimateR. Accessed 11 Nov 2017.

System for Automated Geoscientific Analyses (SAGA). (2018). http://www.saga-gis.org. Accessed 01 May 2018.

Sysuev, V. V. (2014). Basic concepts of the physical and mathematical theory of geosystems. In K. N. Dyakonov, V. M. Kotlyakov, & T. I. Kharitonova (Eds.), *Issues in Geography (Horizons of landscape studies)* (Vol. 138, pp. 65–100). Moscow: Kodeks. (in Russian).

Tomilov, N. A. (2001). *Russians of the lower part of Pritom'ye (the end of XIX – the first quarter of XX century).* Omsk: Omsk Pedagogical University Publishing house.

U.S. Geological Survey (USGS). (2017). http://usgs.gov. Accessed 22 Aug 2017.

Wilson, E. O. (2016). *Half-earth: Our planet's fight for life.* New York: Liveright Publishing Corporation.

Chapter 12
Influence of the Landscape Structure of Watersheds on the Processes of Surface Water Quality Formation (Case Study of Western Siberia)

Vitaly Yu. Khoroshavin, Larisa V. Pereladova, Vladimir M. Kalinin, and Artem D. Sheludkov

Abstract The Authors propose an original system for estimating the runoff of pollutants from different oil-contaminated landscapes of small river catchments in conditions of oil production in the middle-boreal forest of Western Siberia. The system is a set of jointly used methods for determining the quantitative parameters of water and chemical runoff: landscape-hydrological analysis, water-balance flow calculations, and empirical modeling of pollutants from oil-contaminated areas. The water runoff of small rivers in the north of Western Siberia has been very poorly studied, and there are no reliable regime data on water resources. The proposed system requires a minimum of information that is accessible to any specialist who is engaged in assessing the risks of production activities for the environment. Using the methodology, we concluded that the most unfavorable conditions for the formation of the small river water quality are formed on slope landscapes formed on loamy soils and peat bogs when laying the objects of the oil infrastructure, in case of accidents on pipelines. The chapter contains an estimate showing that the removal of petroleum hydrocarbons by one small river with a contamination of 0.25% of the catchment area reaches tens of tons per year.

Keywords Landscape-hydrological analysis of catchment · Small rivers · Pollution runoff · Non-point sources · Water quality · Petroleum hydrocarbons

V. Y. Khoroshavin (✉) · L. V. Pereladova · V. M. Kalinin · A. D. Sheludkov
University of Tyumen, Tyumen, Russia
e-mail: v.y.khoroshavin@utmn.ru

© Springer Nature Switzerland AG 2020
A. V. Khoroshev, K. N. Dyakonov (eds.), *Landscape Patterns in a Range of Spatio-Temporal Scales*, Landscape Series 26,
https://doi.org/10.1007/978-3-030-31185-8_12

12.1   Introduction

The river basin is a complex system of elementary watersheds. Elementary watersheds differ from each other in landscape, hydrological, and geochemical characteristics. Each elementary catchment is a landscape complex with particular relief, spatial patterns of rocks, soils, and plant communities. Soils have different filtration properties, which are determined by their texture. The combination of all these characteristics determines the water regime of elementary watersheds. The interaction of several elementary watersheds influences the water balance and chemical parameters of surface waters within the basin of a higher-order basin. Surface runoff and interflow control the flow of matter within landscape complexes and watersheds as a whole.

The authors assume that catchment area is a set of elementary watersheds. The methodological basis of the work is the landscape-hydrological analysis of the territory, based on this approach. Landscape-hydrological analysis (LHyA) makes it possible to solve the primary task of assessing the risk of sources of pollution in surface waters. It provides the opportunity to identify areas in catchments that contribute most to the formation of quantitative and qualitative flow of small rivers. LHyA allows detecting the watershed complexes that determine the quantitative and qualitative characteristics of small river flow. The method is helpful in assessing the vulnerability of these areas to pollutants as well as in calculating the quantitative parameters of pollutant accumulation and migration in various landscape conditions.

Many factors affect the quality of surface water. The formation of the chemical composition of surface waters is a complex landscape process. Water quality is affected by many factors such as landscape-geochemical conditions in the catchment area, the impact of human activities, and processes within water bodies and watercourses (Komlev 2002; Moiseenko and Gashkina 2010). Anthropogenic factors have been playing an ever-increasing role in regulating the environmental and geochemical parameters of surface waters in recent decades (Fenger 2009; Kulmala 2018). Production activity affects the quality of water directly and through the change in the hydrological functions of landscapes in river catchments.

12.2   Various Aspects of the Relevance of the Issue

The quality of the surface waters has caused concern in the areas of oil and gas production for more than 50 years. The high content of petroleum hydrocarbons (PHC) in the waters is the most troubling phenomenon. The researchers noted cases that the ecological norms of the petroleum products content were repeatedly exceeded in the fishing river waters.

Multiple cases of exceeded norms were observed in 1990–2010 in Nizhnevartovsk, Surgut, and Oktyabrsky municipal districts of the Khanty-Mansi Autonomous Area – Yugra. The most common source of petroleum products in the rivers is the discharge of sewage from industrial enterprises and from storage of fuel and oils. However, discharge of sewage containing oil products often is not in compliance with the official reports. What is the source of oil products entering the surface waters? The analysis of the literature allows drawing a conclusion about the significant influence of non-point sources of water bodies' pollution on water quality (Methods...1973; Novotny and Chesters 1981; Novotny 1988; Mikhailov 2000; Zhuang et al. 2012). For example, 60% of the nitrogen load and 44% of the phosphorus load in the Danube River are due to non-point sources (Environmental... 1994).

The experience of long-term research on Western Siberian oil fields (Gashev and Kazantseva 1998; Kalinin 2010; Khoroshavin and Moiseenko 2014) showed evidence that oil-contaminated watersheds are the main non-point sources of petroleum hydrocarbons input to small rivers. Oil-contaminated areas of watersheds result from accidents at oil and gas production facilities (e.g., oil spills). Oil-contaminated sites were found mainly in the areas that were being developed for more than 20 years. They can occupy up to 2–10% of the area of the old oil deposit. Losses of oil account for 2–3.4% of the extracted volume at the fields of the Tyumen region (Nikanorov et al. 2002; Gendrin 2006). Accidents on the intra-field oil pipelines are the main cause of oil spills (90% of cases). Intra-field pipelines are destroyed due to corrosion of metal. Annually, accidents occur more than 4000 times in the oil fields of Yugra. Accidents lead to a total spill of more than 23,000 tons of oil. This oil pollutes more than 3.84 km^2 of catchment areas (Environmental... 2016). It is important to take into account that not every accident is detected.

Oil and oil product (gasoline, fuel, etc.) spills occur in in various landscapes (forest, swamp, valley, meadow, floodplain, highland, lowland, etc.). The conditions of the local oil field and the water-physical properties of soils determine the subsequent behavior of oil in the catchment area. Determination of the regularities of the petroleum product behavior in particular landscape conditions is one of the actual tasks of forecasting the quality of waters in oil production regions.

There are areas of terrain from which flushing of oil into the riverbed occurs as quickly and fully as possible. On the other hand, in certain areas oil accumulates and is conserved or decomposed. The chapter focuses on revealing landscape and landscape-hydrological differentiation of the PHC flows on the watershed of a small river using the original technique in the conditions of the middle taiga of Western Siberia.

12.3   Methods and Materials

The Model Soil and Water Assessment Tool (SWAT) proposed by the research team under Jeffrey G. Arnold (US Department of Agriculture) leadership proved its relevance for assessing water flow from unexplored catchments. SWAT allows evaluating the hydrological characteristics of catchments taking into account certain natural parameters such as hydrophysical properties of soils, etc. (Jha et al. 2007; Gevaert et al. 2008; Krysanova and Arnold 2008). SWAT has gained international acceptance as a robust interdisciplinary watershed modeling tool as evidenced by international SWAT conferences, hundreds of SWAT-related papers presented at numerous other scientific meetings, and dozens of articles published in peer-reviewed journals. Unfortunately, SWAT is poorly adapted to the conditions of strong waterlogging of the watersheds and was not tested for studying the removal of PHC from the surface of the catchment area. Usually, water quality calculations using SWAT are associated with the removal of nutrients from rural catchments and assessment of erosion processes (Gassman et al. 2007; Williams et al. 2008). SWAT has a lot of limitations. It does not allow taking into account many natural features of Western Siberia, especially the extreme waterlogging of taiga water catchments.

Consideration of a larger set of hydrological functions of landscapes is needed to achieve the study aim. Landscape-hydrological analysis (LHyA) is a methodological basis of our study. Professor Victor Glushkov (Glushkov 1936), one of the outstanding Russian hydrologists of the early twentieth century and the founder of the State Hydrological Institute of Russia, formulated the idea of the impact of the geographical landscape on natural waters. He laid the foundations for a geo-hydrological analysis of the territory, which later developed into a landscape-hydrological method. Professors Aleksandr Antipov and Lev Korytny (Institute for Geography RAS, Siberian branch, Irkutsk) as well as their colleagues were leaders in the development of landscape-hydrological analysis for watersheds of rivers in Western Siberia (Antipov et al. 1989; Kapotov et al. 1992; Kalinin et al. 1998; Antipov et al. 2007; Antipov and Korytny 2012).

Hydrological characteristics of small rivers are very poorly studied in Western Siberia. Hence, the method of water balance is the only appropriate way to obtain quantitative data on the water runoff. The water regime and the water balance depend on the landscape conditions in the catchment area. Analysis of the literature shows that Kalinin's water-balance calculation method is well-adapted for assessing the water balance of the boggy taiga watersheds in Siberia (Kalinin et al. 1998; Kalinin 1999). The method is based on two formulas (Eqs. 12.1 and 12.2) used to calculate the water runoff from swamped and drained (forest, slope) parts of the catchment area. It allows us to calculate only the volume of water flow from the catchment area of a small river, taking into account the landscape-hydrological structure of the territory. The opportunity to take into account the seasonal features of the water flow formation is an important advantage of our model. Spring is the main period of the formation of surface runoff and intra-soil flow in Western Siberia. Usually up to 90% of the water flow of the Western Siberian rivers is formed in spring during snowmelt.

$$\gamma_{dry} = \frac{ax}{\sqrt[3]{1 + \left[\dfrac{WC_{max} - SHS_{max}}{W_s - SHS_{max}}\right]^{3m}}} \qquad (12.1)$$

$$\gamma_{swamp} = x \left[a - \left(a - \eta_{dry} \right) \left[1 + \left(3.7 \frac{h}{h_{crl}} \right)^{-5} \right]^{-\frac{1}{3}} \right] \tag{12.2}$$

where γ_{dry} is the depth of water runoff from forest and slope, mm; γ_{swamp} is the depth of water runoff from swamp and other wetlands, mm; x is the snow storage at the end of winter, mm; WC_{max} is the full water capacity of active layer of soil, mm; SHS_{max} is the maximal soil hygroscopicity of active layer of soil, mm; W_s is the volume of water reserve in soil of snow melting period, mm; h is the average level of intra-soil and groundwater during spring water runoff; and h_{crl} is the critical level of intra-soil and groundwater, when influence of intra-soil water on runoff is finished, m. a, b, m, η_{dry} are parameters, which are determined by meteorological, soil, and relief condition on the part of catchment (Kalinin 2010).

The second part of the assessment system should allow us to calculate the amount of chemicals that are leached from the catchment area. We need a formula that describes the dependence of the removal of pollutants on parameters that can be easily determined by remote observation methods or using reference materials. A large number of methods were proposed that allow evaluating the flow of chemicals along with the water runoff (Novotny 1985; Results...1983; Vompersky and Idson 1986; Vollenweider 1989; Walker et al. 1989). The methods were developed separately for different substances and often for specific geographical (landscape) conditions. The empirical model (Eq. 12.3) can be used to quantify the removal of petroleum hydrocarbons from various landscape-hydrological parts of the catchment area. This empirical model describes the dependence of the concentration of petroleum hydrocarbons in the river waters from the areas of oil spills, the age of their existence, and the moduli of water and oil flow in the area. The model was built for the conditions of the Middle Ob and subsequently adapted to the conditions of the middle taiga and permafrost zone of Western Siberia (Kalinin 2001, Khoroshavin 2005; Khoroshavin and Moiseenko 2014). The specific character of this model is explained by the fact that oil spills are not widespread in the other regions of the world. The model also takes into account the natural features of the landscapes of Western Siberia, in particular high bog occurrence, permafrost, and large seasonal differences in the formation of river runoff.

$$\mu = a_m M_p \left[1 - \exp\left(-a_g \frac{f_p}{F} \right) \right] + a_b M \left(1 - \frac{f_p}{F} \right) \tag{12.3}$$

where μ is the modulus of petroleum hydrocarbon washout, mg s^{-1} km^2; M is the aggregated modulus of water runoff from watershed areas, L s^{-1} km^2; M_p is the modulus of water runoff from oil-contaminated catchment part, L s^{-1} km^2; f_p is the area of oil contaminated on the catchment, km^2; F is the catchment area, km^2; a_b is the parameter equal to oil product concentration in outlet without oil-contaminated areas (background state); a_m is the parameter equal to oil product concentration in outlet when $M_p = 1$ and maximum of oil-contaminated catchment; a_g is the parameter, what show oil components fluidity, it connected with oil spill age.

The value of the parameter a_m is 0.42, and the value of the parameter a_g is 40 for the conditions of the middle taiga landscape of Western Siberia. We obtained the values of the parameters empirically.

To apply the method for assessing the quality and quantity of the water in a small river, the following materials are required: topographic maps at 1:100,000 scale, accessible space images with a resolution at least 30 m, weather and climate, hydrophysical properties of soils, and information on plant communities. We collected the materials for the catchment of the Yeniya River in 2014–2017.

12.4   Study Area

We chose the watershed of the Yeniya River as the study area to verify the quantitative assessment system (Eqs. 12.1, 12.2, and 12.3) for the washout of pollutants from different landscapes of the small catchment area. The case of the Yeniya watershed territory is interesting because the territory is used for oil production and t, at the same time, has the status of regional-level natural protected area. Almost the entire catchment area lies within the borders of the natural park "Kondinskie ozera" which is the lakes in the sources of the Konda River (Fig. 12.1). The Yeniya River is located in the middle taiga; its length is 7.8 km; the catchment area is 190 km². The Yeniya is classified as a small river.

Oil production has been carried out within the catchment area for 18 years. Lukoil Company, the only operating company in the study area, performed most drilling operations and construction of oilfield facilities in 1998–2010. It built a large number of oil and gas production facilities, drilled more than a hundred wells, and laid hundreds of kilometers of infield oil pipelines. Since the first days of the company operating, it has used environmentally friendly technologies. The company collected and removed drill cuttings from the territory and made double-walled oil pipelines and special isolation of the earth embankment for placing well clusters. Oil extraction is conducted in compliance with environmental and safety measures taking into account the special status of the territory. However, the risk of emergencies occurrence still exists despite the precautionary measures. For example, heavy vehicles constantly carry large loads in the territory. Moreover, the Talnikovoe oil field is expected to be expanded.

Petroleum hydrocarbons (PHC) enter the soils during the oil spills. The specific discharge from the catchment area determines the speed and completeness of flushing of soluble and suspended PHC. The distribution of the specific discharge is uneven in time and in the basin space. It depends on the hydrological functions and parameters of the catchment landscapes. Besides, the intensity of oil flushing from the surface of the catchment area might change over a season and thus in the landscape space.

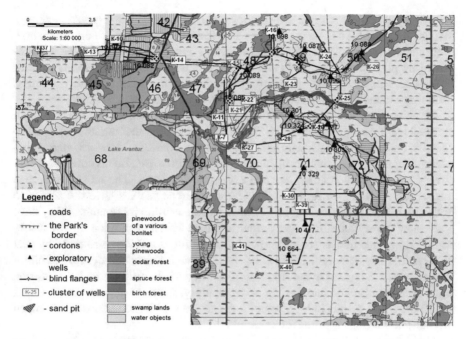

Fig. 12.1 Schematic map of the Yeniya River catchment (the middle taiga of Western Siberia, the left bank of the upper reaches of the Konda River)

The most intensive washout of the PHC from the catchment often takes place during the spring flood when it is formed at all geomorphic levels. In summer, the water runoff is formed during prolonged and intense rainfalls, usually only on wetlands: mires and river floodplains. Analysis of the spatial distribution of the PHC flushing into the riverbed is a more complex task. Understanding the landscape structure of the catchment area is needed to study the role of the landscape in oil pollution of rivers.

Landscape structure of the Yeniya River catchment involves a combination of forests, swamps, and mires. Dominating forest communities are *Pinetum vaccinioso-cladinosum* on sandy Haplic Podzols and *Pinetum ledoso-hylocomiosum* on sandy-loamy Gleyic Albeluvisols. Large areas are occupied by sphagnum bogs on deep (more than 5 m) and moderately deep (1–2 m) peat deposits. *Betuleto-Piceetum caricoso-hylocomiosum* and willow forests grow in the valleys of the Yeniya River and its tributaries.

Raised sphagnum bogs with 2–5 m thickness of peat predominate in the area. They occupy up to 65% of the area of the studied catchment area (Fig. 12.1). Forest stands occupy 23.4% of the catchment area. About half of this area is represented by swamped forests with *Pinetum ledoso-hylocomiosum* and *Pinetum sphagnosum* (with Dicranaceae, *Polytrichum*) on Gleyic Albeluvisols. The rest of the catchment area is covered by floodplain forests. Rare meadow plots cover small areas. The main watershed landscapes of the catchment area belong to two landscape types that are strikingly opposite from a hydrological point of view. The first type is peat bogs that are able to store a large amount of water in spring. Reserves of bog waters flow evenly into riverbeds during the warm season. The second landscape type is represented by pine forests on sandy soils and grounds. These soils have high rate of infiltration, subsurface and ground flow, evaporation, and transpiration. The runoff from forest areas is rapid, and there is no reserve of moisture in the soil.

Pinetum vaccinioso-cladinosum forests on Haplic Podzols are located mainly on elevated landforms at elevations of 70–75 m a.s.l. and on the morainic hills, composed of quartz white sands with small pebbles and boulders. Moisture interception occurs on gentle slopes that occupy lower positions in catena. In this case, the lichen cover is replaced by Dicranaceae and *Polytrichum* mosses. The most widespread landscapes are *Pinetum ledoso-hylocomiosum*. Green mosses have the ability to hold water. Due to this fact, forests with a moss cover are more humid; they have lower flow rates. Basically, water output is realized via transpiration or with a subsurface flow. Pine forests are located on hills and ridges that are surrounded by peat bogs. As a result, mineral-island landscape is formed. *Sphagnum*-dominated oligotrophic bogs are located at elevation of 50–65 m a.s.l. In most cases, the transition from the mineral island into the bog occurs sharply; the slope gradient may account for 10–30°. This results in high velocity of surface runoff and temporarily perched groundwater during snowmelt and rainfall.

The valleys and floodplains of small rivers attract special attention when discussing the migration of pollutants. The Yeniya River has a distinct valley in the middle reaches. The width of the valley is 300 m; the depth is 3 m. The floodplain is left-sided, boggy, and overgrown with deciduous forest (*Salix lapponum, Betula pubescens*). The valley and floodplain are permanently waterlogged and have a close hydraulic link to the riverbed. The bed has a width of 3–5 m and a depth of 0.7–1.5 m.

Each landscape complex has its specific hydrological functions. We have classified all landscapes of the catchment area in five types of runoff-forming complexes (SCs) based on the uniformity of hydrological functions (Fig. 12.2). SCs are parts of the river basin, represented by a combination of natural and man-made complexes and characterized by relative homogeneity of the conditions for the formation of surface, subsoil, and groundwater flow (textures, filtration properties of soils, etc.) (Table 12.1).

We have united all the bog territories into one runoff-forming complex, despite the differences in the vegetation and depths of the peat deposit. Three types of landscapes were grouped in forestry SCs. Flooded SCs were delimited separately.

12.5   Results and Discussion

The landscape-hydrological analysis of the territory of the experimental catchment area made it possible to identify five parts of the territory that differ in the peculiarities of the water runoff formation and the washout of pollutants from the soil surface (Table 12.1). We used the water balance calculations, meteorological information,

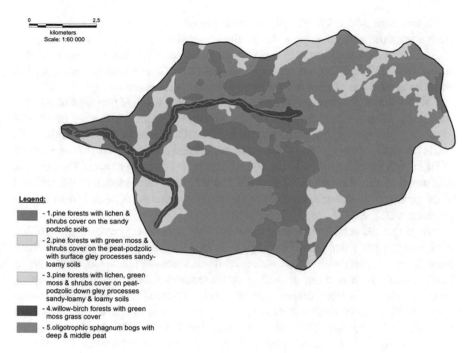

0 ━━━━ 2.5
kilometers
Scale: 1:60 000

Legend:

- 1.pine forests with lichen & shrubs cover on the sandy podzolic soils

- 2.pine forests with green moss & shrubs cover on the peat-podzolic with surface gley processes sandy-loamy soils

- 3.pine forests with lichen, green moss & shrubs cover on peat-podzolic down gley processes sandy-loamy & loamy soils

- 4.willow-birch forests with green moss grass cover

- 5.oligotrophic sphagnum bogs with deep & middle peat

Fig. 12.2   Map of runoff-forming complexes (SCs) of the Yeniya River catchment

Table 12.1   Quantitative characteristics of the runoff-forming complexes of the catchment area of the Yeniya River

Number of SC (on the map)	Plant type	Soils	Texture of soil and sediments	Area, km^2	Steepness of slope, %
1	Forest (*Pinus sylvestris, Pinus sibirica*) with lichen, moss, and shrub cover (different species of Cladoniaceae, *Polytrichum*, Dicranaceae)	Albeluvisols	sandy	8.22	10.60
2		Gleyic Albeluvisols (gley near the surface)	sandy loamy, light loamy	1.31	2.00
3		Gleyic Albeluvisols (gley in deep horizons)	sandy	4.45	4.40
4	Forest in floodplain (*Picea obovata, Salix lapponum, Betula pubescens*) with grassy cover	Complex of Fluvisols and Histosols	sandy	0.96	10.00
5	Bogs with moss and shrubs cover (different species of *Sphagnum, Carex lasiocarpa, Ledum palustre, Cassandra angustifolia, Rubus chamaemorus*, and others)	Histosols	peat	28.62	1.00

and knowledge of the hydrophysical properties of soils to evaluate a specific discharge from the catchment area during the spring flood (70 days) as at average of 15.71 L/s^{-1}*km^2 for a long period (1960–2000). Table 12.2 shows the specific discharge of the PHC from the oil-contaminated areas. These values were obtained with the application of calculations and verified experimentally.

Oil production and transportation at the Talnikovoe oil field can lead to oil contamination of any of the listed SCs. Our task is to determine exactly those SCs (landscape complexes) from which contribute most to the rapid removal of the PHC to the pollution of river waters. Therefore, we carried out a forecast for the removal of PHC with possible oil pollution for all SCs in the catchment area. The forecast was based on calculations using a specially established dependence of the removal of oil products on the area and age of the oil-contaminated area and hydrological parameters (Eq. 12.1).

We performed calculations for various possible areas of oil contamination (0.05, 0.25, and 0.5 km^2) and for different periods of time (during the spring flood and throughout the year) within the boundaries of all watershed SCs. Concentration of PHC is equal to maximum allowable concentrations (MAC) adopted in Russia, given that contamination comprises small areas in each SC (0.05 km^2). The concentration of PHC in the river will be 0.11–0.13 mg*L^{-1} which is 2–2.5 times as high as MAC, given that with contamination comprises 0.5 km^2. This holds true for SC No. 1 and No. 3. The concentration of PHC in the river will be 0.20–0.22 mg L^{-1} (fourfold in excess of MAC) with 0.5 km^2 oil-contaminated areas into SC Nos. 2, 4, and 5. The rate of flushing of PHC varies from 91 to 153 mg*s^{-1}, subject to contamination of large areas (0.5 km^2). These indicators are 2–6 times higher than the background washout of PHC (27 mg*s^{-1}).

Table 12.2 Assessment of possible oil pollution of the Yeniya River in the spring flood period. A background concentration of PHC in the waters of the river is 0.05 mg*L^{-1}

Number of SC, landscape parameters	runoff, mm	M_p, L s^{-1} км^2	f_p, км^2	μ, mg s^{-1} км^2	S, mg s^{-1}	ρ, mg L^{-1}
1. *Pinetum vaccinioso-cladinosum* on the sandy Podzols	57.59	9.52	0.05	0.81	35.16	0.05
			0.25	1.44	62.94	0.09
			0.5	2.09	91.18	0.13
2. *Pinetum ledoso-polytrichosum* on sandy-loamy Gleyic Albeluvisols (gley near the surface)	96.86	16.02	0.05	0.93	40.49	0.06
			0.25	2.00	87.34	0.13
			0.5	3.10	135.00	0.20
3. *Pinetum ledoso-cladinosum* on sandy-loamy and loamy Gleyic Albeluvisols (gley in deep horizons)	44.26	7.32	0.05	0.77	33.35	0.05
			0.25	1.26	54.69	0.08
			0.5	1.75	76.36	0.11
4. *Saliceto-Betuletum caricoso-hylocomiosum* in floodplains on Fluvisols	96.08	15.89	0.05	0.93	40.39	0.06
			0.25	1.99	86.85	0.13
			0.5	3.08	134.10	0.20
5. Oligotrophic sphagnum bogs with deep and moderately deep peat layer	113.51	18.77	0.05	0.98	42.75	0.06
			0.25	2.24	97.66	0.14
			0.5	3.52	153.50	0.22

In general, an analysis of the risks of oil contamination of small river waters using the approach presented showed that the greatest danger to small river water quality of the middle taiga during the spring flood is the discharge of PHC from potentially contaminated bog SC, especially from the areas located on the periphery of the peat bogs. If oil production is carried out in the catchment and emergency situations occur, the annual removal of the PHC by the waters of one small river may amount to 8–24 tons with the concentrations of PHC indicated in Table 12.2.

12.6   Conclusions

The use of the original system of methods for assessing the removal of pollutants from the catchment area allowed us to obtain the following final results. In the taiga zone of Western Siberia, the greatest risk of oil pollution in small rivers is induced by the bog landscapes. This is due to the fact that bog landscapes provide the largest contribution to the formation of water and chemical flow of small rivers. Under equal conditions of oil pollution, the removal of petroleum hydrocarbons from oil-contaminated bog and forest landscapes differs by a factor of 1.5. Projecting oil fields requires special attention to the landscape conditions of the catchment area of the river when placing oil and gas extraction objects. This will reduce the risks of pollution of water bodies in emergency situations. Oil industry companies engaged in construction of oil-producing and oil-transport facilities in the bog landscapes should apply technologies that are capable of reducing the risk of oil contamination to a minimum. One of the possible variants is the placement of oil objects in forest landscapes. However, this option may produce harmful effects as well. It is not always applicable because of the need to protect and rationally use forest resources.

Acknowledgments   The Research was supported by the Ministry of Education and Science of the Russian Federation (Agreement 5.8859.2017/9.10).

References

Antipov, A. N., & Korytny, L. M. (2012). Siberian school for landscape hydrology. In *Issues in Geography* (*Geographical-hydrological research*) (Vol. *133*, pp. 32–48). Moscow: Kodeks. (in Russian)

Antipov, A. N., Vakulin, K. Y., et al. (1989). *Landscape-hydrological characteristics of West Siberia*. Irkutsk: Institute of geography of Siberia and Far East SB AS USSR. (in Russian).

Antipov, A. N., Gagarinova, O. V., & Fedorov, V. N. (2007). Landscape hydrology: Theory, methods, implementation. *Geography and Natural Resources, 3*, 56–67. (in Russian).

Environmental Programme for the Danube River Basin: Danube Integrated Environmental Study. Report Phase 1. (1994). *Commission of the European Communities*.

Environmental situation in the Khanty-Mansi Autonomous Area-Yugra in 2015. Report. (2016). Khanty-Mansysk: Regional Environmental Department. (in Russian)

Fenger, J. (2009). Air pollution in the last 50 years – from local to global. *Atmospheric Environment, 43*(1), 13–22.

Gashev, S. N., & Kazanceva, M. N. (1998). The degree of pollution of catchment areas as an indicator of pollution of aquatic ecosystems in oil production. In *Clean water. Proceedings of 3th All-Russia scientific-practical seminar* (pp. 34–36). Tyumen: Tyumen State University. (in Russian).

Gassman, P. W., Reyes, M. R., Green, C. H., & Arnold, J. G. (2007). The soil and water assessment tool: Historical development, applications and future research direction. *Transactions of the American Society of Agricultural and Biological Engineers, 50*(4), 1211–1250.

Gendrin, A. G. (Ed.). (2006). *Environmental escort for exploration oil and gas deposits. Issue 2. Natural state monitoring on the objects of oil and gas complexes.* Novosibirsk: GPNTB SB RAS. (in Russian).

Gevaert, V., van Griensven, A., Holvoet, K., Seuntjens, P., & Vanrolleghem, P. A. (2008). SWAT developments and recommendations for modeling agricultural pesticide migration measures in river basins. *Hydrological Sciences Journal, 53*(5), 1075–1089.

Glushkov, V. G. (1936). Geographical & hydrological method of research. *Review of Hydrological Institute of USSR, 57–58*, 5–9. (in Russian)

Jha, M., Gassman, P. W., & Arnold, J. G. (2007). Water quality modeling for the Racoon River watershed using SWAT2000. *Transactions of the ASABE, 50*(2), 479–493.

Kalinin, V. M. (1999). *Landscape-hydrological analysis of small river catchments. Tutorial.* Tyumen: Tyumen State University Publishing house. (in Russian).

Kalinin, V. M. (2001). The flow of oil products into the river network from non-point sources (case study of Middle Ob' region). *Proceedings of Tyumen State University, 2*, 11–21. (in Russian).

Kalinin, V. M. (2010). *Water and petroleum (Hydro-ecological issues of Tyumen region).* Tyumen: Tyumen State University Publishing house. (in Russian).

Kalinin, V. M., Larin, S. I., & Romanova, I. M. (1998). *Small rivers in anthropogenic impact (case study of Eastern Zaural'e).* Tyumen: Tyumen State University Publishing house. (in Russian).

Kapotov, A. A., Kravchenko, V. V., Fedorov, V. N., et al. (1992). *Landscape-hydrological analysis of territory.* Novosibirsk: Nauka. (in Russian).

Khoroshavin, V. Y. (2005). *Technogenic transformation of the hydrological regime and quality of small rivers waters on the oil and gas deposits in the Pur River basin* (Ph.D. Thesis). Ekaterinburg: Institute for Integrated Water Use and Water Resources Conservation. (in Russian)

Khoroshavin, V. Y., & Moiseenko, T. I. (2014). Petroleum hydrocarbon runoff in rivers flowing from oil-and-gas-producing regions in Northwestern Siberia. *Water Resources, 41*, 532–542.

Komlev, A. N. (2002). *Regularities of formation and methods of river flow calculations.* Perm': Perm' State University Publishing house. (in Russian).

Krysanova, V., & Arnold, J. G. (2008). Advances in ecohydrological modeling with SWAT – a review. *Hydrological Sciences Journal, 53*(5), 939–947.

Kulmala, M. (2018). Build a global earth observatory. *Nature, 553*(7686), 21–23.

Methods for identifying and evaluating the nature and extent of non-point sources of pollutants. EPA-430/9-73-014. (1973). Washington, DC: U.S. Environment Protection Agency.

Mikhailov, S. A. (2000). *Diffuse pollution of aquatic ecosystems. Assessment methods and mathematical models: Review.* Barnaul: Den'. (in Russian).

Moiseenko, T. I., & Gashkina, N. A. (2010). *The formation of the chemical composition of lake water in a changing environment.* Moskow: Nauka Publishing house. (in Russian).

Nikanorov, A. M., Stradomskaja, A. G., & Ivanik, V. M. (2002). *Local monitoring of pollution of water bodies in areas of high technogenic impacts of the fuel and energy complex* (Book series "water quality"). St. Petersburg: Hydro- and Meteorogical Publishing House. (in Russian).

Novotny, V. (1985). Discussion of "Probability model of stream quality due to runoff" by D.M. Di Toro. *Journal of Environmental Engineering, ASCE, 111*(5), 736–737.

Novotny, V. (1988). Diffuse (non-point) pollution – a political, institutional, and fiscal problem. *Journal of Water Pollution Control Federation, 60*(8), 1404–1413.

Novotny, V., & Chesters, G. (1981). *Handbook of non-point pollution*. New York: Van Nostrand Reinhold.

Results of the Nationwide Urban Runoff Program. Final Report. Volume I. NTIS PB-84-185-552. (1983). U.S. Environmental Agency.

Vollenweider, R. A. (1989). Assessment of mass balance. Principles of lake management. In R. A. Vollenweider & S. E. Jorgensen (Eds.), *Guidelines of lake management* (Vol. 1, pp. 53–69). Shiga: ILEC/UNEP Publ.

Vompersky, S. E., & Idson, P. F. (Eds.). (1986). *Forests impact on water resources*. Moscow: Nauka. (in Russian).

Walker, J. F., Picard, S. A., & Sonzogni, W. C. (1989). Spreadsheet watershed modelling for non-point source pollution management in a Wisconsin area. *Water Resources Bulletin, 25*(1), 139–147.

Williams, J. R., Arnold, J. G., Kiniry, J. R., Gassman, P. W., & Green, C. H. (2008). History of model development at Temple, Texas. *Hydrological Science Journal, 53*(5), 948–960.

Zhuang, Y., Thuminh, N., Beibei, N., Wei, S., & Song, H. (2012). Research trends in non-point source during 1975-2010. *Physics Procedia, 33*, 138–143.

Chapter 13
Comparison of Landscape and Floristic Diversity in Plain Catchments at the Level of Elementary Regions

Dmitry V. Zolotov and Dmitry V. Chernykh

Abstract We examine the interrelations between elementary regional landscape and floristic units (microregions) in the basins of medium-sized rivers and lakes on the example the southeast of West Siberia. In conditions of smoothness of zonal transitions and anthropogenic transformation of the territory, it is proposed to establish the differences between microregions using differential types of geosystems or geosystems-indicators, the occurrence of which is controlled by local features of the abiotic template and basin structure. Qualitative and quantitative measures of similarity were used as an additional criterion of regionalization. In case of zonal homogeneity, the types of geosystems are compared, whereas under zonal heterogeneity the types of geosystems-analogs are applied.

Floristic regionalization is based on regional- and topological-level differential and subdifferential species, which enable us to distinguish the floristic microregions. The differential regional-level species reflect the zonality of the territory, and their difference manifests itself at the border of a common areal or its parts. Topological-level differential species are associated with the peculiarities of the abiotic template and basin structure as well as with the presence of specific ecotopes within the compared microregions, which may be zonally homogeneous.

Floristic and landscape differentiation are directly related through the distribution of differential plant species within the geosystems-indicators. Nonlinearity and ambiguity of links between floristic and landscape diversity is shown based on field data.

Keywords Differential plant species · Geosystems-indicators · Landscape and floristic microregions · Ob plateau · Altai Krai

D. V. Zolotov
Institute for Water and Environmental Problems, Russian Academy of Sciences, Siberian Branch, Barnaul, Russia

D. V. Chernykh (✉)
Institute for Water and Environmental Problems, Russian Academy of Sciences, Siberian Branch, Barnaul, Russia

Altai State University, Barnaul, Russia

13.1   Introduction

The aim of our research was to solve one of the fundamental problems of comparative landscape or ecological floristics (Kozhevnikov 1978, 1996) developing at the
junction of landscape science and botanical geography: the relationships between
landscape structure and spatial organization of flora at the elementary regional level.
Details of the original method for studying landscape and floral diversity in comparison with similar studies in Russia and abroad are seen in Zolotov and Chernykh
(2015, 2016).

Our approach was developed on the example of the Ob plateau (Altai Krai,
Russia), which is characterized by zonal, morpho-lithological (geological and geomorphological) and basin heterogeneity. We consider microregions, subregions, and
regions as regionalization units at the elementary regional level. The methodology
of our landscape studies and regionalization was developed predominantly within
the framework of the achievements of Russian-Soviet tradition taking into account
V.L. Kaganskiy's (2003) main paradigms of regionalization, B.B. Rodoman's
(2010) notion of the *cartoids*, A.Yu. Reteyum's (2006) distinction between notions
of *landscape-area* and *landscape-system*. Our biogeographical (floristic) studies are
more universal and similar to the global traditions (see, e.g., Hengeveld 1990), and
the classics of Russian-Soviet floristics are well-known in the world (e.g., Takhtajan
1986; Malyshev 1991; Yurtsev 1994). One of the most important particular features
of the Russian-Soviet floristics is the preference of the A.I. Tolmachev's method of
specific (elementary) floras and natural units in contrast to the European and
American method of grids or standard areas (e.g., Kottas 2001) and administrative units.

Recently we have shown (Zolotov 2009; Chernykh and Zolotov 2011; Zolotov
and Chernykh 2015, 2016) that the locations of differential and subdifferential species of higher vascular plants and the corresponding partial flora can be linked to
geosystems-indicators (differential types of landscape units) while floristic boundaries – to the boundaries of landscape units, microregions, and catchments. Thus,
the exact floristic boundaries at the lower regional and topological level can be
defined as the boundaries of enclosing landscape and basin units. This study method
can be called *landscape-and-basin approach* to floristic regionalization or landscape-
basin-floristic (landscape-floristic) regionalization.

13.2   Study Area

Since 1995, the authors have been engaged in floristic and landscape research at the
Ob plateau in the Altai Krai, where the Barnaulka river basin (hereinafter BB) firstly
began to be studied as the model area, and then (since 2008), the study was extended
to the adjacent Kasmalinsky basin (hereinafter KB). The study area (Fig. 13.1) is a

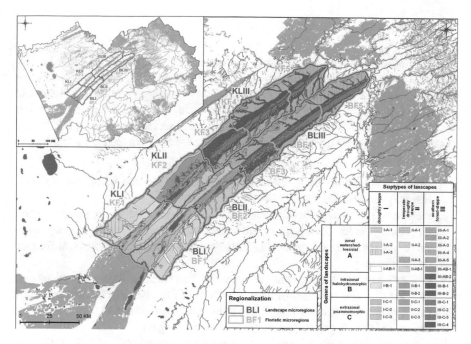

Fig. 13.1 Landscape map of KB (K) and BB (B) at the level of types of *mestnost* groups (TTG), landscape (L), and floristic (F) regionalization. Landscape microregions: Kasmalinsky, droughty-steppe (KLI), temperate-droughty-steppe (KLII), and south-forest-steppe (KLIII); Barnaulsky, droughty-steppe (BLI), temperate-droughty-steppe (BLII), and south-forest-steppe (BLIII). Floristic microregions, Kasmalinsky – Seliverstovsky (KF1), Guseletovsky (KF2), Kadnikovsky (KF3), Rebrikhinsky (KF4), and Pavlovsky (KF5); Barnaulsky, Novichikhinsky (BF1), Zerkalsky (BF2), Serebrennikovsky (BF3), Ziminsky (BF4), and Cheremnovsky (BF5). Territory ratio of landscape and floristic microregions, KLI ≈ KF1, KLII≈KF2, KLIII≈KF3 + KF4 + KF5, BLI ≈ BF1, BLII≈BF2, and BLIII≈BF3 + BF4 + BF5

river and lake basins that were formed as a result of erosion and neotectonic processing of *ancient flow gullies* (hereinafter *AFG*). Thus, the Barnaulka AFG was transformed into BB (5862.6 km², the left tributary of the Ob river) and the Kasmalinskaya AFG – into the Kasmala river basin (the left tributary of the Ob river) and the lake Gorkoye drainless basin into which the other Kasmala river flows. Together, these two basins with an unclear water divide are below referred to as KB (6623.7 km²). AFG stretch southwestward (i.e., from the Ob river to the Irtysh river), but in the contemporary basins, their flow direction has mainly changed to the opposite. Thus, in the Kasmala and Barnaulka river basins, the flow is oriented from SW to NE to the Ob river. In the lake Gorkoye basin, which is located approximately in the middle of drainless fragment of Kasmalinskaya AFG, the runoff has two main directions – from SW (a system of temporary streams and channels between lakes) and from the NE (Kasmala river) to the lake Gorkoye. KB and BB cross two natural zones and three subzones (droughty-steppe, temperate-droughty-steppe, and southern-forest-steppe). This enables us to use them as unique model objects for comparative-geographical research.

13.3   Methods and Approaches

The methodical basis of the research consists of a set of traditional and original methods of landscape science (large-scale and medium-scale mapping, landscape structure analysis, identification of geosystems-indicators, regionalization) and botanical geography and comparative floristics (chorology of differential species, analysis of taxonomic and typological structure of elementary regional and partial floras, their relationships, regionalization) with necessary corrections according to the authors' approach:

1. Analysis of publications while elaborating and substantiating the authors' research methodology and collecting field data.
2. Field research: collection of herbarium material, floristic and landscape descriptions using routes, transects, and key areas, and laying of reference geographical points for the inventory of Yu.R. Shelyag-Sosonko local floras, B.A. Yurtsev partial floras.
3. Geo-referencing floristic and landscape descriptions to electronic maps and field identification of units of interest with the GPS navigator Garmin Montana 650 t, which allowed working in the field with geo-referenced raster images of any scale (topographic maps, aerial and space images).
4. Visual interpretation of remote sensing data (aerial and satellite imagery), including automated unsupervised and supervised classification, for electronic mapping based on collected field materials.
5. Electronic mapping using the ArcGIS 9.2 software for creating landscape maps at the levels of types of complex *urochishche* groups, *mestnost,*[1] *mestnost* groups, and species of landscapes.
6. Use of electronic databases Microsoft Office Access 2013 to systematize and analyze the floristic and landscape information. Unified floristic database for the Kasmalinsky and Barnaulka river basins consists of six tables the largest of which currently has 17,395 lines, where the line is a description of the particular species location in specific geographical point based on floristic lists or herbarium data.
7. Studying the chorology of differential and subdifferential species to delimit floristic microregions using the author's methodology developed on the basis of A.I. Tolmachev's specific floras, B.A. Yurtsev's differential species and partial floras, Yu.P. Kozhevnikov's landscape floristics and biogeographical method of "thickening of areal boundaries", and V.B. Sochava's partial geosystems.
8. Standard methods of structural taxonomic and typological analysis for the comparative study of elementary regional and partial floras and their relationships (inclusion, similarity, etc.).

---

[1] *Urochishche* and *mestnost* are the Russian terms for hierarchical levels of landscape morphological units. See Glossary and Chap. 1 for definitions.

9. Original technique for identifying geosystems-indicators and analysis of their role in the landscape structure (Chernykh and Zolotov 2011).
10. Original technique of complex comparative study of landscape typological and floristic taxonomic diversity (Chernykh and Zolotov 2011).

13.4 Results

Qualitative analysis of landscape structure and regionalization using differential types of units of landscapes and landscapes-analogs (geosystems-indicators). We created landscape maps of the Kasmalinsky and Barnaulka river basins (scale 1:100000–500,000) at the levels of types of complex *urochishche* groups, *mestnost*, *mestnost* groups (Fig. 13.1), and species of landscapes. A unified landscape regionalization of adjacent basins at the level of microregions was elaborated (Table 13.1, Fig. 13.1). We delimited three landscape microregions in each basin, respectively, Kasmalinsky droughty-steppe (KLI), temperate-droughty-steppe (KLII), south-forest-steppe (KLIII), Barnaulsky droughty-steppe (BLI), temperate-droughty-steppe (BLII), and south-forest-steppe (BLIII). Each microregion is characterized by an individual ratio and set of landscape units at all considered levels of differentiation. At the level of types of *mestnost* groups or kinds of landscapes and at the lower levels, there are specific differential types of units that provide opportunity to distinguish adjacent microregions belonging to the same natural subzone (subtype of landscapes).

Our study showed evidence that, similar to the distribution of differential plant species in microregions, it is possible to analyze the presence of differential types of landscape units (geosystems-indicators). To compare the microregions within the same subzone, this can be done using common types of units, and for different subzones, it is necessary to use landscapes-analogs (Table 13.1), i.e., types of spatial units that have similar abiotic template but differ in bioclimatic parameters. As a rule, differences between microregions decrease shifting from the lower to the higher hierarchical level and increase in the opposite direction. In other words, the probability that a differential type of *mestnost* group (or kind of landscape) occurs is lower than that of type of *mestnost*, and the probability of existence of the latter is lower than that of a type of complex *urochishche* group (Chernykh and Zolotov 2011). Table 13.1 shows the ratio of the types of *mestnost* groups and their analogs in landscape microregions and subregions of the study area. For example, the landscape microregion BLII differs from KLII by the presence of differential types of *mestnost* groups II-A-3 and II-C-3 and from BLI and KLI due to the differential types of *mestnost*-analog groups A-e and B-b.

Using this data, it is convenient to illustrate the representativeness of the elementary region in relation to a higher rank order. So, although the microregion KLII does not include all types of *mestnost* groups occurring in the subregion KBLII, namely, 78% (7 of 9), it covers all the characteristic genera of landscapes (A, B, C).

Table 13.1 Analogs of types of *mestnost* groups (TTG) and their distribution in droughty-steppe (I), temperate-droughty-steppe (II), and south-forest-steppe (II) subzones of KB (K) and BB (B). For particular TTG (I-A-1 ... III-C-4) below or in parentheses, there are the areas (km²) in landscape microregions (KLI ... BLIII) and subregions (KBLI, KBLII, KBLIII)

Analogs of TTG	K – Kasmalinsky basin			B – Barnaulka river basin			KB – territory as a whole		
	KLI	KLII	KLIII	BLI	BLII	BLIII	KBLI	KBLII	KBLIII
A-a	I-A-1 169.3	II-A-1 129.9	III-A-1 283.4	I-A-1 161.3	II-A-1 146.9	III-A-1 168.7	I-A-1 330.6	II-A-1 276.9	III-A-1 452.1
A-b	–	–	III-A-2 23.8	–	–	III-A-2 153.8	–	–	III-A-2 177.7
A-c	I-A-2 574.8	II-A-2 478.7	III-A-3 838.3	I-A-2 474.9	II-A-2 513.8	III-A-3 828.3	I-A-2 1049.6	II-A-2 992.5	III-A-3 1666.6
A-d	I-A-3 229.6	–	III-A-4 418.5	I-A-3 105.7	–	III-A-4 383.1	I-A-3 335.3	–	III-A-4 801.6
A-e	–	–	–	–	II-A-3 9.8	III-A-5 61.9	–	II-A-3 9.8	III-A-5 61.9
AB-a	I-AB-1 92.0	II-AB-1 33.9	III-AB-1 145.9	I-AB-1 138.2	II-AB-1 105.5	III-AB-1 95.3	I-AB-1 230.2	II-AB-1 139.4	III-AB-1 241.1
AB-b	–	–	III-AB-2 266.6			III-AB-2 136.6	–	–	III-AB-2 403.2
B-a	I-B-1 341.7	II-B-1 387.0	III-B-1 580.5	I-B-1 96.3	II-B-1 80.2	III-B-1 264.1	I-B-1 438.0	II-B-1 467.2	III-B-1 844.5
B-b	–	II-B-2 158.5	III-B-2 27.7		II-B-2 8.6	III-B-2 27.2	–	II-B-2 167.1	III-B-2 54.9
C-a	I-C-1 239.6	II-C-1 99.6	III-C-1 407.8	I-C-1 322.8	II-C-1 182.3	III-C-1 500.5	I-C-1 562.4	II-C-1 281.9	III-C-1 908.3
C-b	I-C-2 141.3	II-C-2 179.9	III-C-2 68.7	I-C-2 230.5	II-C-2 109.9	III-C-2 122.0	I-C-2 371.8	II-C-2 289.8	III-C-2 190.7

C-c	I-C-3 87.1	–	III-C-3 124.6	I-C-3 196.9	II-C-3 51.6	III-C-3 127.3	I-C-3 283.9	II-C-3 51.6	III-C-3 252.0
C-d	–	–	III-C-4 95.0	–	–	III-C-4 58.6	–	–	III-C-4 153.6
Types of TTG in total:	8 (1875.4)	7 (1467.5)	12 (3280.8)	8 (1726.6)	9 (1208.6)	13 (2927.4)	8 (3602.0)	9 (2676.1)	13 (6208.2)
	9 (3342.9)			10 (2935.2)			10 (6278.1)		
	27 (6623.7)			30 (5862.6)			30 (12486.3)		

Analogs of TTG on genera of landscapes (Fig. 13.1): A – Zonal watershed covered with loess. A-a – Plano-convex gently rolling ridge (*ouval*) tops. A-b – Wide flat interfluves locally with depressions. A-c – Rarely dissected gently sloping (less often sloping) slopes of the ridges (*ouvals*). A-d – Densely dissected sloping and gently sloping slopes of the ridges (*ouvals*). A-e – Remnant flat surfaces of AFG bottoms with depressions, hollows, and shallow gullies. AB – Valley-*balka* systems and small river valleys in A and B. AB-a – Wide shallowly incised with vaguely formed bottoms. AB-b – Deeply incised with clearly formed bottoms. B – Intrazonal halohydromorphic. B-a – Flat and slightly inclined rolling, sometimes hillock-rolling AFG terraces. B-b – Remnant flat surfaces of AFG bottoms with depressions, hollows, and lakes. C – Extrazonal psammomorphic. C-a – AFG bottoms with hillocks, crests, and hollows. C-b – Deeply incised flat-rolling sometimes hillock sites of AFG bottoms and deltas with lakes. C-c – Inclined hillock and crest hollow peripheral part of AFG bottoms. C-d – Valleys of medium and small river inheriting AFG bottoms

On the other hand, the microregion BLII contains all types of *mestnost* groups in the enclosing subregion.

Quantitative analysis of landscape structure and regionalization. The landscape regionalization at the lowest regional level aimed at the "bottom-up" hierarchy building can rely upon the comparative study of the landscape structure using quantitative measures of similarity. Landscape microregions initially identified by morphological and auxiliary features can be combined into subregions and regions based on linkage density.

We applied the Sorensen-Czekanowski coefficient (Table 13.1), to evaluate the density of linkages between the landscape structure of microregions and their aggregates according to zonal and basin principles (Fig. 13.2). Area of type of *mestnost* group or its analog was taken as a weight while calculating the coefficient. Note that inter-basin intra-subzonal landscape connections are often stronger than intra-basin inter-subzonal ones. This serves as a basis for combining microregions into subregions (or parts thereof) according to the subzonal rather than basin principle. Intra-basin connections between the microregions of the steppe zone are stronger than that between adjacent microregions of the steppe and forest-steppe zones, and the connection of the southern-forest-steppe microregions with the remote droughty-steppe ones is stronger than with the adjacent temperate-droughty-steppe ones but weaker than among themselves (Fig. 13.2a). Inter-basin intra-subzonal aggregates

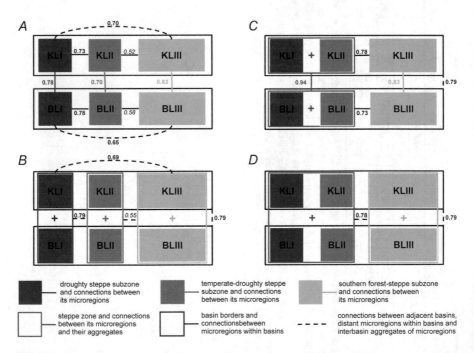

Fig. 13.2 Landscape structure relationships of landscape microregions and their aggregates in the Kasmalinsky (K) and Barnaulka river (B) basins. Quantitative Sorensen-Czekanowski coefficient (as the weight of type of *mestnost* group or its analog their area) is used

of steppe microregions (subregions) are also more closely related to each other than to the southern-forest-steppe aggregate (subregion) (Fig. 13.2b), which gives grounds for referring them to different landscape regions, i.e., the boundary between KLII-KLIII and BLII-BLIII of higher rank order than KLI-KLII and BLI-BLII. Intra-basin inter-subzonal (intrazonal) steppe aggregates are most densely linked (0.94), while the connections of the southern-forest-steppe microregions are weaker with each other but stronger than with steppe intra-basin aggregates (Fig. 13.2c). This finding also supports the abovementioned version of delimiting regions. Steppe and forest-steppe parts of the territory are connected almost as much as the basins under consideration (Fig. 13.2d), which is the result of the sum of strong inter-basin intra-subzonal connections between the KB and BB microregions and illustrates their extreme similarity in landscape structure as similar samples, rather than as a single *chorion* with a homogeneous internal structure.

Analysis of distribution of differential species at regional (Chernykh and Zolotov 2011) **and topological levels for landscape-floristic regionalization.** Differential species of topological level are associated with specific ecotopes or geosystems-indicators that make one microregion differ from another, their genesis being associated with local rather than regional factors. For example, under same or similar climate and lithology, the microregions may differ in basin organization. Thus, the landscape microregion KLII (Table 13.1, Figs. 13.1 and 13.2) corresponding to floristic KF2 (Fig. 13.1) is the poorest in number of types of *mestnost* groups, but the type II-B-2 occupies the largest area among analogs (B-b – remnant surfaces of AFG bottoms). Here the presence of drainless depressions with saline lakes determines the existence of a number of stenotopic halophytes, for example, *Halocnemum strobilaceum*, *Limonium suffruticosum*, etc. Both these species were found only in this microregion within the model territory. Moreover, for the boundaries KF2-KF1,[2] KF2-BF2, and KF2-BF1, they act as differential species of topological level, because in all adjacent to KF2 microregions climatic conditions allow these species to occur. In all the microregions adjacent to KF2, these species are absent due to specific ecotopological structure determined by differing basin organization, greater drainage in particular. For the boundaries KF2-KF3 and KF2-BF3, these species are treated as regional-level differential species, as they reach the northeastern border of their area.

Differential species can be divided into the actual differential and subdifferential ones according to the degree of reliability. The degree of reliability is primarily determined by the probability of detection of these species beyond the limits of *chorion* boundaries associated with them. Naturally, this is largely determined by the stenotopic nature of species. For example, eurytopic steppe species can be encountered far beyond the steppe zone limits on steep southern slopes with suitable moistening conditions. Whereas for calciphytes specific rocks (limestones) are also needed, so it is less likely to occur outside the corresponding *chorion* limits; hence,

---

[2] Here and below in the designation of the boundary, the microregion with presence of differential species stands on the first place and the one with absence on the second place.

the degree of reliability is higher. Therefore, the former (eurytopic) species will be treated as subdifferential, and the latter (stenotopic one) will be differential. For example, in microregion KLI (≈KF1) within zonal watershed-loess landscapes (A) at burial mounds (Zolotov and Biryukov 2009), a number of steppe species occur (*Paeonia hybrida, Iris glaucescens, Allium lineare, Stipa lessingiana, S. praecapillata*), which despite long-term targeted search were not found in the adjacent microregion BLI (≈BF1) in the same subzone but belonging to another basin (Table 13.2). This fact cannot be fully explained by the absence of known groups of burial mounds in BLI, because all these species can be easily found in suitable ecotopes outside of these natural-anthropogenic objects where anthropophobic elements of native flora in the form of flora-isolates are preserved. For example, *Paeonia hybrida* in KLI occurs outside the burial mounds on the untilled field borders, etc. Even more interesting is the fact that *Iris glaucescens, Allium lineare, Stipa lessingiana*, and *S. prae-*

Table 13.2 Specific species of native fraction of KB flora (absent in BB) and their distribution by floristic (KF) and landscape (KL) microregions

No.	Species of higher vascular plants	KLI	KLII	KLIII		
		KF1	KF2	KF3	KF4	KF5
1.	*Ophioglossum vulgatum* L.	–	–	–	+	–
2.	*Gymnocarpium dryopteris* (L.) Newm.	–	–	–	+	–
3.	*Pulsatilla turczaninovii* Kryl. et Serg.	–	+	–	–	–
4.	*Halocnemum strobilaceum* (Pall.) M.Bieb.	–	+	–	–	–
5.	*Polygonum gracilius* (Ledeb.) Klok.	+	–	–	–	–
6.	*Limonium suffruticosum* (L.) O.Kuntze	–	+	–	–	–
7.	*Elatine alsinastrum* L.	–	–	–	+	–
8.	*Trientalis europaea* L.	–	–	–	+	–
9.	*Urtica galeopsifolia* Wierzb. ex Opiz	–	–	–	+	–
10.	*Hedysarum gmelinii* Ledeb.	–	+	–	+	+
11.	*Lotus krylovii* Schischk. et Serg.	+	–	–	–	–
12.	*Vicia megalotropis* Ledeb.	–	–	–	+	–
13.	*Rhinanthus vernalis* (N.Zing.) Schischk. et Serg.	–	–	–	+	–
14.	*Ziziphora clinopodioides* Lam.	–	–	–	–	+
15.	*Artemisia sericea* Web. ex Stechm.	–	+	+	+	–
16.	*Cirsium canum* (L.) All.	–	–	–	–	+
17.	*Pilosella caespitosa* (Dumort.) P.D.Sell et C.West	–	–	–	+	–
18.	*Potamogeton alpinus* Balb.	–	–	–	+	–
19.	*Iris glaucescens* Bunge	+	+	+	+	–
20.	*Allium clathratum* Ledeb.	–	–	–	+	–
21.	*Allium lineare* L.	+	–	–	–	–
22.	*Malaxis monophyllos* (L.) Sw.	–	–	–	+	–
23.	*Melica altissima* L.	–	–	+	–	–
24.	*Stipa lessingiana* Trin. et Rupr.	+	–	–	+	+
25.	*Stipa praecapillata* Alechin	+	–	–	–	–
Total specific species in the microregion:		6	6	3	15	4
				18		

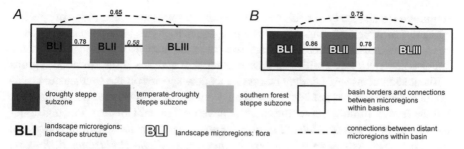

Fig. 13.3 Ratio of landscape (*A*) and floristic (*B*) relationships between landscape microregions (the example of the Barnaulka river basin (BB)). *A* – quantitative Sorensen-Czekanowski coefficient, *B* – qualitative Sorensen-Czekanowski coefficient

capillata were not found in the Barnaulka river basin, although, for example, *S. lessingiana* occurs even in KF5. Of these species, only *S. praecapillata* occur in KF1 close to the northeastern limit of the area, marking the boundary KF1-KF2 and being here a regional-level differential species. Thus, for the boundary KF1-BF1, all listed steppe species are subdifferential for the topological level, since suitable ecotopes occur in BF1 with rather high probability of detection.

Analysis of landscape-flora relationships at the elementary region level. The floristic diversity changes more smoothly and regularly than the landscape one. This follows from the assessment of density of linkages between the landscape structure and floras of landscape microregions. Thus, landscape connections between remote microregions are often stronger than between adjacent ones, whereas floristic connections between adjacent microregions are always stronger than between remote ones (Fig. 13.3). However, landscape differences are more obvious than differences in floras, because in the first case we deal with formalized characteristics of particular differential types of landscape units and in the second case – with specific species, the differential ability of which is often questionable due to the uncertainty of the factors limiting their distribution.

Not all the specific species of KB (Table 13.2) can be considered as differential and subdifferential for topological level (Zolotov and Chernykh 2016) since they are not reliable enough to distinguish the microregion floras in the same subzone. Some species (*Polygonum gracilius, Urtica galeopsifolia, Lotus krylovii, Rhinanthus vernalis, Pilosella caespitosa*) are taxonomically complicated or have been distinguished from closely related ones quite recently. Such species require a targeted search in the BB with good chances to be found there. In BB there are suitable ecotopes for riparian-aquatic and aquatic sporadically encountered species (*Elatine alsinastrum, Potamogeton alpinus*), and they could be just missed by chance. This holds true for rare meadow-forest (*Vicia megalotropis*) and meadow-steppe (*Melica altissima*) species as well. The distribution of mesophytic forest species being rare for the Ob plateau (*Ophioglossum vulgatum, Gymnocarpium dryopteris, Trientalis europaea, Malaxis monophyllos*) could be explained by the rarity of corresponding forest ecotopes in the southern-forest-steppe, but in the BB visually similar communities without these species occur. The identification of differential features requires a special study of ecology of these communities and the biology of species. The most reliable differential species – obligate halophytes (*Halocnemum strobila-ceum*, *Limonium suffruticosum*) – are strictly associated with well-manifested *sor* solonchaks[3] (less often with meadow solonchaks) and are absent in microregions without such ecotopes. *Cirsium canum* – meadow species on the southeastern margin of the area – could not be missed in adjacent territories due to a specific habitus, but the factors limiting its southeastward movement remain unclear. The multiplicity of steppe species (*Pulsatilla turczaninovii*, *Hedysarum gmelinii*, *Ziziphora clinopodioides*, *Artemisia sericea*, *Iris glaucescens*, *Allium clathratum*, *A. lineare*, *Stipa lessingiana*, *S. praecapillata*) can be explained by greater areas and dissection of zonal landscapes in KB and the presence of groups of burial mounds (Zolotov and Biryukov 2009).

Analysis of richness of partial floras of the genera of landscapes (mega-ecotopes) and their determination by the landscape structure. At the moment according to 1995–2017 data and literary sources (Zolotov 2009), native flora of the Kasmalinsky and Barnaulka river basins has 870 species (85.8%) of 373 genera (85.7%) and 103 families (92.0%); alien fraction of flora – 144 species (14.2%) from 104 genera (23.9%) and 38 families (33.9%); and anthropogenically transformed flora, 1014 species (100.0%) of 435 genera (100.0%) and 112 families (100.0%).

The entire native flora of study area (870 species) consists of partial floras of genera of landscapes (Table 13.3): zonal watershed loess (A), intrazonal halohydromorphic (B), and extrazonal psammomorphic (C). The partial flora of the zonal watershed-loess landscapes (A) unites 552 species, of which 23 species (4.2%) are specific. The partial flora of intrazonal halohydromorphic (B) despite the smaller area (Table 13.1) is slightly larger – 566 species, of which 17 (3.0%) species are specific. The largest is the flora of the extrazonal psammomorphic landscapes (C) – 793 species, of which 226 species (28.8%) are specific. Hence, it is obvious that the flora of extrazonal psammomorphic landscapes (C) is the most original and rich, and contribution of these landscapes to the flora richness in the study area is the greatest. The association of partial floras in aggregates confirms these conclusions. The joint flora of the zonal watershed-loess (A) and intrazonal halohydromorphic (B) landscapes includes only 642 species in which all specific species constitute 75 species (11.7%) and the mutual specific – 35 species (5.5%). The joint flora of intrazonal halohydromorphic (B) and extrazonal psammomorphic (C) landscapes consists of 845 species, in which all specific species are 316 species (37.4%) and mutual specific species – 73 species (8.6%). The joint flora of zonal watershed-loessial (A) and extrazonal psammomorphic (C) landscapes unites 851 species, in which all specific species are 302 species (35.5%) and mutual specific species – 53 species (6.2%). The joint flora of zonal watershed-loess (A) and extrazonal psammomorphic (C) landscapes is slightly richer, but the joint flora of intrazonal halohydromorphic (B) and extrazonal psammomorphic (C) landscapes is much more original,

---

[3] *Sor* solonchak has dense or loose salt crust, unlike meadow solonchak.

Table 13.3 Distribution of native flora species of KB and BB by partial flora of genera of landscapes (A, B, C)

Distribution types of species by genera of landscapes	Number of species in partial flora	Share of partial floras and their aggregates, %	Share of the total flora, %
A (A, AB, AC, ABC)	552	100.0	63.4
Only A	23	4.2	2.6
B (B, AB, BC, ABC)	566	100.0	65.1
Only B	17	3.0	2.0
C (C, BC, AC, ABC)	793	100.0	91.1
Only C	226	28.8	26.0
AB (A, AB, AC, B, BC, ABC)	642	100.0	73.8
Limited AB (A, AB, B)	75	11.7	8.6
Only AB	35	5.5	4.0
BC (B, AB, BC, C, AC, ABC)	845	100.0	97.1
Limited BC (B, BC, C)	316	37.4	36.3
Only BC	73	8.6	8.4
AC (A, AB, AC, BC, C, ABC)	851	100.0	97.8
Limited AC (A, AC, C)	302	35.5	34.7
Only AC	53	6.2	6.1
ABC	870	100.0	100.0
Only ABC	441	50.7	50.7

which brings together the last two genera of landscapes closer than the first two. The share of species noted in all genera of landscapes (ABC) is also very significant – 441 species (50.7%), which indicates the unity of flora of the study area and Ob plateau as a whole.

13.5   Conclusion

The chapter briefly outlines the most important results and opportunities for comparative study of landscape and floristic diversity at the level of elementary regions:

1. Qualitative analysis of landscape structure and regionalization using differential types of units of landscapes and landscapes-analogs (geosystems-indicators)
2. Quantitative analysis of landscape structure and regionalization
3. Analysis of distribution of differential species of regional and topological levels for landscape-floristic regionalization
4. Analysis of ratio of landscape and floristic relationships at the elementary region level
5. Analysis of richness of partial floras of the genera of landscapes (mega-ecotopes) and their determination by the landscape structure

Acknowledgments The research was carried out in the framework of State Assignment of IWEP SB RAS № 0383-2019-0004.

References

Chernykh, D. V., & Zolotov, D. V. (2011). *Spatial organization of landscapes of the Barnaulka river basin*. Novosibirsk: SB RAS Publishers. (in Russian).

Hengeveld, R. (1990). *Dynamic biogeography*. Cambridge, UK: Cambridge University Press.

Kagansky, V. L. (2003). Basic practices and paradigms of zoning. *Regional Research, 1*(2), 17–30. (in Russian).

Kottas, K. L. (2001). Comparative floristic diversity of Spring Creek and nine-mile prairies, Nebraska. *Transactions of the Nebraska Academy of Sciences and Affiliated Societies, 27*, 31–59.

Kozhevnikov, Y. P. (1978). On development of a new branch of phytogeography – environmental floristics. In *Abstracts of VIth congress of all-union botanical society* (pp. 222–223). Leningrad: Nauka. (in Russian).

Kozhevnikov, Y. P. (1996). *Vegetation cover of Northern Asia in historical perspective*. St. Petersburg: World and Family-95 Publishers. (in Russian).

Malyshev, L. I. (1991). Some quantitative approaches to problems of comparative floristics. In P. L. Nimis & T. J. Crovello (Eds.), *Quantitative approaches to phytogeography. Tasks for vegetation science* (Vol. 24, pp. 15–33). Dordrecht: Springer.

Reteyum, A. Y. (2006). Research installations of landscape science. In K. N. Dyakonov (Ed.), *Landscape science: Theory, methods, regional studies, practice. Proceedings of XIth international landscape conference* (pp. 46–49). Moscow: MSU Publishing House. (in Russian).

Rodoman, B. B. (2010). The scientific geographical cartoids (geographical schemes). *Geographical Bulletin, 2*(13), 88–92. (in Russian).

Takhtajan, A. (1986). *Floristic regions of the world*. Berkeley: University of California Press.

Yurtsev, B. A. (1994). Floristic division of the Arctic. *Journal of Vegetation Science, 5*, 765–776.

Zolotov, D. V. (2009). *Checklist of the Barnaulka river basin flora*. Novosibirsk: Nauka. (in Russian).

Zolotov, D. V., & Biryukov, R. Y. (2009). Florae-isolates of burial mounds as natural-anthropogenic elements of steppe landscapes. In *Problems of Botany of South Siberia and Mongolia: Proceedings of the VIII*[th] *International scientific-practical conference* (pp. 401–404). Barnaul. (in Russian).

Zolotov, D. V., & Chernykh, D. V. (2015). Landscape-basin approach to the study of floristic diversity (heterogeneous catchments of steppe and forest-steppe zones of Altai Krai, Russia, as a case study). *Acta Biologica Universitatis Daugavpiliensis, 15*(2), 383–392.

Zolotov, D. V., & Chernykh, D. V. (2016). Interrelation of floristic and landscape regionalization at lower hierarchical level (Ob' plateau, Altai Krai). *Bulletin of Udmurt University, series Biology and Earth. Sciences, 26*(2), 35–44. (in Russian).

Part IV
How Patterns Indicate Genesis and Influence Future Evolution Trends

Chapter 14
Altitudinal Landscape Complexes of the Central Russian Forest–Steppe

Anatoly S. Gorbunov, Vladimir B. Mikhno, Olga P. Bykovskaya, and Valery N. Bevz

Abstract The chapter considers vertical differentiation of plain landscapes, which is the universal feature of geosystem sequence, determined by the relief morphometry, mainly absolute and relative altitudes. Intrazonal nature of vertical differentiation is reflected in the structural organization of geosystems, the nature and intensity of the landscape-forming processes, and the specificity of the landscape pattern. We analyze vertical differentiation in terms of geocomponents (geology, topography, climatic and hydrological parameters, soils, and vegetation), landscape-forming processes (erosion, karst, suffosion, and landslides), various categories of geosystems (forest, steppe, meadow, marsh, and field), structure of types of landscapes, the stages of relief and landscape development, and the specificity of the landscape patterns. Based on this, the term "altitudinal landscape complex" is suggested as a special category, which defines a paradynamic system of landscapes interrelated by a general direction of the physico-geographical process. Wide range of elevations in the south of the East-European Plain determined the vertical differentiation of landscapes of different taxons (section, district, landtype association, landtype, and site). In turn, this provided the reason for establishing taxonomy of altitudinal landscape complexes, which includes four levels: step, variant, storey, and layer.

Keywords Vertical differentiation · Plain landscapes · Altitudinal landscape complex · East-European Plain

A. S. Gorbunov (✉) · V. B. Mikhno · O. P. Bykovskaya · V. N. Bevz
Voronezh State University, Voronezh, Russia

© Springer Nature Switzerland AG 2020
A. V. Khoroshev, K. N. Dyakonov (eds.), *Landscape Patterns in a Range of Spatio-Temporal Scales*, Landscape Series 26,
https://doi.org/10.1007/978-3-030-31185-8_14

14.1 Vertical Differentiation of Plain Landscapes: Research Issues

The problem of vertical differentiation of plain landscapes has been attracting attention of many scientists for a long time (Milkov 1947; Beloselskaia 1969; Rowe and Sheard 1981; Zonneveld 1989; Nikolaev 1999; Baily 2002; Berezhnoi et al. 2007; Divíšek et al. 2014). However, in spite of its timeliness, the issue in landscape geography still remains poorly studied. This is explained by both rather complex character of the phenomena manifestation and insufficiently developed methodological basis. In particular, there is still a lack of understanding in the issue of relationships between zonal, azonal, and provincial features in plain regions. On the other hand, factors driving vertical differentiation of geosystems differ from region to region. In some cases, relief serves as a key driver; in others, paragenetic[1] connections or qualitative state of landscape sphere matter.

System-based concept of paradynamic complexes[2] is believed to be a relevant new methodological approach that could enable us to find proper solution to the problem of vertical differentiation in plains. As an object of research, the territory of the Central Russian Upland was chosen. In this chapter, we made an attempt to substantiate an existence of unique zonal–azonal–altitudinal holistic complexes, to classify them, and to analyze opportunities to indicate this landscape category. For this purpose, the authors considered the following guidelines.

1. A vertical differentiation of plain landscapes is a universal property of geosystems' qualitative change depending on differences in relief and specifically its absolute and relative altitudes. Unlike altitudinal zonation in mountains, which is predetermined by significant contrasts in absolute elevations, plain landscapes are not commonly subject to drastic changes in landscapes. Therefore, vertical differentiation manifests itself only within intrazonal differences. Nonetheless, it can be detected well in a structure and dynamics of geosystems. Consequently, it acts as an indicator of zonal landscapes transformation.
2. A key reason for a vertical differentiation of landscapes is change of vertical matter and energy flow intensity in a landscape sphere—gravitational energy in particular.
3. Terrain elevation is responsible for important invariant properties of landscape structure and by this determines a direction of landscape-forming process (Khoroshev 2013).
4. Manifestation of a vertical differentiation of landscapes has a regional-specific nature and is common for highland and lowland plains.
5. Altitudinal landscape complexes (ALC), resulting from vertical transformation of relief, are original paradynamic systems of landscapes formed due to a joint manifestation of two main directions of the physico-geographical process—

---

[1] Paragenetic system: a set of contrasting landscape units linked by common process.

[2] Paradynamic complexes: a set of contrasting landscape units linked by matter flow.

horizontal and vertical ones. A diversity and intensity of backbone flows act as the main drivers of ALC. They are closely related to an interaction of contrast environments and involve both endogenous and exogenous factors. Anthropogenic factors also contribute much to the formation of the structure and development of ALC. All this allows us to consider ALCs as original geosystems. Their adjacent structural elements have close interrelation by direct and reverse flows of matter and energy. They are characterized by a certain range of elevations, particular lithologic and geomorphic frame, genesis, age, structure, and direction of development. Differences in these components result in differentiation of ALCs.

6. ALCs can be well observed in a landscape texture (Victorov 2006), especially in types of local landscapes. Differences in types of local landscapes can be traced at all levels of vertical differentiation of landscapes.

14.2   Classification Categories of Altitudinal Landscape Complexes

The main classification categories of ALCs of plains have a following subordination: step, variant, storey, and layer (Mikhno and Gorbunov 2001) (Fig. 14.1). This classification takes into account elevation, age, genesis, lithologic–geomorphic structure, landscape genesis specifics, nature of dynamic relationships, direction of development, and interactions in geosystems.

Altitudinal landscape steps are intrazonal paradynamic landscape systems of a regional level. They emerge as a result of conditions inherent for the particular absolute altitudes. They share common features of elevation, genetic unity, age, structure of landscapes, relative homogeneity of abiotic template, intensity of neotectonic movements, and landscape genesis. According to F.N. Milkov (1947), within forest–steppe zone of East-European Plain three altitudinal steps can be distinguished: the upper (250–300 m), the middle (150–250 m), and the lower (up to 150 m) ones. The territory of the Central Russian Upland is completely located within the middle altitudinal landscape step. This is a genetically uniform Neogene–Quaternary high plain shaped by erosion and denudation. It is distinguished by a ubiquitous development of gullies and balkas.[3] A density of a gully-and-balka dissection sometimes can account for 2.5–3 km/km². A relative elevation difference between interfluves and valley bottoms reaches 150 m. Upper Cretaceous carbonates form a lithogenic base. Differentiation is clearly manifested in contrast to upland and slope types of terrains, which comprise approximately equal areas. A negligible thickness of the Neogene–Quaternary mantle (8–10 m) over carbonate rocks and marl determined the widespread development of erosion-chalk, karst, and karst-and-suffosion land-

---

[3] *Balka* is the Russian term for flat-bottom erosional landforms sometimes with ephemeral water courses.

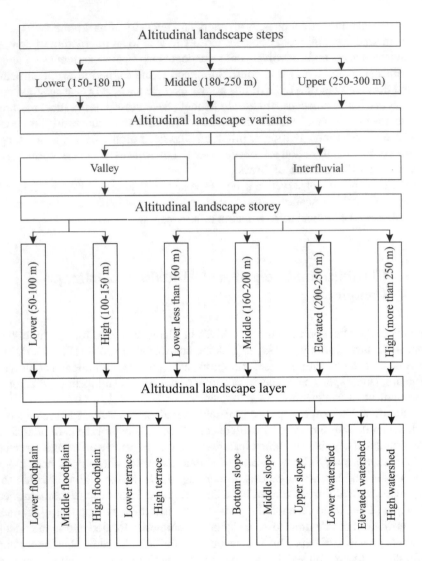

Fig. 14.1 The classification categories of altitudinal landscape complexes of plains

forms. Elevations of interfluves varying from 150 to 200–250 m a.s.l. and more, good drainage conditions, and predominance of loess-like loams on flat interfluves favored the development of typical zonal landscapes such as oak forest–steppe within the middle altitudinal landscape step of the Central Russian Upland. In its natural state, it was characterized by a predominance of gramineous–forbs steppes on Chernic Chernozems with fragments of oak forests on Phaeozems in ravines and uplands. Some remnants of a virgin forest–steppe have been preserved only in strict

Table 14.1 The change of the coefficients of the ratio between heat and moisture and the irrigation norms of forage grasses within the altitudinal landscape stories of the south of the Central Russian Upland

Interfluve altitudinal landscape stories	Elevation a.s.l. (m)	Hydrothermal coefficient[a]	Humidity coefficient[a]	Norm of irrigation m³/ha
Typical forest–steppe (east of the Central Russian Upland)				
High	> 250	1.26	1.19	2400–2450
Elevated	200–250	1.16	1.05	2700–2750
Middle	160–200	0.98	0.96	3000–3050
Typical forest–steppe (west of the Central Russian Upland)				
High	> 250	1.26	1.19	2400–2450
Elevated	200–250	1.24	1.15	2500–2550
Middle	160–200	1.20	1.12	2700–2750
Southern forest–steppe (east of the Central Russian Upland)				
Elevated	200–250	0.95	0.90	3350–3400
Middle	160–200	0.85	0.81	3650–3700
Lower	< 160	0.81	0.79	3800–3850

[a]Hydrothermal coefficient (HTC) is calculated as:

$$HTC = \frac{P_{>10}}{0.1 \Sigma t_{>10}}$$

where $P_{>10}$ is total precipitation for the period with mean daily air temperature above 10 °C (mm) and $\Sigma t_{>10}$ is the sum of mean daily air temperatures for the same period (°C)
Humidity coefficient (HC) is calculated as:

$$HC = \frac{P}{E}$$

where P is total annual precipitation (mm) and E is annual evaporation capacity (mm)

nature reserves. Nowadays, landscape structure is characterized by dominance of field upland type of terrain and meadow–steppe slope type of terrain. Parametric differences between landscapes of the lower and middle altitudinal steps are given in Table 14.1.

Altitudinal landscape variants are paradynamic systems of landscape complexes with specific relative elevation range, unique nature of matter and energy flows, and a set of landscape-forming processes. Landscape is divided into interfluvial and valley variants. In this case, we define interfluvial landscapes as flat interfluve areas and genetically associated slopes of an erosion network. In valley landscapes, we include floodplain and terrace geosystems. Low terraces and floodplains of the Don river have an elevation range rarely exceeding 30 m. This results in low energy potential of surface runoff (294 J/m²/L), and, hence, a weak erosion-induced dissection (0.05 km/km²). On the contrary, neighboring interfluvial landscapes have a deeply dissected topography (up to 100 m), surface runoff reaches 980 J/m², and results in a dense erosional pattern (1.5–2 km/km²). The categories of valley and interfluvial landscapes are fundamentally different from each other by a set of attributes as follows:

- Relative topographic position: all valleys are located lower than the interfluves.
- Dominant landscape-forming process: denudation on interfluves; accumulation in valleys.
- Landscape structure: rarity of landscapes being close to the zonal type in valleys but their prevalence on interfluves (up to 60–70% of an area).
- Type of landscape genesis: most valley landscapes are hydromorphic and sandy lithogenic ones formed with a direct involvement of an underlying substrate and the influence of groundwater, while the key attribute of interfluves is occurrence of autonomous, climate-dependent landscapes.

The main landscape unit we used to disclose an internal structure of variants is district.

Altitudinal landscape stories appear in the form of geographically disparate paradynamic systems of landscapes (districts), united by a similarity of elevations, geological and geomorphic structure, groundwater depth, relative homogeneity of soil and vegetation cover, microclimate, direction of landscape-forming processes, manifestations of dynamics, and landscape-forming matter flows. The main distinctive feature of district is the confinement to neotectonic structures, the same geological and tectonic structures as well as one type of relief. Districts with similar structure form altitudinal landscape stories (Fig. 14.2).

For example, a group of the valley landscapes of the largest rivers, including a floodplain and a lower terrace, with absolute altitudes of 60–100 m, represents the valley low altitudinal landscape storey. It is characterized by a predominance of hydrogenous landscapes, floodplain gramineous–forbs meadows on Arenic Fluvisols and moderately moist floodplain forests on stratified soils, and sedge-dominated fens. The Don river floodplain and the first terrace are the typical examples of this landscape storey. High altitudinal landscape stories are clearly distinguished by the prevailing relief outside river valleys. The interfluve low altitudinal landscape storey includes districts with elevations up to 160 m a.s.l. It is dominated by forest-and-field flat loamy plains with numerous suffosion depressions with gentle and medium-steep slopes. They have a widespread occurrence on left banks of the rivers Potudan, Tikhaya Sosna, and Chernaya Kalitva. Elevations of 160–200 m a.s.l. are typical for the middle landscape storey. Landscapes are dominated by forest–field–steppe loamy plains with a net of gullies with deep or medium-deep downcutting in chalk rocks. This storey is widespread on the chalk-dominated south of the Central Russian Upland. It is well pronounced, in particular, at the left-bank part of the basins of the Svapa, Seversky Donets, and Aidar rivers as well as at the part of the Kalach upland and the southern Pooskolye region. The elevated altitudinal landscape storey occurs at the altitudes of 200–250 m. Its main feature is a dominance of forest–field–steppe undulating loamy plains with a gully net deeply cut into chalk rocks. The storey includes high sections of the Timo-Schigrovskaya low ridge, the interfluve of the Oskol and the Don rivers, etc. Finally, the highest topographic positions are occupied by the high altitudinal landscape storey

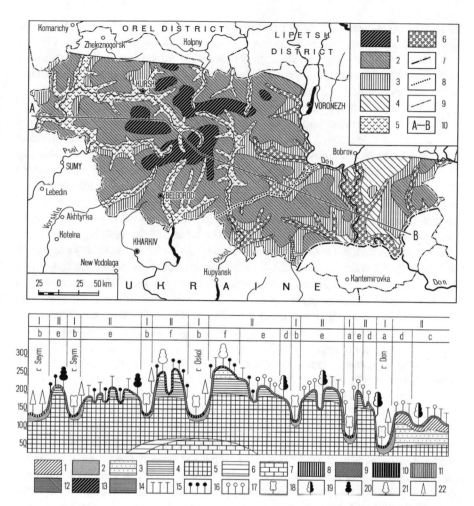

Fig. 14.2 A structure of altitudinal landscape stories of the Chalk south of the Central Russian Upland forest–steppe zone

Map legend: Interfluvial variant, altitudinal landscape storey: 1, high (more than 250 m); 2, elevated (200–250 m); 3, middle (160–200 м); 4, low (less than 160 m). Valley variant, altitudinal landscape storey: 5, high (more than 100 m); 6, low (less than 100 m); 7, boundaries of forest–steppe zone; 8, boundaries of typical and southern forest–steppe subzones; 9, boundaries of municipal districts; 10, line of a transect

Transect legend: 1, quaternary clay loam; 2, quaternary sandy deposits; 3, Neogenic sand and clay deposits; 4, Paleogenic clay; 5, chalk deposits; 6, Jurassic sediments; 7, Devonian limestones; 8, Greyi-Luvic Phaeozems (dark gray forest soils); 9, Albic Luvisols (light gray forest soils); 10, Arenic Fluvisols (stratified alluvial soils); 11, Greyi-Luvic Chernozem soils; 12, Chernic Chernozem soils; 13, Luvic Chernozem soils; 14, Haplic Chernozem soils; 15, agrophytocoenosis; 16, grassland steppe; 17, gramineous–grassland steppe; 18, floodplain forest; 19, ravine oak forest; 20, upland forest; 21, interfluve oak forests; 22, pine forest. Altitudinal variants: I, valley; II, interfluve. Altitudinal landscape storey: a, valley low; b, valley high; c, interfluve low; d, interfluve middle; e, interfluve elevated; f, high elevated

(250–276 m). It includes the most elevated areas of Oboyansk, Fatezh, Schigrov, Seym–Donetsk, and Olimo–Veduzhskoye ridges. This landscape storey is dominated by forest–field–steppe plains and plateaus with typical and Luvic Chernozems with a gully net deeply cut in chalk rocks. An internal structure of altitudinal landscape stories is revealed through the district and landtype associations.

Altitudinal landscape layers are paradynamic systems of *urochishches*[4] or, less commonly, monotonous areas with similar elevations, one type of location, and uniform abiotic template. In the chalk-dominated south of the Central Russian Upland, 11 altitudinal landscape layers were distinguished. The low floodplain (up to 4 m above the river level), middle floodplain (4–8 m), and high floodplain (more than 8 m) layers were formed in Holocene. The aquatic, fen, and meadow landscapes dominate in these layers. Low terrace layer represented by first and second terrace were formed in Late Pleistocene. At present, super-aqual forest landscapes are developing. The high terrace layer unites the Middle Pleistocene high terraces of rivers, covered mainly by pine forests and sandy steppes. The lower slope layer (up to 120 m a.s.l.) was formed in the Pleistocene–Holocene. It is characterized by a weak dissection (up to 30 m) and alternation of steppe areas with oak forests in ravines. The middle slope layer (up to 160 m a.s.l.) develops within a wide range of elevations (30–50 m). Its invariant emerged in the Upper Pliocene now dominated by chalky slopes with low-density calciferous vegetation cover. The upper slope altitudinal landscape layer (up to 200 m a.s.l.) has the Early Pliocene age. It has a large range of elevations (more than 50 m) with predominance of chalky landscapes. The low flat interfluves layer (120–160 m a.s.l.) lies within the limits of the Late Pliocene peneplain. It includes interfluve margins with oak forests on Phaeozems (gray forest soils). The elevated flat interfluve layer (160–200 m a.s.l.) lies within the limits of the Miocene–Pliocene peneplain. In the past, gramineous–forbs steppes and upland oak forests on Chernic Chernozems dominated in its landscapes. The high flat interfluves layer (more than 200 m) has the longest development history in the Central Russian Upland. Its age was established as early Miocene (Shapkinskaya peneplain). At present, the oak forest–steppe here has been replaced by agrophytocoenoses on Luvic Chernozems.

14.3 Issues of Indication of Altitudinal Landscape Complexes

Identification of ALCs faces certain problems. First of all, we have to take into account multiplicity of factors of vertical landscape differentiation. The implementation of comprehensive cartographic and instrumental tools is needed. Field data analysis allowed us to conclude that indication of ALCs is possible using less labor-intensive landscape indication methods. They provide opportunity to get insight into

---

[4] *Urochishche*: morphological unit of landscape adopted in Russian terminology; see Glossary and Chap. 1 for definitions.

missing pieces of the puzzle through the analysis of well-studied ones. This method is based on a strong interconnection and interdependence of natural processes and events, components, and complexes.

Our research showed evidence that in some cases erosional and karst geosystems serve as relevant indicators of ALC. Within the chalk-dominated south of the Central Russian Upland, the following main genetic varieties of karst landscapes occur: karst per se, karst-and-suffosion, karst-and-erosion, and karst-and-landslide. In most cases, altitudinal landscape layer turned out to correspond to a certain genetic variant of karst formations. For example, karst-and-erosion forms predominate in the slope layers, karst-and-suffosion forms predominate on terraces of a low layer, and the karst itself predominates on flat interfluves layers. However, sometimes application of landscape layers indication on the basis of chalky landscapes classification face difficulties due to a complexity of choosing indicators. In particular, it is not easy to distinguish genesis of visually similar concave landforms, for example, suffosion depressions, karst funnels, and karst-and-suffosion landforms. For this purpose, along with methods of engineering geology, landscape indication tools can be useful. By so doing, it became possible to determine genesis of underdeveloped funnels in some parts of the chalk-dominated south of the Central Russian Upland based on analysis of relationships between funnel morphology and volume of underground cavities containing fallen rocks of roofs (Mikhno 1976).

Field observations indicated that each high-elevation landscape storey has its peculiar structural features of chalk–karst geosystems. Hence, landscape spatial structure of karst formations can be used as an appropriate indicator of landscape-forming processes at these landscape stories.

Consider the example of karst landscapes, chosen as indicators of the altitudinal landscape layers in the chalk-dominated south of the Central Russian Upland. In this case, objects of indication are the lower, middle, elevated, and the high landscape stories.

Lower altitudinal landscape storey includes river floodplains, as well as the first and, rarely, the second terraces. These landscapes were influenced by the neotectonic joints and shears, which favored water accumulation. This is evidenced by thick strata of alluvial sandy and clayey deposits (15–50 m). Naturally, such conditions are not suitable for the development of karst landscapes that occur on the floodplains quite rarely. Karst funnel-shaped sinkholes are located in the river beds or at the foot slopes of river valleys, where fissure-karst waters emerge (the Kisliai, the Potudan, and the Oskol rivers). The floodplain of the Sejm river near Solntsevo town of the Kursk region is unique in this respect. This region is distinguished by a strong shear and a series of tectonic faults. This causes crack development in the chalk–marl rocks. Abundant surface water and groundwater (seven rivers merge in this region) contribute to the formation of karst cavities, which in turn lead to the forming of karst landforms on the surface. Most of them have a karst-and-suffosion genesis. Density of similar objects in this region exceeds $50/km^2$, waterlogging being a distinctive feature. Approximately nine out of ten cavities are occupied by fens or lakes, sometimes covered by bulrush and reeds. On low terraces, the proportion of lacustrine-fen karst-and-suffosion landscapes does not exceed 50–60% of the total area. In addition to the described region, karst-and-suffosion geosystems seldom occur in the floodplains of the Oskol, the Don, the Potudan, and the Seversky Donets rivers. In general, the main feature of the low altitudinal landscape storey is the poor development of karst geosystems and the predominance of hydromorphic karst formations with wetland vegetation in their structure.

The middle altitudinal landscape storey includes a territory with elevations up to 160 m a.s.l. The thickness of the Neogene–Quaternary sediments ranges from 10 to 60 m. Most of them overlie the chalk–marl rocks and thereby slow down karst development. The widespread karst landscapes of this storey occur on the interfluves of Kleven–Obista–Amonka rivers (Kursk region), as well as in the valleys of large rivers in the southern part of the Central Russian Upland. Within the boundaries of those valleys, the hydromorphic karst sedge-dominated hummocky formations are the predominant karst landscapes on high floodplains and low flat interfluves, though with low occurrence.

The conditions for karst development within the terrace terrain type are relatively good due to two reasons. First, the sands composing terraces have a good filtering capacity, and, second, the underlying chalk–marl rocks are strongly cracked. The main distinctive features of the structure of the karst geosystems are their hydromorphic properties and dense vegetation cover. This is especially typical for the karst landforms on the interfluves of the Cleveni–Obista–Amonka rivers, where up to 80% of the karst landscapes are represented by large valleys, mostly overgrown by willow. A similar picture can be observed on the left bank of the Donskaya Seimitsa river, and on the terraces of the Seimas, Oskol, and Black Kalitva rivers.

The elevated altitudinal landscape storey comprises the territories with elevations of 160–200 m a.s.l. The surface is densely dissected by ravines and gullies (1–1.5 km/km^2) with outcroppings of the chalk–marl rocks on the slopes. Karst landscapes of the elevated level are distinguished by the domination of the karst funnel-shaped sinkholes overgrown by vegetation on the watersheds, as well as the step-like erosive karst forms on the slopes. A typical example is the interfluve of the Rzhava and Donskaya Seimitsa rivers, where the density of the karst landforms accounts for 80/km^2.

The high altitudinal landscape storey was formed in areas with elevations of watersheds up to 200–250 m a.s.l. and more. Most of these territories belong to the central parts of the Voronezh crystalline massif—the area of the neotectonic uplifts and elevated chalk–marl rocks. This storey is indicated by the landforms with deep cavities overgrown by forests and meadow–steppe, classified as the covered chalk–karst type. The thickness of the Neogene–Quaternary sediments rarely exceeds 10 m, and they are underlain by thick strata of chalk–marl rocks. In the areas of tectonic uplifts they are strongly fractured, which encourages karst. Conditions of the upper landscape storey are the best for the karst landforms development. First, upper areas of the storey receive 5% as much precipitation as at lower stories, and, second, a thin layer of clay and loam has a weak screening effect for the penetration of water into the chalk. As a result, the density of karst landforms here is maximal in comparison with other stories, in some regions accounting for more than 100 landforms per square kilometer. For example, on the Ubli-Kotla rivers interfluve 107 karst landforms were observed, 90% of which had karst origin. A vast majority of them are covered with forests. This is one of the main features of karst landscapes of the upper storey. Lakes and mires occur rarely. There are karst funnel-shaped sinkholes overgrown by forests, along with the ones overgrown by shrubs, steppified and plowed. Another example is the karst landscapes of the Sejm–Mlodat rivers interfluve (Streletskaya steppe) and Mlodat–Polnaya rivers interfluve (Cossack steppe), which have a wide occurrence of the deep funnel-shaped sinkholes with active ponors, covered by forests, or funnel-shaped sinkholes with meadows and meadow–steppes, which are classified as the covered karst type.

14.4   Vertical Differentiation of Landscapes and Landscape–Typological Mapping

Altitudinal differences of flat landscapes are critical to the content of traditional landscape maps, particularly for the division to high altitudinal geomorphologic variations of types of landtype association, which are equivalent in terms of economic use. A total of seven basic types were identified within the region. The interfluve loamy type corresponds to the elevated plain areas with sufficient drainage, without signs of erosion, with steepness up to 3°, and with a depth of groundwater more than 5 m. This type appears as flat, inclined, undulating interfluves characterized by zonal soils that are formed in the covering loess-like loams. The interfluves sandy type is associated with the glaciofluvial sandy sediments, which are common for the East-European Plain on the areas of the Don (Gunz) glaciation epoch. This type has increased forest cover and hummocky-and-pitted relief. The interfluve nondrained type occurs along most rills with ephemeral stream courses. Slow drainage and the level of groundwater that is close to the surface (3–5 m) favor development of Gleyic Chernozem, which is replaced by Gleyic Solonetz and Albi-Mollic Solonetz in hollows and cavities. The interfluve hilly type is formed with the participation of sandy-argillaceous deposits of the Paleogene, including interlayers of dense sandstone, more resistant to denudation. As a result, an interfluve becomes an uneven surface, where the network of troughs is interspersed by buttes shaped by erosion. According to the elevations, the interfluve types of the terrain are divided into altitudinal variations: elevated (more than 200 m a.s.l.), middle (200–160 m a.s.l.), and lower (less than 160 m a.s.l.).

The slope type of land association includes surfaces with a steepness of over 3°, mainly a network of ravines and rills and slopes of river valleys. Landscapes are characterized by outcrops of bedrock, considerable dissection, active soil erosion, and sparse vegetation cover. The localization of the variations is based on local differences of elevations. Elevation values are used to delimit slope landscapes with network of a deep (more than 30 m), medium (10–30 m), and weak (less than 10 m) erosion. The terrace type of land association was formed within the boundaries of sandy left banks of rivers, in the structure of which two altitudinal–geomorphic variations can be delimited: low-elevation variation, including the first and second terraces, and the high-elevation variation, which includes the third terrace. The floodplain type covers the river valleys bottoms flooded during high water, which have the lowest elevations. The basis for the type of terrain differentiation is the height of the surface above the water's edge in the river. Three altitudinal-geomorphologic variations can be identified using this indicator: low (height above the water level is less than 4 m), lowered (4–8 m), and high (more than 8 m). A fragment of the map of the landtype association is shown in Fig. 14.3.

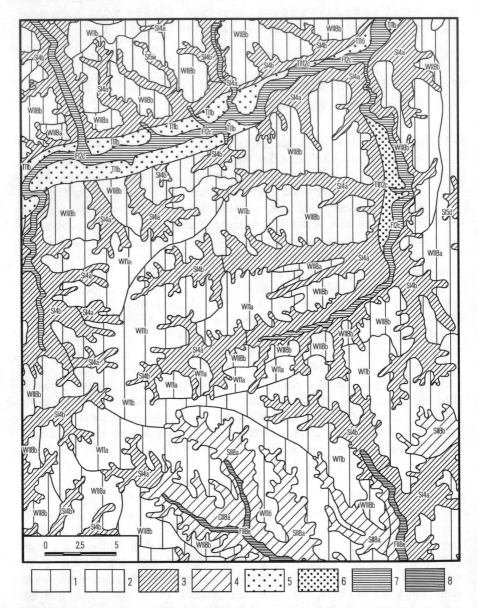

Fig. 14.3 The landtype association of the Central Russian Upland (fragment of the map)
Legend: Landtype association: 1, elevated watershed loamy; 2, middle watershed loamy; 3, slope deep erosion; 4, slope middle erosion; 5, high terrace; 6, lower terrace; 7, high floodplain; 8, middle floodplain; WI1a, SI4b ... , index of landtype

14.5   Vertical Differentiation of Landscapes and Optimization of Landscape–Ecological Situation

The theory of vertical differentiation of plain landscapes provides some new opportunities for solving practical problems. In particular, the specificity of altitudinal landscapes can be taken into account for the purposes of the landscape optimization and rational spatial organization, in processes of landscape planning and design, and in industrial and civil construction (Gorbunov and Bykovskaya 2012). Nowadays, landscape optimization involves the ways to increase the socio-economic functions, provided by landscapes. Successful implementation of such optimization is possible with the help of environment-transforming measures, which require considering regional, typological, dynamic, geochemical, and ecological features of landscapes. Such features strongly depend on the terrain elevation. Therefore, when solving the optimization-related problems, landscape vertical differentiation must be taken into account. For example, the flat interfluve type of terrain at elevations of more than 250 m a.s.l. is distinguished by a convex profile, a deep level of groundwater, a predominance of Luvic Chernozems, and relatively well-developed forest cover. At elevations of 120–160 m a.s.l., the same flat interfluves appear as spruce forests on the plains with groundwater near the surface, nutrient-rich Chernic Chernozems, and aspen coppice. These facts should be taken into account during designing agrolandscapes to calculate irrigation standards and input of fertilizers, to create forest belts and ponds. The most important role in optimizing the landscape and ecological situation in the region is assumed to belong to water reclamations. Land reclamation in this case is expedient due to the lack of natural moisture, which is needed to ensure successful growth of cultivated plants and to achieve larger crops. Calculations show that the ratio of heat and moisture in the south of the Central Russian Upland with chalk rocks is not always optimal. It happens because main climatic characteristics are changing. They depend on the latitude, elevation, and slope aspect. Study of ALCs provides basis for calculating irrigation norms (Table 14.1).

The obtained results allow us to conclude that vertical differentiation of landscapes is a crucially important phenomenon in the south of the Central Russian Upland with chalk rocks. Each step, storey, and variation of altitudinal landscapes has its peculiar structural dynamics. Consideration for such features is essential for solving problems related to optimization of landscape–ecological environment and to rational use of natural resources of the region.

References

Baily, R. G. (2002). *Ecoregion – Based design for sustainability*. New York: Springer.
Beloselskaia, G. A. (1969). Basic issues of landscapes vertical differentiation in a central forest-steppe. In *Issues in landscape geography* (pp. 16–24). Voronezh: Voronezh State University Press. (in Russian).

Berezhnoi, A. V., Gorbunov, A. S., & Berezhnaia, T. V. (2007). *A vertical differentiation of land-scapes of the Central Russian forest-steppe*. Voronezh: Voronezh State University Press. (in Russian).

Divíšek, J., Chytrý, M., Grulich, V., et al. (2014). Landscape classification of the Czech Republic based on the distribution of natural habitats. *Preslia, 86*, 209–231.

Gorbunov, A. S., & Bykovskaya, O. P. (2012). Issues on optimizing the ecological situation and vertical differentiation of landscapes of the forest-steppe zone of the Chalk south of Central Russian upland. *Arid Ecosystems, 2*(2), 91–97. https://doi.org/10.1134/s2079096112020035.

Khoroshev, A. V. (2013). Problems of studying the landscape polystructurality. In I. I. Mamay (Ed.), *Landscape collection* (pp. 170–195). Moscow-Smolensk: Oikumena. (in Russian).

Mikhno, V. B. (1976). A landscape-indicating method of The Central Russian upland karst funnel genesis indication. *Zemlevedenie (Earth science), 11*, 208–210. (in Russian).

Mikhno, V. B., & Gorbunov, A. S. (2001). High-altitude landscape complexes of the Chalk south of the Central Russian upland. *Proceedings of Voronezh State University, Series Geography and Geoecology, 1*, 6–24. (in Russian).

Milkov, F. N. (1947). About a phenomenon of landscapes vertical differentiation on the Russian plane. *Problems of geography, 3*, 35–41. (in Russian).

Nikolaev, V. A. (1999). *Landscapes of Asian steppes*. Moscow: Moscow University Press. (in Russian).

Rowe, J. S., & Sheard, J. W. (1981). Ecological land classification: A survey approach. *Environmental Management, 5*, 451–464.

Victorov, A. S. (2006). *Basic problems of mathematical morphology of landscape*. Moscow: Nauka. (in Russian).

Zonneveld, I. S. (1989). The land unit – A fundamental concept in landscape ecology, and its appli-cations. *Landscape Ecology, 3*(2), 67–86.

Chapter 15
Landscape Structure as Indicator of Debris Flow and Avalanche Activity in the Russian Caucasus Mountains

Marina N. Petrushina

Abstract The chapter focuses on landscape structure of paragenetic geosystems shaped by snow avalanches and debris flows with main attention to the stopping zone. The research was performed in the mountains of Western and Central Caucasus based on the long-term field observations, landscape mapping, interpretation of remote sensing data, and phytoindication. Landscape structure in the zones of natural processes activity is polystructural and depends on the type of processes, their frequency, and the internal features of the affected landscapes. Wet avalanches are followed by the strongest changes. A single influence, especially accompanied by the airwave, often results in relatively rapid recovery of zonal vegetation. As a result of prolonged avalanche releases, nature complexes of neighboring zonal types or subtypes may develop. A series of complexes indicating the frequency and duration of avalanches emerges in the impact zones. We distinguished several types of landscape patterns indicating the avalanche activity. Intensification of landscape changes in the late twentieth and the early twenty-first century due to the large avalanche and debris flow releases was detected.

Keywords Landscape structure · Debris flows · Avalanches · Landscape indication

15.1 Introduction

Mountain regions are distinguished by metachronous and polystructural landscape pattern as a result of well-pronounced superposition of three types of geographical space differentiation according to V.N. Solnetsev (1997)—geostationary,

M. N. Petrushina (✉)
Lomonosov Moscow State University, Moscow, Russia

biocirculation, and geocirculation ones.[1] The first one is associated with the abiotic substrate, which in mountains is characterized by complex geological stratification, the high terrain ruggedness, the lithological contrasts, and the high occurrence of rocky slopes that form the frame of the landscape structure. The formation of hierarchical biocirculation structures is manifested in the mosaics of the soil-vegetation cover resulting from the variety in solar energy input. The altitudinal distribution of heat and moisture generate significant climatic gradients. The geocirculation type of structure is primarily associated with high gravitational gradients. It is shaped by exogenous processes, often catastrophic, such as debris flows, snow avalanches, rock falls, and landslides. As a result, we can identify specific *paragenetic*, or functionally holistic, *geosystems*. They are integrated by a variety of lateral flows and consist of natural territorial complexes (NTC) of three zones—initiation, transit, and stopping (or accumulation). Paragenetic geosystems complicate landscape structure of mountainous regions and play significant role in their dynamics (Gvozdetskiy 1979; Samoylova et al. 2004; Kovalev 2009).

The study of paragenetic geosystems as indicators of natural processes activity is urgently needed since snow avalanches and debris flows are the most dangerous hazards in the mountains. The effects of these processes are reflected in the structure and ecological state of landscapes and their components. This allows using structure as an indicator of debris flow and avalanche activity. Most studies are devoted to their impact on vegetation (Potter 1969; Tushinskiy and Turmanina 1971; Turmanina 1980; Rapp and Nyberg 1981; Malanson and Butler 1986; Hupp et al. 1987; Strunk 1989; Rapp et al. 1991; Rixen et al. 2007; Aleinikova et al. 2005). Some studies, described landscape structure in the zones of their release (Fedina 1977; Akifyeva et al. 1978; Khapaev 1978; Ishankulov 1982; Oliferov 1982; Petrushina 2001, 2007; Khoroshev 2005), with strong emphasis on its complexity and dynamism. Considering the specifics of snow avalanche and debris flows paths, Butler (2001) proposes to consider them as geomorphic process corridors dissecting the matrix and performing five functions of a corridor: habitat, conduit, filter, source, and sink. He proposed to include them in the definition of corridor accepted in landscape ecology (Forman and Godron 1986).

In addition, studies in the areas affected by these processes make it possible to study exodynamic landscape successions as a particular case of the long-term state of NTC, which is important for improving indication techniques and revealing the features of mountain dynamics.

The study of debris flows and avalanches becomes more vital at present due to an increase of their activity under changing climatic conditions and anthropogenic impact. The end of the twentieth century and the beginning of the twenty-first century are distinguished by large and disastrous debris flows and avalanches in different mountain regions (Pebetez et al. 1997; Gruber and Margreth 2001; Szymczak

---

[1] See Chap. 1 for details.

et al. 2010; Stoffel et al. 2014; Perov et al. 2017), including the Caucasus mountains of Russia (Seynova et al. 1998; Zalikhanov 2001; Oleinikov 2002; Zaporozhchenko and Kamenyev 2011). Their effect on landscapes became a good basis for studying landscape dynamics.

The aim of our investigation is to reveal landscape structure and dynamics in the zones of snow avalanches and debris flows release in the Russian sector of the Caucasus and use them as indicators of processes activity.

15.2   Study Area and Methods

The investigation has been carried out in the Baksan, Chegem, and Cherek river basins, the highest parts of the Central Caucasus, and in the Teberda river basin within the Western Caucasus (Fig. 15.1). The region is favorable for active snow avalanches and debris flows especially in the Central Caucasus. In the study area, mountain ridges are higher than 3000–3500 m a.s.l. with some peaks above 5000 m a.s.l., Elbrus (5642 m) being the highest one. Valleys with rocky steep slopes are deeply dissected with elevation range up to 1000–1900 m. Palaeo–glacial landforms are typical for the region including different kars with intensive snow accumulation and numerous erosional landforms (more than 1 km/km²) serving as the paths for avalanches and debris flows. Widespread glaciers being in the retreat stage promote the formation of new periglacial landforms, including ice-cored moraines and damped lakes, which act as sources of the glacial outburst floods with large destructive effects, as witnessed in the years 2000, 2002, 2006, and 2017 (Perov et al. 2017; Chernomoretz et al. 2018).

The climate of the area is moderately continental and humid in the Western Caucasus and dry in the Central Caucasus. Mean winter air temperature ranges from −7° (2150 m) to −10° (3000 m). Solid precipitation varies from 400 to 600 mm. The maximum annual precipitation can vary in wide range and may exceed the

Fig. 15.1 The location of the study area

average two or three times, which is the common reason for the debris flow and avalanche release. During prolonged snowfalls (some times for 5 or even for 21 days), the precipitation can range from 84.6 to 228.5 mm. The distribution of the snow cover is irregular due to the strong winds. Mean snow depth ranges from 0.6–0.8 m (2150 m) to 2 m (2800–3000 m) reaching the maximum of 10–12 m on the slopes near the tops of ridges. The number of days with avalanches varies from 26 to 50 in a year (Troshkina 1992). The last decades are characterized by increasing temperature and precipitation contrasts between years and months. The maximum precipitation occurs in March exceeding in some years the average month amount in three times (Petrushina 2015). Rains including rainstorms with intensity of 50–85 mm per day contribute to 55–75% of annual precipitation. The precipitation of 20–25 mm per day and more is usually enough for starting of the debris flow (Seynova and Tatyan 1977).

The region is characterized by altitudinal zonation, exposure-induced landscape contrasts, especially in the Central Caucasus, curvy character of zone boundaries affected by topography, and intensive geomorphic processes. Glacio–nival, meadow (subnival, alpine, and subalpine), forest landscapes with pine (*Pinus hamata*), pine–birch (*Betula pendula*), and birch (*B. pendula, Betula Litwiinovii*) predominate in the valleys of the Central Caucasus. The Teberda river basin, located to the west in more humid climate, is characterized by deciduous (*Fagus orientalis, Carpinus caucasuca*), mixed, and coniferous forests with fir tree (*Abies nordmanniana*), spruce (*Picea orientalis*), and pine (*P. hamata*). Crooked birch and acer (*Acer trautvetterii*) stands dominate at higher altitudes than coniferous forests. The high occurrence of avalanche and debris flows geosystems is the specific feature of the study region (Petrushina 1992). Landscapes shaped by palaeo- and modern debris flows and avalanches occupy approximately a half of the valley floors in the Central Caucasus, thus forming the complex landscape structure.

To reconstruct the avalanche and debris flow activity and landscape dynamics we used fine-scale landscape mapping (1:50,000, 1:25,000 and 1:10,000), field observations since 1977, phytoindication including phytocenology, lichenometry (using *Rhizocarpon geographicum*), and dendrochronology, and examination of historical maps, aerial images, satellite data (TERRA (ASTER), World View-1, LANDSAT), and on-land photos of different years. Landscapes have been examined with particular focus on the zone of avalanche and debris flow stop-on. To determine the rate and character of succession stages, we conducted repetitive observations of vegetation and soils on plots including terrains after disaster or with known data of debris flow and avalanche releases. Particular attention has been paid to the initial stages of succession. In the zones of avalanche releases, emphasis has been made on the ecological state of vegetation, particularly trees and shrubs (diversity, height, live forms and character of damage), which are a good indicator of their impact (Akifyeva 1980; Vlasov et al. 1980; Burrows and Burrows 1976; Erschbamer 1989; Butler and Sawyer 2008; Bullschweiler and Stoffel 2010; Simonson et al. 2010). Detailed investigation along the transect, using a 10 m spacing, was also carried out to study interrelations between the components of the geosystems in the model plots.

Fig. 15.2 Avalanche geosystems in the upper part of Baksan river

Most detailed research of avalanches has been performed in the upper part of the Baksan Valley, characterized by regular and catastrophic avalanches of different types (Fig. 15.2). Since the area was well studied, mainly in the 1960s and 1970s (Akifyeva et al. 1970; Zolotarev 1980; Troshkina 1992), it has been chosen as one of a test sites for landscape monitoring. It stretches for 5 km along the valley with slopes rising from 2200 to 3700 m a.s.l. Several of 19 avalanche tracks reach the valley floor. The area is famous for large fans in the zone of stop-on. It has been changed strongly by intense avalanche release from the late 1950s until today. The last decades are known for particular large avalanches in winters of 1967/68, 1968/69, 1973/74, 1975/76, 1978/79, 1986/87, 1992/93, and 2001/2002.

15.3   Results and Discussion

Debris flows and avalanches cause either simplification or complication of landscape structure in three zones—initiation, transit, and stopping—with large-scale disturbance in the latter zone. Each of them has its own spatial structure, the most dynamic in the stopping zone, where NTC of different ages are formed. They can be regarded as original spatial models of temporal changes in landscapes (Petrushina 2001). Debris flows usually destruct the whole NTC of lower hierarchical levels (including landforms and deposits) while avalanches affect, as a rule, vegetation only. Various landscape spatial patterns were revealed in the zones of avalanche and debris flows of several types and frequency.

15.3.1   Landscape Structure of the Debris Flow Geosystems

The debris flow fans are characterized by a segmented fan—in some sites, a lobed and streamed pattern—determined by flow character and debris accumulation, such as the accumulation of coarse sediments in the upper part of the fans and along stream channels, and the development of ridge–hollow relief. In the Central Caucasus, five main types and several subtypes of cones were distinguished based on the characteristics of the landscape structure and its dynamics (Petrushina 2007). The first type of cones usually of large size with well-incised channel (up to 7 m deep and more) is characterized by the formation of ridge–hollow relief with well-developed zonal vegetation and soils (steppe, forest) or meadow and forest–meadow (if they are used for haymaking). These fans are not subject to impact of debris flows nowadays but have a lot of traces of their high activity in the past. The second type of fans is distinguished by rather stable and homogeneous pattern in the marginal parts not affected by current debris flows. A series of nested fans with age less than 100 years occurs in the lower central sector. These fans with a well-incised channel are usually rather large and old. They are often formed along the south-facing slopes of ranges in the steppe altitudinal belt. They are particularly well distinguished in contrasting structure of old steppe or meadow NTC and young small-leaved forests with shrubs, since they are usually located at lower altitudes (below 1800 m). The landscape pattern of the younger parts of the fans is often heterogeneous.

The structure of the fans of the third type is more complex but with poorly developed main channel. The debris flows here are usually divided into a number of channels with a large stream moving along the shortest path to the main river. The presence of such channels is a reliable indicator of large debris flows, often of glacial genesis, typical for regions with active glaciation degradation. Coarse gravel deposits are quite typical, complicated by stone spots (especially from water-stone debris flows). The largest deposits accumulate in the upper sectors where *Pinetum cladinosum* forests on primitive soils are formed. The lower sectors are dominated by *Pinetum fruticuloso-hylocomiosum* and *Pinetum herbosum* forests on Umbrisols. The characteristic attributes are as follows: the inversion of microrelief formed by the newly formed levees within the older parts of fans, buried soils, and mosaic combination of natural complexes of different ages often with the poorly interrelated vegetation and soils. Most fans in the upper parts of the river basins fall into this type. As the main channel gradually deepens, this type of fans may evolve to the second type. The third type of fans are often affected by snow avalanches, which disturb some regularities of the landscape differentiation. In the early twenty-first century, several debris flows induced by snow avalanches occurred in the sites where they were not observed for the past 40–50 years if ever.

The fourth type of fans is characterized by intermediate features of the first two types. The fifth type, with a simpler internal structure, is typical for fans with the annual release of the so-called "man-made" debris flows. These types of fans can be divided into several subtypes according to their size, frequency of the debris flow events, magnitude, and position in the different altitudinal belts with more subtype variations in the third type of the fans.

A decrease of young debris flow fans in comparison with the ancient ones was observed in all valleys in the area. The largest fans are typical for the basins of the Central Caucasus. Different age of fans is represented by four evolution stages—initiation, formation, quasi-stable state, and destruction—that form specific chrono series. The character of landscape vertical structure, mainly of vegetation and soils, is the criteria for identification of the stages. These stages can be considered as landscape succession stages within the existence of one zonal type of landscape (Petrushina 2001, 2007). The rate of the succession stages on debris flow fans depends on the position in the altitudinal belt, texture and thickness of the recent deposits, frequency of the flows and effect of other natural processes (snow avalanches, erosion, flooding, etc.), anthropogenic impact, and the distance to the well-developed complexes. The succession rate depends on the comprised area and intensity of the event, which is in agreement with the findings of other researchers (Pickett and White 1985; Connell and Slatyer 1977). Numerous fans at the initial stages of NTC formation indicate the activity of debris flows in the study area.

15.3.2 Landscape Structure of the Avalanche Geosystem

It is well known that avalanches are determined by terrain and climate (Mc Clung and Schaerer 1993). The landscape features of the initiation and transit zones, relief and vegetation in particular, often predetermine the character of the avalanche release. Avalanches of various types, frequency, and magnitude also cause different landscape changes. The episodic impact (often catastrophic) of powder avalanche as well as snow airwave usually destructs or affects only vegetation (the species diversity, density of forests, height and forms of trees and shrubs, and phenological phases), while wet avalanche with large quantities of deposits (snow, rocks debris, and organic matter) results in transformation of the whole NTC of the low hierarchic level including deposits, relief, and soils. Distinct boundaries of the NTC were observed in the areas of regular avalanche releases of similar sizes.

Changes in the NTC boundaries and spatial landscape pattern are typical for the zones of avalanche release inducing the development of various landscape patterns (Table 15.1).

In the subnival and alpine belts, the results of avalanching are not so easily detected as in the forest belt. However, uneven distribution of snow cover due to topography and different time of avalanche release causes the changes in the type of functioning, microcomplexity of soils, and vegetation. This results in emergence of the spotty landscape patterns. The sites of avalanche release are often indicated by

Table 15.1 Landscape patterns in the zones of avalanche release (according to literature and original results)

Zone of avalanche release	Altitudinal belt (zonal landscape type and subtype)	Landscape pattern
Initiation	Subnival, alpine (subalpine)	Spotty
Transit	Alpine, subalpine, forest	Longitudinal-stripped (channeled avalanche), wide longitudinal-stripped (sluff avalanche)
Stop-on	Subalpine, forest, steppe	Fan-shaped (fans), finger-shaped (snow airwave), polyconcentric, spotty, with rectilinear boundaries (avalanche of fresh snow)

snowpack laying until June or even July, active nivation and erosion, and formation of nival niches, erosional landforms, etc. They are also indicated by colored alpine carpets with *Sibbaldia procumbens*, *Taraxacum stevenii*, and *Pedicularis nordmanniana*, which usually cover the central sectors of niches. These plants with short growth period often indicate the places with winter snow depth of 3–5 m. The peripheral sectors are often covered by *Campanula sp.* These carpets form beautiful colored aspects in contrast with adjacent herbal communities. If avalanches release in spring, in places of their initiation the snow is melting more rapidly and plants appear earlier and use longer period of vegetation. *Festuca sp.* can be observed in these sites as well.

The transit zone of avalanches is usually located in subalpine and forest belts. The releases of channeled avalanches of different magnitude and frequency are indicated by linear hollows of different width with either tall-grass subalpine meadows or with low-grass alpine meadows or with different leaf-bearing krummholz or even without soils and vegetation (Mears 1992; Patten and Knight 1994; Butler 2001). On slopes, these communities alternate with undisturbed coniferous landscapes with spruce, fir, or pine forests. Longitudinal-stripped landscape patterns are characteristic of such sites. Spotty-longitudinal-stripped patterns can be observed in the areas with joint action of channeled and sluff avalanches. The complexes with prostrate shrubs (willows, junipers, and rhododendron) or sweetbriers with practically unchanged height often appear in the zones of initiation and transit of sluff avalanches in subalpine and forest belts.

As the subalpine belt can be the zone of initiation, transit, and stopping at the same time, its landscape structure is more complex. The inversion alpine meadows dominated by *Sibbaldia semiglabra*, *Ranunculus sp.*, *Gentiana sp.*, and some other species occur here often. The meadows with *Nonea ventricosa*, *Trollius ranunculinus*, *Anemona fasciculate*, and *Pulsatilla aurea* as indicators of snowy sites are characteristic of the lower part of this belt. The south-facing slopes are commonly covered by mesophilous meadows with *Calamogrostis caucasica*, *Cephalaria gigantea*, *Betonica grandiflora*, *Euphorbia iberica*, and some other species. Since the zone of avalanche stopping often is located in the lower part of the subalpine meadow belt, the tree line in these areas is usually indented due to the avalanche release as it was observed in different regions (Gorchakovskiy and Shiyatov 1985; Suffling 1993; Holtmeier 2009). In comparison with the climate-determined tree line, the real one can decrease by 200–300 m and in some cases even more.

Avalanche fans with fan-shaped, polyconcentric, or spotty landscape patterns often correspond to the zone of stopping. This zone usually penetrates the forest landscapes in the valley bottoms and foot slopes, north facing as a rule. The sites with avalanches of rather similar frequency and magnitude (small and medium size) are indicated by polyconcentric patterns, while the sites of snow airwave impact have finger-shaped landscape patterns (Akifyeva 1980; Petrushina 2001). The episodic release of powder snow avalanches usually results in the rectilinear character of boundaries. When the impact of avalanches of different power and type, coming from neighboring avalanches, is imposed, the landscape pattern of the territory becomes more complicated and the boundaries are less pronounced.

A series of nature complexes indicating the frequency and duration of avalanches emerges in the release sites as well (Fig. 15.3). The extreme patterns of the series indicate the most frequent impact of avalanches and their practical absence. The corresponding NTCs may belong to various zonal types or subtypes. Peripheral

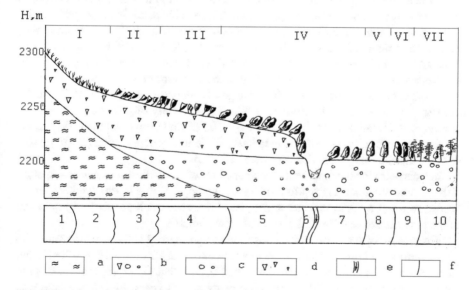

Fig. 15.3 Natural complexes in the zone of avalanche impact
Bedrock: 1, crystalline schists. Deposits: 2, of mixed genesis; 3, avalanche rock debris; 4, alluvium. Soils: 5, primitive; 6, primitive on buried Umbrisol; 7, sod; 8, meadow–forest; 9, Cambisol; 10, forest with podzolization features. Vegetation: 11, primitive plant aggregation; 12, sparse subalpine herb–grass meadow with pioneer species. 13, willow stand; 14, willow elfin woodland; 15, herb–grass meadow; 16, birch–elvin woodland; 17, birch-crooked stem forest; 18, mature birch forest; 19, light birch–pine forest; 20, pine forest with bilberries.
Lines—Frequency of avalanche release: 1, annually; 2, annually or once in 2–3 years; 3, once in 3–(5)–10 years; 4, once in 10–20 years; 5, once in 20–30 years; 6, once in 30–50 years; 7, once in more than 50 years.

complexes with coniferous forests are the least dynamic, but subject to the most dramatic changes after the large avalanches. The central sectors of the fans (as a core) represent peculiar stable dynamical invariants of avalanche geosystems. They are manifested as the azonal subalpine meadows with sparse vegetation or with tall herbs, in some places with rare willow shrubs within the forest altitudinal belt. Subalpine complexes prevail in the sites subject to regular powder avalanches. Quite rarely these sites are occupied by complexes with the dominance of alpine species *S. semiglabra*, *Alopecurus sericeus*, etc. Mesophilic tall-grass meadow communities on avalanche cones commonly indicate fine-textured substrate. Sites of ground avalanche release contain a larger amount of rocky material and, hence, are most often distinguished by a sparse herbal community dominated by *Polygonum panjutinii*, *Achillea grandiflora*, etc. On stony clastic deposits (especially on south-facing slopes), the typical communities include combination of species from various ecological groups, namely xerophytes and mesophytes.

Complexes with krummholz birch communities are indicative of the avalanching with the interval of 5–10 years, while those with tall birch forests indicate the avalanching of once in 10–30 years. Mature birch–pine forests in the Central Caucasus and birch–spruce or birch–fir in the Western Caucasus occur in the places with avalanche release approximately once in 30–50 years and with coniferous forests not more often than once in 50 years or even rare. This result is in agreement with the previous studies in the Central Caucasus (Akifyeva et al. 1978; Turmanina 1980) and close to data of Mears (1992). The nature complexes of different ages, of contrast zonal subtypes (forest coniferous, subalpine birch and meadow), can replace each other at a short distance making landscape structure more complex. The inversions of plants and their communities are good indicators of the zone of avalanche impact (Vlasov et al. 1980; Suffling 1993). The diversity of plants can increase in birch-crooked stands up to 45–60 species on 100 m^2 and even more than 100 species on fans with steppified subalpine meadows, while in adjacent coniferous (pine) forests the number of species does not exceed 15–25.

In areas with the development of a stepped relief, cascading geosystems are often formed, and the lenticular pattern of landscape appears due to the combination of transit zones of track avalanches and zones of their accumulation that may be at the same time the zones of initiation of another avalanches.

The release of the large avalanches in the upper part of Baksan Valley since the mid-twentieth century, including the recent events in 2002 and 2018, resulted in the destruction of landscapes, which rarely had been affected before: the age of pine trees accounts for 150–200 years, some of them 320 years. Pine forest area was reduced by six times accompanied by fragmentation into separate patches. In this period, landscape structure of avalanche fans was simplified while in the valley bottom it became more complex. The reduction of complexes with pine and willow scrubs and expansion of aspen stands, post-forest, and subalpine meadows were observed in the study areas. The birch-crooked stem forests were reduced on the fans and expanded twice in the valley bottoms (Petrushina 2001).

Single avalanche release, especially accompanied by the snow airwave, often results only in vegetation change and relatively rapid recovery of zonal vegetation without soil modification due to higher stability of soils and their longer characteristic time scale. This is true in most cases for the avalanche fans along the south-facing slopes. Vegetation regrowth near the north-facing slopes is slower and the recovery goes through succession stages — the formation of birch and birch–pine forests. In the sites experiencing intensification of avalanching followed by vegetation changes, the soil begins to reconstruct its structure and physical and chemical properties. This process after destruction of forest vegetation typically involves the gradual increase of pH, organic matter, and exchange cations contents, and more intense accumulation of biogenic chemical elements (P, Mo, Zn, Mn, etc.). Soils in avalanche stopping zones are often in discordance with vegetation, which is manifested in high occurrence of poorly differentiated soils and soils with buried horizons. For example, the soils under the birch forest may resemble either forest Cambisols or meadow Umbrisols (Petrushina 2001). In the area of avalanches, internal landscape self-regulation is manifested as a pattern transformation according to the changing inner conditions. Nowadays, the rate of the forest recovery in the upper part of Baksan Valley is slowed down due to the frequent avalanching and anthropogenic impact.

Numerous observations in the study area showed evidence that in recent decades debris flows and snow avalanches occur in places where they were not observed for 50–60 years or were never observed. Besides that, debris flows sometimes follow avalanches in the same year along common channels, which leads to a complication of the landscape structure in the sites of their releases.

Based on the study results, we suggested classification of paragenetic geosystems according to the following set of attributes: (1) genesis (debris flow, scree–debris flow, debris flow–avalanche, complex mixed genesis); (2) the temporal mode of action (permanent, periodic (with various frequencies), rare, and very rare); (3) the time of action (year-round, winter, winter–early spring, spring–early summer, summer, autumn, autumn–winter); (4) size and impact area (small, medium, large, very large), intensity of action; (5) the position of the zones of matter mobilization in high-altitude belts (nival–meadow–forest, subalpine meadow–forest–steppe); (6) the complexity of structure in the areas of matter mobilization (simple, complex with a combined transit and accumulation zone, complex with a series of combined transit and accumulation zones of the substance); (7) the nature of the internal structure (homogeneous, heterogeneous, contrasting); and (8) the dynamics of the structure in the stopping zone (constantly dynamic, with periodic changes in nature components, with periodic changes in the landscape structure at different frequencies).

15.4   Conclusions

Metachronous and complex landscape structure of avalanche and debris flow geosystems evidences the dynamism of mountain landscapes and relative permanence of these processes in time. The character of landscape changes depends on the type of debris flows and avalanches, their activity, and the internal features of the affected landscapes. We distinguished several types of landscape spatial patterns in the zones of initiation, transit, and stopping of avalanches of different types and magnitudes with the most dramatic changes in the latter zone. The stopping zone is characterized by the diversity of complexes of various ages, often with contrast type of vegetation, with disparity of plants and soils, the recurrent change of their boundaries in time, and the inversion of plant communities and natural complexes of different altitudinal belts.

The avalanches and debris flows including extreme catastrophic events during the recent decades caused elevated dynamism of landscape patterns. As a result of avalanching, the landscape structure becomes more heterogeneous at the valley bottoms and more homogeneous on the avalanche fans.

The growth of snowiness, with maximum precipitation in spring in particular, favors the increased avalanche impact, mainly of wet type, on landscapes and their functionality. The complication of landscape structure on the fans due to the common effect of debris flows and avalanches is typical for the current state of the study area.

In modern climatic conditions, there is a high probability of permanency in debris flow and avalanche activity. This necessitates further study of their impact on landscapes, monitoring of their activity, and assessment of territories by hazard level.

Acknowledgments This research was conducted according to the State target for Lomonosov Moscow State University "Structure, functioning and evolution of natural and natural-anthropogenic geosystems" (project no. AAAA-A16-116032810081-9).

References

Akifyeva, K. V. (1980). *Methodical tool for deciphering aerial photographs in the study of avalanches*. Leningrad: Gidrometeoizdat. (in Russian).

Akifyeva, K. V., Kravzova, V. I., & Turmanina, V. I. (1970). Large-scale complex investigations of avalanche cones as objects of on colorgraghic materials. *Informatsionny Sbornik MGG, 15*, 55–72. (in Russian).

Akifyeva, K. V., Volodicheva, N. A., Troshkina, E. S., et al. (1978). Avalanches of the USSR and their influence on the formation of natural-territory complexes. *Arctic and Alpine Research, 10*(2), 223–233.

Aleinikova, A. M., Petrushina, M. N., Sukhoruchkina, E. S., et al. (2005). Phytoindication of the age of moraine and mud-flow deposits of the Central Caucasus (the Gerkhozhan-su river basin). *Proceedings of Moscow University, series 5 Geography, 4*, 40–47. (in Russian).

Bullschweiler, M., & Stoffel, M. (2010). Tree-ring and debris flow: Recent developments, future directions. *Progress in Physical Geography, 34*(5), 625–645.

Burrows, C. J., & Burrows, V. L. (1976). *Procedures for the study of snow avalanche chronology using growth layers of woody plants*. University of Colorado Institute of Arctic and Alpine Research, Occasional Paper 23.

Butler, D. R. (2001). Geomorphic process–disturbance corridors: A variation on a principle of landscape ecology. *Progress in Physical Geography, 25*(2), 237–248.

Butler, D. R., & Sawyer, C. F. (2008). Dendrogeomorphology and high magnitude snow avalanches. A review and case study. *Natural Hazards Earth System Science, 8*, 303–309.

Chernomoretz, S. S., Petrakov, D. A., Aleynikov, A. A., et al. (2018). The outburst of Bashkara glacier lake (Central Caucasus, Russia) on September 1, 2017. *Earth's Cryosphere, 22*(2), 70–80.

Connell, J. H., & Slatyer, R. O. (1977). Mechanisms of succession in natural communities and their role in community stability and organization. *American Naturalist, 111*, 1119–1144. http://www.columbia.edu/cu/e3bgrads/JC/Connell_1977_AmNat.pdf.

Erschbamer, B. (1989). Vegetation on avalanche paths in the Alps. *Vegetatio, 80*, 139–146.

Fedina, A. E. (1977). Dynamics of mountain landscapes. In *Relief and landscapes* (pp. 200–207). Moscow: MSU Publishing House. (in Russian).

Forman, R. T. T., & Godron, M. (1986). *Landscape ecology*. New York: Wiley.

Gorchakovskiy, P. L., & Shiyatov, S. G. (1985). *Phytoindication of environmental conditions and natural processes in high mountains*. Moscow: Nauka. (in Russian).

Gruber, U., & Margreth, S. (2001). Winter 1999: A valuable test of the avalanche-hazard mapping procedure in Switzerland. *Annals of Glaciology, 28*, 328–332.

Gvozdetskiy, N. A. (1979). *Principal problems of physical geography*. Moscow: Vysshaya Shkola. (in Russian).

Holtmeier, F.-K. (2009). *Mountain timberlines: Ecology, patchiness, and dynamics* (Advances in global change research, 36). New York: Springer.

Hupp, C. R., Osterkamp, W. R., & Thornton, J. L. (1987). Dendrogeomorphic evidence and dating of recent debris flows on Mount Shasta, Northern California. *U.S. Geological Survey Professional Paper, 1396–B*, 1–30.

Ishankulov, M. Sh. (1982). Landscapes of cones of removal and formation of their morphological structure. *Issues in Geography, 121*. (in Russian).

Khapaev, S. A. (1978). Dynamics of avalanche natural complexes: An example from the high-mountain Teberda State Reserve, Caucasus Mountains, USSR. *Arctic and Alpine Research, 10*(2), 335–344.

Khoroshev, A. V. (2005). Effect of avalanche and mudflows on the structure of the components links in the high-mountain landscapes. *Proceedings of Congress of Russian Geographical Society, 2*, 95–100. Sankt-Peterburg. (in Russian).

Kovalev, A. P. (2009). *Landscape as itself and for man*. Khar'kov: Burun Kniga. (in Russian).

Malanson, G. P., & Butler, D. R. (1986). Floristic patterns on avalanche paths in the northern Rocky Mountains, USA. *Physical Geography, 7*, 231–238.

Mc Clung, D. M., & Schaerer, P. A. (1993). *The avalanche handbook*. Seattle: The Mountaineers.

Mears, A. I. (1992). *Snow-avalanche hazard analysis for land use planning and engineering* (Bulletin 49). Denver: Colorado Geological Survey.

Oleinikov, A. D. (2002). Snow avalanches on the Great Caucasus in the conditions of climate warming. *Materials of glacilogical research, 93*, 67–72. (in Russian).

Oliferov, A. (1982). *Geographical aspects of melioration of mudflow landscapes*. Simferopol: Publishing house of Simferopol University. (in Russian).

Patten, R. S., & Knight, D. H. (1994). Snow avalanches and vegetation pattern in Cascade Canyon, Grand Teton National Park, Wyoming, USA. *Arctic and Alpine Research, 26*(1), 35–41.

Pebetez, M., Lugan, R., & Raeriswyl, P. A. A. (1997). Climatic change and debris flows in high mountain regions: The case study of the Ritigraben torrent (Swiss Alps). *Climate Change, 36*, 371–380.

Perov, V., Chernomorets, S., Budarina, O., et al. (2017). Debris flow hazards for mountain regions of Russia: Regional features and key events. *Natural Hazards, 88*(1), 199–235.

Petrushina, M. N. (1992). Landscapes of the Baksan valley. In G. I. Rychagov & I. B. Seinova (Eds.), *Nature use of the Elbrus area* (pp. 120–152). Moscow: MSU Publishing House. (in Russian).

Petrushina, M. N. (2001). Impact of debris flows and snow avalanches on the high mountain landscapes. *Materials of glacilogical research, 91*, 96–104. (in Russian).

Petrushina, M. N. (2007). Effect of debris flow activity on the landscapes of the Central Caucasus. In *Proceedings of the international conference on debris-flow hazards mitigation: Mechanics, prediction, and assessment, proceedings* (pp. 67–76). Rotterdam: Millpress.

Petrushina, M. N. (2015). Influence of avalanche and debris flow activity on the current state of landscapes of the Western Caucasus. *Proceedings of Moscow University, series Socio-ecological technologies, 1–2*, 111–126. (in Russian).

Pickett, S. T. A., & White, P. S. (Eds.). (1985). *The ecology of natural disturbance and patch dynamics*. New York: Academic.

Potter, N. Jr. (1969). Tree-ring dating of snow avalanches tracks and geomorphic activity of avalanches, northern Absaroka Mountains. Wyoming. In S. A. Schumn & W. C. Bradley (Eds.), *Contribution to quaternary research. Geological Society of America, Special Paper123* (pp. 141–165). United States.

Rapp, A., & Nyberg, R. (1981). Alpine debris flows in Northern Scandinavia. Morphology and dating by lichenometry. *Geografiska Annaler Series A, Physical Geography, 63*(3/4), 183–196.

Rapp, A., Li, J., & Nyberg, R. (1991). Mudflow disasters in mountainous areas. *Ambio, 20*(6), 210–218.

Rixen, C., Haag, S., Kulakowski, D., et al. (2007). Natural avalanche disturbance shapes plant diversity and species composition in subalpine forest belt. *Journal of Vegetation Science, 18*, 735–742.

Samoylova, G. S., Avessalomova, I. A., & Petrushina, M. N. (2004). Mountain landscapes. Levels of space organization. In K. N. Dyakonov & E. P. Romanova (Eds.), *Geography, society and environment. Vol. II. Functioning and present-day state of landscapes* (pp. 84–100). Moscow: Gorodets. (in Russian).

Seynova, I. B., & Tatyan, L. V. (1977). The critical meaning of meteorological parameters of debris flow hazard situations in the high mountainous region of the Central Caucasus. *Meteorology and hydrology, 12*, 74–82. (in Russian).

Seynova, I. B., Malneva, I. V., & Kononova, N. K. (1998). Dynamics of and forecasting of glacial debris flows of the Central Caucasus. *Materials of glacilogical research, 84*, 114–120. (in Russian).

Simonson, S. E., Greene, E. M., Fassnacht, S. R., et al. (2010). Practical methods for using vegetation patterns to estimate avalanche frequency and magnitude. *Proceeding of the International Snow Science Workshop, 548–555*.

Solnetsev, V. N. (1997). *Structural landscape studies*. Moscow: MSU Publishing House. (in Russian).

Stoffel, M., Tiranti, D., & Huggel, C. (2014). Climate change impacts on mass movements – Case studies from the European Alps. *Science of the Total Environment, 493*, 1255–1266. https://doi.org/10.1016/j.scitotenv.2014.02.102.

Strunk, H. (1989). Dendrogeomorphology of debris flow. *Dendrochronoligia, 7*, 15–24.

Suffling, R. (1993). Induction of vertical zones in sub-alpine valley forests by avalanche-formed fuel breaks. *Landscape Ecology, 8*, 127–138.

Szymczak, S., Bollschweiler, M., Stoffel, M., et al. (2010). Debris-flow activity and snow avalanches in a steep watershed of the Valais Alps (Switzerland): Dendrogeomorphic event reconstruction and identification of triggers. *Geomorphology, 116*, 107–114.

Troshkina, Y. S. (1992). Avalanches in the Elbrus region. In G. I. Rychagov & I. B. Seinova (Eds.), *Nature use of the Elbrus area* (pp. 64–85). Moscow: MSU Publishing House. (in Russian).

Turmanina, V. I. (1980). Influence of avalanche activity on the vegetation. In *Avalanches of Priel'brus'ye* (pp. 47–62). Moscow: MSU Publishing House. (in Russian).

Tushinskiy, G. K., & Turmanina, V. I. (1971). Phytoindication of glacial-debris flow dynamics of the last millennium. In G. K. Tushinskiy (Ed.), *Phytoindication methods in glaciology* (pp. 142–153). Moscow: MSU Publishing House. (in Russian).

Vlasov, V. P., Khanbekov, I. I., & Chuenkov, V. S. (1980). *Forest and snow avalanches*. Moscow: Lesnya promyshlennost'. (in Russian).

Zalikhanov, M. C. (Ed.). (2001). *Inventory of the avalanche and debris flow hazards in the North Caucasus*. Saint Petersburg: Gidrometeoizdat. (in Russian).

Zaporozhchenko, E. V., & Kamenyev, N. S. (2011). Debris flow dangers of the 21st century in the Northern Caucasus (Russia). In R. Genevois, D. L. Hamilton, & A. Prestininzi (Eds.), *Debris-flow hazards mitigation: Mechanics, prediction, and assessment* (pp. 813–822).

Zolotarev, A. E. (1980). Study of snow and avalanches in Elbrus region by fotogrammetry method. In *Avalanches of Priel'brus'ye* (pp. 47–62). Moscow: MSU Publishing House. (in Russian).

Chapter 16
Multiscale Analysis of Landscape Structure

Alexander V. Khoroshev

Abstract Dominant landscape-ecological models either focus on hierarchical organization of one natural geocomponent (in most cases plant cover, land use, or relief) or describe relations at a single hierarchical level. In contrast, we propose the tool to reveal multiple independent hierarchies based on *interactions* between properties of natural geocomponents. We use multidimensional scaling and response surface regression models to examine emergent effects imposed by combination of elements of the higher-scale system to the focus-level system. We show the effect of topographically induced landscape pattern on present-day processes manifesting in variety of soil–plant relationships on the example of the middle taiga of East European plain. The research confirmed the hypothesis that the combined effect of several higher-order geosystems provides emergent effect for the lower-order landscape unit. Probabilistic landscape mapping showed the areas with perfect adaptation of soils and vegetation to abiotic environment as well as areas with high probability of several stable states. Increase in relief dissection makes inter-level relations more complex due to growing effect of fine-scale processes. Combination of landforms serves as indicator of the processes which were acting in Pleistocene and resulted in various outcroppings of carbonate-rich sediments depending on relief dissection.

Keywords Scale · Hierarchy · Geocomponent · Soil–plant relationships · Response surface regression · Emergent effects

A. V. Khoroshev (✉)
Department of Physical Geography & Landscape Science,
Lomonosov Moscow State University, Moscow, Russia

© Springer Nature Switzerland AG 2020
A. V. Khoroshev, K. N. Dyakonov (eds.), *Landscape Patterns in a Range of Spatio-Temporal Scales*, Landscape Series 26,
https://doi.org/10.1007/978-3-030-31185-8_16

16.1   Introduction

Obvious growing interest to multiscale organization of landscapes is encouraged by
the necessity to translate information among hierarchy levels (O'Neill 1988; Wu
and David 2002; Oline and Grant 2002; Burnett and Blaschke 2003; McGarigal
et al. 2016; Malanson et al. 2017). The key idea of the concept is that each property
of a landscape reflects superposition of effects generated at various hierarchical
levels. Hierarchy theory suggests that multiple scales of pattern will exist in landscapes because of the multiple scales at which processes are acting (Turner and
Gardner 2015). Since landscape-ecological studies traditionally focus on linkages
both between natural geocomponents (e.g., substrate, soil, water, air, vegetation,
and animals) and between spatial units, it is crucial to reveal interactions between
spatial and nonspatial effects. Limitations upon interactions between geocomponents in a landscape system are imposed by differences both in characteristic space
scale and time scale inherent to bodies and processes.

The most critical issue in multiscale landscape studies is the quantitative evaluation of contributions from each scale level to spatial variability of landscape attributes (Borcard and Legendre 2002; Cushman and McGarigal 2002; Jin Yao et al.
2006). The most commonly used concept that relates pattern to process in landscape
ecology, the patch-corridor-matrix model, perceives the landscape as a planimetric
surface (Hoechstetter et al. 2008). The need to include vertical dimension to landscape pattern models requires consideration for the role of topographic and geological structures (Bolstad et al. 1998; Dorner et al. 2002; Sebastiá 2004; Khoroshev
and Aleshchenko 2008; Dragut et al. 2010; Cushman and Huetmann 2010; Bastian
et al. 2015). The focus on the critical significance of abiotic environment in shaping
landscape structure has been instrumental for the Central and East European schools
in landscape ecology and physical geography, even since the earliest stages (Berg
1915; Solnetsev 1948; Neef 1967; Angelstam et al. 2013; Bastian et al. 2015). The
landscape's relief can be interpreted in various ways: (i) as a legacy of former time
processes (e.g., sedimentation in Pleistocene), (ii) as an indicator of geological
structure affecting nutrient supply, and (iii) as a binding factor for the present-day
matter flows (e.g., erosion, water migration, and seeds dispersion). The patterns in a
landscape surface that are of interest to landscape ecologists may also be interpreted
as emergent properties of particular combinations of surface heights and slopes
across the study area (McGarigal and Cushman 2005). Weaver and Perera (2004)
criticized the models simulating the fate of each pixel independently (termed pseu-
dospatially explicit by Malanson 1996) and argued that accounting for spatial
dependence creates more reliable output for analyzing spatial patterns and relating
those patterns to ecological processes.

Dominant landscape-ecological models focus either on hierarchical organization
of one natural geocomponent (in most cases, plant cover or land use which can be
easily detected from satellite imagery or relief detected from digital elevation model
[DEM]) or describe relations at a single hierarchical level. In contrast, we propose
the tool to reveal multiple independent hierarchies based on *interactions* between

properties of natural geocomponents. The challenge is to distinguish contributions of effects generated by *several* higher-order geosystems that can impose constraints on properties of the focus-level system. In this chapter, we examine emergent effects imposed by combination of elements of the higher-scale geosystem to the focus-level geosystem. A set of correlating properties of soil and vegetation is hypothesized to vary in space within the constraints imposed by combination of landforms in some neighborhood.

Hierarchical levels are not postulated a priori but are induced based on evaluation of linkages between the properties of the focus unit and spatial emergent properties of higher-order geosystem. Geosystems of different rank orders are believed to be generated by processes acting at different space and time scales. Soil and vegetation can reflect constraints from the higher-order geosystem by certain groups of attributes, and not necessarily by the whole set of attributes. Each attribute can receive signals from several rank-orders of geosystems simultaneously. The group of attributes governed by the same higher-order geosystem forms partial geocomplex that indicates manifestation of a single ecological process. We show the effect of topographically induced landscape pattern on present-day processes manifesting in variety of soil–plant relationships. In other words, we hypothesize that remedy of palaeo-processes manifesting in relief morphology determines present-day processes.

16.2   Material and Methods

The research was performed in the middle taiga of East European plain (the southern Arkhangelsk region of Russia). The study area (the Zayachya River basin, 154 km^2) is located within the Ustyanskoye plateau composed of Permian sedimentary rocks (the Sukhona formation). Elevations range from 100 to 175 m a.s.l. Physical environment of the landscape was shaped by morainic and limnoglacial accumulation in Riss period of Pleistocene. The Zayachya River belongs to the Severnaya Dvina basin. Flow directions of the first- and second-order streams are affected strongly by the system of lineaments stretching northeastward and northwestward (Khoroshev 2003). Development of the Zayachya terraces with alluvial deposits overlaying Permian marls and morainic loams dates back to the late Würm. Terraces and floodplains of the small rivers were formed in Holocene (Avessalomova et al. 2016).

The relatively high humus content in soils indicates high concentrations of Ca and Mg due to close to surface carbonate morainic loams or Permian marls, in most cases on valley slopes. The most typical zonal soil process—podzolization—is inherent for substrates where loams are covered by 30–70 cm thick sandy layer inherited from Pleistocene glacial lakes. Oligotrophic mires occupy the central sections of the flat interfluve areas. Transect stretching from the bog to the valley slopes shows gradual substitution of *Pinetum eriophoro-sphagnosum* communities on Histic Gleysols or Histosols by *Piceetum myrtilletosum* forests on Haplic Podzols and further—by *Piceetum oxalidosum* forests on Umbrisols or Rhendzic Leptosols. In secondary forests after clear-cutting aspen dominates on the most nutrient-rich soils, birch—on soils with medium nutrients supply, and pine—on the poorest soils with thick sandy layer. During recovery succession all of them are replaced by *Picea abies*, except for the units on sandy terraces where pine preserves its domination.

Field data were collected at 184 sample plots included description of plant cover, soils, landforms (genesis, shape, slope angle, and aspect). Phytocoenosis was described by five groups of attributes: abundance of species in the tree storey (10 variables), shrubs (10 variables), low shrubs (9 variables), herbs (50 variables), and mosses (4 variables). Soils were characterized by three groups of attributes: thickness of genetic horizons (11 variables), Munsell color measured over the interval 5 cm (30 variables), texture measured over the interval 5 cm (10 variables).

We used topographic map (1:50,000) to compose digital elevation model (DEM) with resolution 400 m using triangulation technique. We deliberately focused on low-resolution DEM to omit from examination fine details of relief (small rills, oxbows, and debris slopes, etc.) and to concentrate on mesoscale landforms (e.g., ravines, floodplains, and terraces) and related contrasts in soil and plant cover. Each sample plot was considered to be representative for the square operational territorial unit (OTU) with linear dimension 400 m. ArcView 3.2a and Fracdim software were applied as tools to calculate four morphometric attributes of relief at a range of spatial extents centered on the focal site provided by field data: standard deviation of elevations (vertical dissection, VD below), total length of valleys (horizontal dissection, HD below), vertical curvature (VC), and horizontal curvature (HC). We calculated landscape variables in a moving window with linear dimension ranging from 1200 to 6000 m with step size 800 m. These indices are believed to characterize intensity of lateral transfer at the site and general drainage conditions.

The following procedure below referred to as "multiscale analysis of landscape structure" (MALS) includes several steps (Khoroshev 2019).

At step 1 we performed reduction of dimensionality critically needed because of a great number of landscape properties. Strong deviation of raw data from normal distribution required to apply nonparametric techniques and nonmetric multidimensional scaling (NMDS) as a method with no restrictions for normality and nonlinearity. For each group of attributes, we calculated nonparametric Gamma correlations and converted correlation (r) matrix to distance (d) matrix by equation $d = 1 - r$. We applied multidimensional scaling to distance matrix to derive sensitivity ($a_{1i} \ldots a_{4i}$) of each variable to axes (*dimensions* in terms of NMDS). To derive appropriate number of axes we plotted the stress value against numbers of dimensions and analyzed the scree plot obtained. For most groups of variables, smooth decrease in stress values appeared to level off to the right of number of dimensions "4." Therefore, four-dimensional configuration was assumed to be appropriate. Then we calculated coordinates of sample plots (i.e., OTU provided by field descriptions) on the axes of ecological factors by solving equations system:

$$y_i^{\,j} = a_{1i} x_1^{\,j} + a_{2i} x_2^{\,j} + a_{3i} x_3^{\,j} + a_{4i} x_4^{\,j} \tag{16.1}$$

where y_i^j is the known value of variable i measured in sample plot j; $a_{1i}...a_{4i}$ are the sensitivity coefficients for variable i in relation to axes 1, 2, 3, or 4; and $x_1^j...x_4^j$ is the coordinate value for the sample plot j.

The axes were rationalized as sensitivity to ecological factors according to ecological requirements of species or horizons. For example, if the thickness of humus horizon is scoring low on the axis and that of podzolic horizon scoring high, the axis was interpreted as base saturation depending strongly on properties of soil-forming deposits. The opposite scores for podzolic and peat horizons indicate degree of water mobility.

At step 2 unified data base was composed. It included attributes of the studied landscape units (expressed by coordinates of each sample plot on the NDMS axes) and morphometric relief attributes in a hypothetic higher-order geosystem. Normal distribution of NDMS axes values allowed performing Principal Components Analysis (PCA) to distinguish groups of landscape attributes with various contributions of internal (focus-level) and external (higher-level) factors of variability. PCA enabled us to derive orthogonal "super-factors." Suppose, the first "super-factor" controls attributes varying in concordance with the relief properties of, say, the A higher-order geosystems. The second "super-factor" controls the group of attributes that ignore properties of the A geosystem but obey properties of the higher-order geosystem of another size B. It is also possible that some factors reflect combined emergent effect induced by geosystems of several rank orders. The abovementioned group of super-factors controls attributes that are sensitive to *external* influences generated at the higher level (i.e., *inter-level interactions*). The second group of "super-factors" controls attributes of various geocomponents that vary in concordance with each other but independently of the higher-order geosystems. This testifies the result of *internal* intercomponent relations at the focus scale level (i.e., *intra-level interactions*) and reflects self-development in an ecosystem (soil–vegetation and sediments–soils relationships, etc.). The third group involves attributes that describe single geocomponent (e.g., vegetation) and are related to each other (e.g., layers of phytocoenoses subject to simultaneous changes during succession) but insensitive neither to other geocomponents (e.g., soil) nor to higher-order geosystems. Of course, each attribute of landscape unit to some extent can be sensitive to any of three groups. However, PCA provides the opportunity to range factors by their contribution to spatial variability of geocomponents as well as to distinguish properties insensitive to this or that group of factors. Each factor reflects separate effect of some ecological process.

At step 3 for each attribute we composed two response surface regression (RSR) models [Eq. 16.2.1, 16.2.2] and compared their quality in order to evaluate contributions of internal and external factors.

$$y = a + b_1 x_1 + b_2 x_1^2 + b_3 x_2 + b_4 x_2^2 + b_5 x_1 x_2 + ... + b_m x_n x_k \pm \varepsilon \quad (16.2.1, 16.2.2)$$

where y is the coordinate of a sample plot on the NMDS axis, x_n is the value of internal [Eq. 16.2.1] or external [Eq. 16.2.2] PCA "super-factors."

The equation (16.2.1) relates the attribute to the combined effect of all the internal factors independent on the higher-order geosystems. The equation (16.2.2) relates the attribute (e.g., sensitivity of herbs to water supply) to the combined effect of all the external factors depending on combinations of landforms in the higher-order geosystems. In each equation coefficient of determination r^2 describes proportion of explained variance. By comparing coefficients and statistical significance (p-value) of equations we obtained the opportunity to answer the question: whether it is necessary to consider broad-scale information (e.g., relief of heterogeneous landscape) in order to explain spatial variability of attribute of interest. For example, the possible result can be as follows: to predict abundance of shrubs it is enough to have information about composition of tree layer; but in order to predict thickness of podzolic horizon one needs information on terrain ruggedness in surroundings which is critical for soil water regime and cations leaching. By this we separated contributions of external and internal factors.

Step 4 is aimed at determination of relevant scale level to explain variability of topography-driven attributes. NMDS coordinates were related to morphometric attributes of relief in a square neighborhood of each pixel provided by field description. RSR equations were composed separately for each linear dimension of a hypothetic higher-order geosystem (3, 5, ..., 15 times as large as the linear dimension of OTU):

$$y = a + b_1 x_1 + b_2 x_1^2 + b_3 x_2 + b_4 x_2^2 + b_5 x_1 x_2 + \ldots + b_m x_n x_k \pm \varepsilon, \qquad (16.3)$$

where y is the value of attribute (NMDS coordinate), x_n is the morphometric attribute of relief in the higher-order geosystem (VD, HD, VC, and HC).

Comparison of equations composed for various size of a higher-order geosystem provides the following opportunities:

- To choose the equation with the highest coefficient of determination and, hence, to determine "*resonance scale level*" of the higher-order geosystem that affects the OTU state
- To clarify whether one or several scale levels are critical for the OTU state
- To identify the set of topographic variables that serve as statistically significant predictors for the OTU state
- To determine whether the predictor correlates with the dependent attribute positively or negatively

Note that to compose regression models we did not use raw field data but coordinates on MDS axes. They are used as surrogates for the values of ecological factor that exerts similar influence on a number of correlating initial raw variables. It follows that selection of the "resonance scale level" with the better explanatory power enables us to reveal relevant characteristic space scale for the ecological process which is critical for the corresponding group of landscape attributes.

At step 5 we tested the hypothesis that the combined effect of several higher-order geosystems (step 3) contributes to spatial variability of OTU attributes more effectively than individual effects of each higher-order geosystem (step 4). It was assumed that we inevitably lost some amount of information while calculating factor values at step 2. In contrast, we preserve information while composing equations at step 4 separately for each hypothetical "resonance scale level." Suppose that equation for the combined effect of several hierarchical orders (step 3) explains more variability of the attribute as compared to any of the equations explaining the individual influence of higher-order geosystem (number of predictors being the same—four in each equation). This testifies that, despite loss of information, *combined* effect of several higher-order geosystems is more significant than *individual* contribution of any higher-order geosystem. For example, abundances of nutrient-sensitive species in OTUs depend both on redistribution of nutrients among micro-landforms and on redistribution between the interfluve area and the river valley. Alternatively, regression equation for a single higher-order geosystem can have better quality than equation for the combined effect. By so doing, we conclude that the corresponding group of attributes obeys to ecological processes on either a single or several hierarchical levels.

At step 6, the purpose was to create a series of cartographic models that would take into consideration both concordances in spatial distribution of landscape attributes and multiplicity of higher-order geosystems. At previous steps we identified relevant spatial parameters of higher-order geosystems for each group of attributes. Hence, it became possible to use these parameters in identifying topographical units that could be treated as the higher-order geosystems for OTU as well. We performed classification of pixels (OTUs) by a set of four abovementioned morphometric relief attributes in a "resonance" square environment determined at step 4. Multistructural organization of the landscape forces us to compose a series of maps. Each of these maps shows *partial geosystems, or geocomplexes* (e.g., water sensitive and nutrient sensitive) shaped by similar specific system-forming process with varying degree of manifestation across a landscape. Partial geocomplexes reflect the variability of individual geocomponents with due consideration given to relations with the remaining geocomponents depending on integrating function (Solon 1999).

However, even for a single type of partial geocomplexes with a uniform integrating function it is necessary to consider multiplicity of governing scale levels. Hence, classification of OTUs could be based on a set of morphometric attributes calculated for all the significant scale levels. For example, suppose that water-dependent properties of soil and vegetation in OTU correlate with relief at two scale levels: in the closest environment with linear dimension (LD) 400 m and in the distant environment with LD 2000 m. Then we use attributes of both environments to classify pixels. Statistically significant attributes were selected using forward stepwise method. Discriminant analysis was applied to determine relevant number of classes which should distinguish the maximum possible percentage of OTUs characterized by soil and vegetation properties.

Naturally, it is almost impossible to provide field description for every OTU within the study area. Therefore, we classified OTUs provided by field description by those attributes of soils and vegetation that appeared to be sensitive to the same ecological factor. Separate classifications were performed for water-sensitive and nutrient-sensitive attributes. Then number of classes was used in discriminant analysis as a grouping variable. Morphometric attributes of relief in significant environment were taken as response variables. Statistically significant morphometric attributes were selected using forward stepwise method. By this, we predicted the soil-vegetation class with the highest membership probability for each OTU with known combination of relief properties. Finally, we derived relatively homogeneous areals with similar attributes of soils and vegetation. Homogeneity within each areal is determined by similar combination of landforms in a set of higher-order geosystems.

Step 7 included quantitative evaluation of uncertainty in delineated landscape units. At the previous step we followed the probabilistic logic: "under the given combination of landforms soil-vegetation class X has the highest probability of occurrence." However, one can easily suppose local landform combinations that permit occurrence of several soil-vegetation classes with similar probabilities. To calculate uncertainty of class prediction we applied Shannon formula:

$$H = -\Sigma p_i * \log(p_i) \tag{16.4}$$

where H is the uncertainty of class membership and p_i is the probability that class i will occur under current combination of landforms in significant neighborhoods.

High uncertainty in OTU can be interpreted as a high potential of multiple stable states. By so doing we delineate both areas with deterministic relations between abiotic and biotic geocomponents and ecotone areas with the highest probability of expansion for this or that soil-vegetation class.

16.3 Results

For each group of variables, we derived four MDS axes to ensure the same number of degrees of freedom in response regression surface equations. Similarly, four morphometric attributes of relief were calculated for square neighborhood. The first four axes explained 60–80% of variance for most raw field data. The first and second axes which are responsible for most variance were interpreted as sensitivity to water supply and nutrient supply. We found evidence that nutrient supply has the highest significance for trees, shrubs, and herbs, while water supply has for low shrubs, mosses, thickness of soil horizons, and soil colors. As an example, we illustrate the physical content of axes for trees and soils. Variables scoring high on one of these axes included *Pinus sylvestris* and those scoring low included *Picea abies*. This corresponds to the gradient from the nutrient-poor sandy soils to the nutrient-rich loamy ones. The values of the first axis for trees correlate with the second axis for soil horizons which shows opposite spatial patterns for thickness of humus and podzolic horizons. The second axis for trees represents water supply gradient which is detected by abundance of *Alnus incana* and *Pinus sylvestris* on poorly drained soils and dominance of *Picea abies* or *Populus tremula* on well-drained soils. This axis has good correlation with the first axis for soil horizons (peat vs. humus and podzolic horizons).

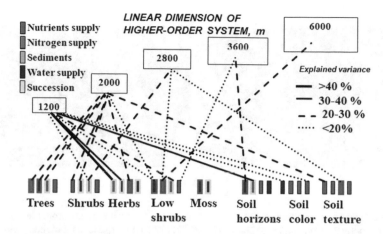

Fig. 16.1 External effects of linkages between soil-vegetation properties and morphometric characteristics of the higher-order geosystems relief. Percentage of variance explained by response regression surface Eq. (16.3)

To transfer information from the hypothetical higher-order scale levels to the focus level we compared quality of RSR models [Eq. 16.2.1and 16.2.2]. VD, HD, VC, and HC were repeatedly computed at a range of spatial extents around each location using a moving window with linear dimensions 1200, 2000, 2800, 3600, and 6000 m. Values were assigned to the central pixel of the square and used as explanatory variables for each NMDS axis coordinates in RSR model. Coefficient of determinations R^2 in equations for each size see in Fig. 16.1. By this we managed to identify peaks of R^2 corresponding to the supposed sizes of hypothetical higher-order geosystems. Similar approach was applied by Cushman and Huetmann (2010), Miguet et al. (2016) to explain the relationship between the biological response and landscape structure for each spatial extent and to find the scale at which the variable most strongly responds.

Figure 16.1 shows evidence that most attributes are scale sensitive and exhibit the best response to the geosystems with linear dimension either 1200 m or 2000 m. Higher-order geosystems with linear dimension 1200 m provide constraints for attributes of herbs, shrubs, low shrubs, thickness, and color of soil horizons (Fig. 16.2a). Note that the list of attributes sensitive to the "1200 m" geosystem involves first axes for herbs, low shrubs, soils colors which are indicative of either water or nutrient supply. It means that the higher terrain ruggedness results in decrease of groundwater level and less occurrence of temporary perched soil water. Herbs and low shrubs respond to water supply by substitution of communities with *Orchis maculata, Luzula pilosa, Linnaea borealis* by those with *Asarum europaem, Gymnocarpium dryopteris, Vaccinium vitis-idaea*, etc. Trees and mosses ignore processes inherent for this level. HD is close to zero in the central sectors of the interfluves, but high in the belt of runoff formation in the upper sectors of gentle slopes. HD decreases as the individual small streams converge and form the deeper and wider side valleys. The increase of VD from the water divides toward the valleys

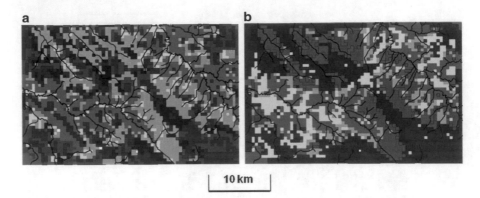

Fig. 16.2 Maps of geosystems based on deterministic model assuming that properties of geosystems are strictly determined by the landforms combination in the square neighborhood with linear dimensions: (**a**) 1200 m (properties of herb, low shrub, shrub layers, soil horizons), (**b**) 2000 m (properties of tree and shrub layers)

and from the upper reaches of the rivers toward the lower ones indicates growing density of joints induced by current neotectonic uplift. In the upper reaches, marls are hidden below the morainic and limnoglacial sediments resulting in lower curvatures, the soils being less base saturated. Commonly, higher occurrence of rich soils on marl and carbonate loams indicates better drainage conditions. Deep dissection favors frequent emergence of carbonate-rich groundwater and development of wet and nutrient-rich habitats with dominance of *Filipendula ulmaria, Aconitum septentrionale, Aegopodium podagraria, Trollius europaeus*. Thus, "1200 m" geosystems correspond to patterns generated by erosion which favored drainage and increased contrast between nutrient-poor and nutrient-rich habitats.

Tree and shrub layers demonstrated the most obvious response to the properties of the larger "2000 m" geosystems which show explicitly step-like organization of relief (Fig. 16.2b). It is induced by southwestward inclination of Permian strata transformed by joints. The lower blocks in the southwestern sector, in contrast to the northern and eastern sectors, were undergone to the more well-manifested influence of the dammed lakes during Würm glaciation and later by fluvioglacial and alluvial accumulation of sand. Poorer nutrient supply is indicated by dominance of pine over spruce, high occurrence of *Juniperus communis* and better development of podzolic process without gleyization. Landscape pattern formed by "2000 m" geosystems can be referred to as strip-like while "1200 m" geosystems generate dendritic pattern.

Now we turn to the results of separating contributions of *internal* and *external* effects. Eight mutually independent "super-factors" explained 54% of variance for the whole set of NMDS axes. Among them the first, second, sixth, and seventh super-factors (referred to as "external" below) turned out to be highly sensitive to relief surrounding focal location (i.e., to higher-order geosystems). Super-factor 1 reflects variability of VD and HD in the distant environment (2800–6000 m) and indicates the high-order neotectonic blocks. Super-factor 2 involves curvatures in the closest environment (1200–2800 m) while super-factor 6 involves curvature in the wider environment. Super-factor 7 reflects correlating variability of VD and VC in the closest environment with linear dimension 1200 m. The same super-factors describe variability of some properties of plant and soil cover, for example, the first and the second axes for herb layer being indicative, respectively, to nutrient and water supply.

In contrast, four other super-factors (nos. 3, 4, 5, and 8) were insensitive to relief of surroundings. They have high factor loadings for the first axis for trees, the second axis for low shrubs, the first axis for mosses, the third axis for herbs. From this fact we conclude that internal interactions between these attributes of vegetation layers in situ have the higher significance than broad-scale processes. These super-factors are referred to as "internal" below. Several attributes demonstrated sensitivity both to external and internal super-factors.

For each axis we compared proportion of variance explained by four relief-sensitive super-factors vs. four relief-insensitive ones (Table 16.1). Each external super-factor integrates emergent effects of constraints translated from the higher scale levels. Coefficient of determination R^2 shows contributions of external factors to the total variability of landscape unit attributes. Similarly, summary contribution of relief-insensitive internal super-factors was evaluated.

For example, percentage of variance explained by relief-sensitive external super-factors for the second axis for trees (sensitivity to water supply) accounted for 31% while relief-insensitive super-factors independently explained 40%. For most axes, super-factors explain much less than 100% of variance (although the first axis for herbs accounts for 86%). Percentage of unexplained variance in the range of 20–40% for most axes evidences that many processes are operating at other scales. Most likely, it is critical to consider fine-scale processes. Hence, it is needed to analyze DEM with higher resolution that would reflect matter redistribution among micro-landforms. This issue is beyond the scope of this chapter and effects of spatial resolution for the evaluation of landscape variability have been addressed (Khoroshev et al. 2013).

Percentage of variance of the first axis for herbs (sensitivity to nutrient supply) explained by properties of higher-order geosystems with linear dimensions 1200, 2000, 2800, 3600, 6000 m accounted for 40, 34, 28, 25, 26%, respectively. However, the combined emergent effect of the external super-factors accounted for 59%. Thus, greater figure integrates not only influence of broad-scale processes (e.g., relief-dependent water migration) but also nonlinear effect of relief-dependent properties of soil and plant cover. Hence, proportion of oligotrophic and megatrophic herb species in the landscape depends on nutrients redistribution governed simultaneously by the geosystems of several scale levels. The largest higher-order geosystems (3600–6000 m) were shaped by different rates of Quaternary cover removal depending on rate of neotectonic uplift and resulting in soil enrichment by base cations from underlying marl. The lower-level embracing geosystems (1200–2000 m) contributed by means of the present-day migration of nutrients with surface and subsurface waters along the slopes and thalwegs.

The internal super-factors proved to produce high contribution to the spatial variability of properties linked with feedbacks. For example, the first axis for herbs

Table 16.1 Contributions of external and internal effects on spatial variability of geocomponents properties expressed by axes of multidimensional scaling (MDS)

Properties (MDS axes)	Percentage of variance explained by		
	4 Relief-sensitive (external) superfactors	4 Relief-insensitive (internal) superfactors	Totally 8 external and internal superfactors
D1 herbs	**59**	27	86
D2 herbs	**45**	20	65
D3 herbs	18	**44**	62
D4 herbs	**24**	15	39
D5 herbs	14	21	35
D6 herbs	23	18	41
D7 herbs	**22**	**21**	43
D8 herbs	13	15	28
D1 trees	16	**34**	52
D2 trees	**31**	**40**	71
D3 trees	**39**	20	59
D4 trees	13	9	22
D1 low shrubs	**25**	**39**	64
D2 low shrubs	11	**37**	48
D3 low shrubs	**32**	38	70
D4 low shrubs	12	17	29
D1 shrubs	**39**	37	76
D2 shrubs	**29**	18	47
D3 shrubs	**33**	15	48
D4 shrubs	12	28	40
D1 mosses	7	**24**	31
D2 mosses	**27**	21	48
D1 soil color	14	21	35
D2 soil color	8	19	27
D3 soil color	**22**	9	31
D4 soil color	19	13	32
D5 soil color	11	10	21
D6 soil color	20	15	35
D1 soil horizons	17	**39**	56
D2 soil horizons	**39**	26	65
D3 soil horizons	10	9	19
D4 soil horizons	12	18	30
D1 soil texture	15	13	28
D2 soil texture	3	15	18
D3 soil texture	12	*21*	33
D4 soil texture	5	11	16

Significant equations are in bold letters

(*Majanthemum bifolium, Oxalis acetosella, Linnaea borealis* vs. *Heracleum sibiricum, Carex canescens, Veronica officinalis, Equisetum sylvaticum*) is slightly sensitive to relief of the "2000 m" geosystem but with no any emergent effect of several higher-order geosystems. At the same time emergent effect of the four internal super-factors is rather large and significant. The attributes have poor correlation with various properties of abiotic environment such as groundwater level and soil properties. However, simple explanation was offered by strong linkage with the forest stand age, canopy cover, canopy height, and tree diameter. Obviously, the abundance of this group of herbs is related to the stage of recovery succession and is adapting to the environment conditions formed by evolving tree layer. Surprising correlation with the attributes of embracing "2000 m" geosystems is easily explained by the pattern of cutting in the study area: mature and overmature stands were preserved on the steep dissected slopes with low access, while young and middle-aged stands dominate on the accessible flat areas and the gentle slopes. Hence, we concluded that self-regulation in phytocoenosis is one of the mechanisms independent of higher-order geosystems but responsible for spatial variability.

To compose series of map of partial water-sensitive and nutrient-sensitive geocomplexes we applied probabilistic approach. We performed classification of units by water-sensitive properties linked by strong Spearman rank correlation. Eight soil-vegetation classes were identified. Then we used discriminant analysis to test the hypothesis that each class occurs under specific combination of relief attributes imposed by the higher-order geosystems as well as by slope gradient in the OTU. Forward stepwise method ($F = 1$) was used to identify significant relief morphometric attributes to distinguish eight soil-vegetation classes. The list of significant attributes used as independent variables included VD in the geosystems of several scale levels as well as HC in the "2000 m" geosystem. Only 32% of the units provided with field descriptions were correctly classified (Wilks' Lambda = 0.41441). This is not a surprise since most water-sensitive properties exhibited high contribution of the internal factors but low contribution of the external ones. The map in Fig. 16.3 shows the location of eight soil–vegetation classes that were predicted based on the highest probability of occurrence under the given relief conditions. However, it shows location more or less perfectly only for classes 2, 6, 7, 8 that exhibited the highest percentage of correct discrimination. In other words, each of these classes of water-sensitive soil-vegetation properties can occur under specific relief conditions that are not inherent for any other class.

Class 2 (Fig. 16.3) is typical for the areas with the highest vertical dissection due to the outcroppings of Permian marls close to water streams. In well-drained habitats secondary aspen forests dominate, soils have thick humus and very thin podzolic horizons. Absence of *Vaccinium myrtillus* (commonly typical for the taiga) indicates low level of groundwater. Hence, water regime resembles the southern taiga rather than in the middle taiga.

Class 6 (Fig. 16.3) with elevated abundance of *Alnus incana* without moss cover occurs in similar geological conditions but with less vertical dissection. Relatively well-drained and nutrient-rich soils were appropriate for plowing which is evidenced by frequent occurrence of Anthric horizons in Anthri-Umbric Albeluvisols.

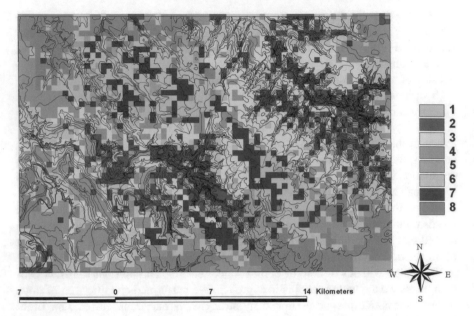

Fig. 16.3 Multiscale cartographic model of the most probable soil-vegetation classes based on water-sensitive properties. Classes 1–8—see in text

Class 7 (Fig. 16.3), in contrast, occurs at large distances from the valleys in the areas with low vertical dissection. It is dominated by *Betuleto-Piceetum myrtilloso-sphagnosum* or *Betuleto-Piceetum herboso-sphagnosum* forests on Gleyi-Histic Albeluvisols.

Class 8 (Fig. 16.3) is located within the wide interfluves with very low vertical dissection in the upper reaches of the basin. Permian marls are covered by morainic loams 20–30 m thick and by peat up to 1–2 m thick. Draining effect of valleys is not manifested. *Pinetum sphagnosum, Pinetum ledoso-sphagnosum, Pinetum myrtilloso-sphagnosum* communities on Histic Gleysols or Histosols dominate.

Analogously, we identified eight soil-vegetation classes based on a set of nutrient-sensitive attributes and performed discriminant analysis in relation to relief properties. Percentage of correct classification accounted for 43% (Wilks' Lambda = 0.25127). Since most attributes are governed by external factors, the quality of discrimination was higher than that for the water-sensitive attributes. Probabilities of each class occurrence are given in Fig. 16.4. The most probable class for each pixel is shown in Fig. 16.5. However, it is obvious that the current DEM resolution (400 m) cannot catch the influence of fine-scale processes on the nutrients redistribution in the landscape as well as heritage of fine-scale Pleistocene processes that were responsible for sedimentation pattern of nutrient-poor sands over morainic loams. Classes 1, 3, 6, 7 exhibited the highest percentage of correct discrimination.

Class 1 (Fig. 16.5) occupies almost the same group of landforms as the class 2 at Fig. 16.3 for water-sensitive properties. However, its areal is less perforated by the

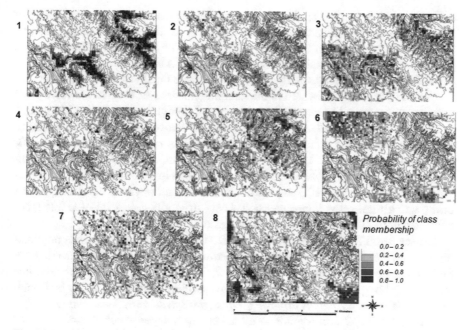

Fig. 16.4 Probability of soil-vegetation classes 1–8 occurrence for the nutrient-sensitive properties at a given combination of landforms in significant square neighborhoods

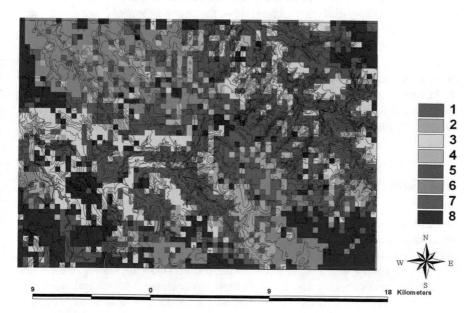

Fig. 16.5 Multiscale cartographic model of the most probable soil-vegetation classes based on nutrient-sensitive properties. Classes 1–8—see in text

other classes. The highest level of the nutrient supply determines high abundance of *Populus tremula* and *Alnus incana* in the tree layer as well as herbs and shrubs that are more characteristic for broad-leaved forests (e.g., *Pulmonaria officinalis, Paris quadrifolia*) or for floodplains and fens (*Filipendula ulmaria, Aconitum septentrionale, Ribes rubrum, Lonycera xylosteum*). Soils have the thickest humus horizons.

Class 3 (Fig. 16.5) is located in the marginal portions of the flat interfluve areas with low dissection. The cover of nutrient-poor limnoglacial sands is thick enough to support *Pinetum myrtillosum* and *Pinetum vaccinioso-hylocomiosum* communities with prevalence of typical boreal herbs (*Trientalis europaea, Linnaea borealis, Luzula pilosa, Lycopodium clavatum, Pyrola rotundifolia, Melampyrum sylvaticum*), shrubs (*Juniperus communis, Salix caprea*) and low shrubs (*Vaccinium vitiasidaea, Vaccinium myrtillus*). Gleyization on the contact of sandy and loamy layers limits agricultural activity despite good access from villages. A lot of arable lands cultivated until 1980s were abandoned because of low efficiency.

Classes 6 and 7 (Fig. 16.5) occur on the flat interfluve areas, the former dominating in the wider sectors, the latter in the narrower ones. Poor drainage favors accumulation of peat layer 10–25 cm thick which isolates the herb roots from morainic loams underlying sands. Boreal herbs and low shrubs dominate including hydrophilous ones (*Carex sp., Juncus filiformis, Equisetim sylvaticum*). *Vaccinium myrtillus* is abundant. Class 7 corresponds to a thinner sandy layer which favors increased aspen abundance in secondary forests instead of birch.

It is worth noting that in the interior interfluve areas nutrient-induced patterns (Fig. 16.5) are more heterogeneous than water-induced ones. Water-induced patterns form a series of more or less monotonous belts from the water divides toward valleys. In the deeply dissected sectors, vice versa, nutrient-induced patterns are less diverse. It follows that neither water supply nor nutrients supply can have the higher importance for landscape mapping.

The map of uncertainty (Fig. 16.6) shows in which locations a set of internal and external factors allows occurrence of several possible soil-vegetation classes. Less uncertainty was detected at the map of nutrient-induced patterns, the lowest uncertainty occurring in the most dissected and the least dissected areas. These locations correspond, respectively, to the maximum and minimum possible marls influence.

16.4 Discussion

Our findings show evidence that we have no rationales to confirm the hypothesis about strictly deterministic relations between abiotic environment, soils, and plant cover as it was supposed at the earliest stages of the landscape science development. The results testify that similar relief conditions and corresponding water- and nutrient-induced patterns allow multiplicity of combinations of soil and vegetation properties. Moreover, the model obviously fails to reflect fine-scale mosaics of landforms and sediments. Therefore, evaluation of patterns uncertainty was performed. The explanation of the detected zones of high uncertainty involves several aspects.

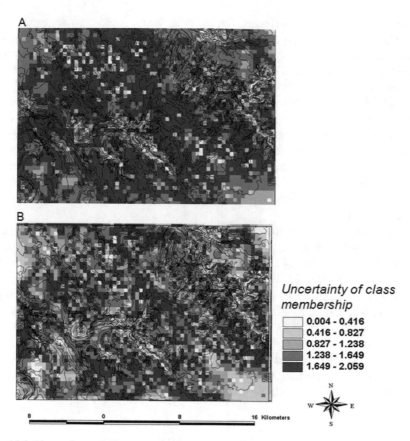

Fig. 16.6 Uncertainty of class membership for the multiscale cartographic models of soil-vegetation classes based on water-sensitive (**a**) and nutrient-sensitive (**b**) properties

First, strict boundary of a landform (e.g., abrupt transition from flat surface to steep slope) not necessarily causes abrupt change in the whole set of properties. One can easily imagine that groundwater level on a flat surface "feels" the neighborhood of a slope not exactly on the boundary but at some distance from it within the marginal portion of a flat area. Therefore, we can expect gradual transition in soil and vegetation properties as we approach the well-manifested topographic boundary. It follows that in a transition zone some water-sensitive properties can resemble slope conditions. At the same time, the other water-sensitive properties look similar to that of a flat area. Hence, high uncertainty can be interpreted as an ecotone where environment (external factor) is more significant than own properties of a unit (internal factors).

Second, types of relations between abiotic environment, soils, and vegetation can vary within the area which shows nonstationarity of intercomponent relations. This phenomenon is of particular importance if relations are expressed by parabolic pattern. In the study area abundance of pine increases in two opposite directions. On the one hand, pine dominates in the wettest habitats (*Pinetum sphagnosum*) in the interior sections of the interfluves. On the other hand, it can also dominate in the driest habitats on the sandy river terraces (*Pinetum cladinosum*) in the deep valleys. Hence, if one tries to relate pine abundance to vertical or horizontal dissection he will obtain zero correlation because the sign of dependence is the opposite in the left and right sections of the "dissection" axis. Therefore, it is necessary to reveal areas with the different types of intercomponent relations. The relevant procedure can use classification of units by regression coefficients in equation linking properties of the focus unit with attribute calculated for its environment using moving window technique (Khoroshev and Aleshchenko 2008). To verify the constancy of relations type under varying grain size it is possible to reveal self-similarity of relations across scales (Khoroshev et al. 2007). Fine-scale mosaics or anthropogenic disturbance can cause important effects which mask broad-scale effects shown by our cartographic model.

Third, high uncertainty indicates the area where bifurcation of landscape state is the most probable if some powerful external factor (e.g., climatic change) exerts influence. The unit can split into several contrast ones because several stable states are possible under the given combination of relief conditions. Moreover, depending on occasional fluctuations (e.g., weather anomalies, anthropogenic disturbance, pest outbreak, and windfall) such a unit can switch to this or that equally possible state. Finally, areas with high uncertainty are the privileged candidates for expansion of communities from their typical topographically determined habitats to the habitats where their survival is also possible but limited by competitors with similar claims. If external conditions tend to become more favorable for one of competitors change of the unit state is highly probable.

16.5 Conclusion

The research showed the efficiency of distinguishing landscape attributes that are governed by internal interactions between geocomponents and those governed mainly by broad-scale processes in higher-order geosystems. Real landscape structure is a superposition of both internal and external factors. Since topography-based landscape mapping is convenient and informative, it is critically needed to identify properly topography-sensitive properties and to determine the relevant scale level of their manifestation.

We identified three principal ways of subordination for the properties of soil and plant cover. First, the property can undergo priority influence of the other geocomponents independently on external factors. Second, the property can be influenced by ecological processes of a single higher-order geosystem that impose strict constraints on the possible range of values. Third, the property can be influenced by emergent effect of several higher-order geosystems, that is, to a set of broad-scale processes.

Our research confirmed the hypothesis that the combined effect of several higher-order geosystems provides the emergent effect for the lower-order landscape unit.

This is in agreement with the idea that cross-scale interactions can generate emergent behavior that cannot be predicted based on observations at single or multiple, independent scales, and the interactions may produce nonlinear dynamics with thresholds (Peters et al. 2004). More generally, this confirms one of the most important attributes of the complex adaptive systems that contains many adaptive components and subsystems nested within each other, giving rise to emergent properties (Messier and Puettmann 2011).

In the study area the information translated from the higher-order geosystems is manifested mainly in drainage conditions that depend on landforms pattern in certain neighborhood and control soil and vegetation properties. Most soil and vegetation properties are governed by not a single scale level. This fact testifies that landscape reflects simultaneous action of a variety of processes with different spatial and temporal scales. The quantitative evaluation of linkages between abiotic environment, soils, and vegetation is highly dependent on size of operational territorial unit. The OTU size chosen in our study proved to satisfy the model of interlevel interactions better for nutrient-induced properties than that for water-induced properties. In other words, our multiscale cartographic landscape model is more relevant to describe nutrient redistribution rather than water supply. Geological and topographic patterns which are responsible for nutrient supply were detected properly enough. To a great extent combination of landforms serves as indicator of the processes which were acting in Pleistocene and resulted in various outcroppings of carbonate-rich sediments depending on relief dissection. To detect water redistribution in details the chosen OTU size is obviously too coarse. Most likely, it means that present-day water redistribution follow to lesser extent the geological patterns but to a much greater extent to fine-scale flows among micro-landforms.

Probabilistic landscape mapping showed the areas with perfect adaptation of soils and vegetation to abiotic environment as well as areas with high probability of several stable states. This way of mapping provides the opportunity to flatten the old contradiction between discrete and continual concept of nature organization. Since both abrupt and gradual transitions between landscape units are possible, mapping uncertainty is the appropriate technique to show both kinds of boundaries. Gradual change of soil or plant cover in space can indicate change in spatial environment (i.e., higher-order geosystem) though own topographic and geological features remain the same. However, our study demonstrated its efficiency for topography-dependent properties only. Equilibrium relations between the focus landscape units and the higher-order geosystems dominate in the areas with low relief dissection which is indicated by high uncertainty of soil-vegetation class membership. Increase in relief dissection makes interlevel relations more complex due to the growing effect of fine-scale processes. Whether the proposed mapping technique can be applied to the topography-independent patterns is a matter of further investigation.

Acknowledgments The study was supported by Russian Foundation for Basic Research (RFBR projects 17-05-00447, 14-05-00170).

References

Angelstam, P., Grodzynskyi, M., Andersson, K., Axelsson, R., Elbakidze, M., Khoroshev, A., Kruhlov, I., & Naumov, V. (2013). Measurement, collaborative learning and research for sustainable use of ecosystem services: Landscape concepts and Europe as laboratory. *Ambio, 42*, 129–145.

Avessalomova, I. A., Khoroshev, A. V., & Savenko, A. V. (2016). Barrier function of floodplain and riparian landscapes in river runoff formation. In O. S. Pokrovsky (Ed.), *Riparian zones. Characteristics, management practices, and ecological impacts* (pp. 181–210). New York: Nova Science Publishers.

Bastian, O., Grunewald, K., & Khoroshev, A. V. (2015). The significance of geosystem and landscape concepts for the assessment of ecosystem services: Exemplified on a case study in Russia. *Landscape Ecology, 30*(7), 1145–1164.

Berg, L. S. (1915). The objectives and tasks of geography. *Proceedings of the Imperial Russian Geographical Society, 51*(9), 463–475 (in Russian). See also in Wiens, J. A., Moss, M. R., Turner, M. G., & Mladenoff, D. J (Eds.) (2006). *Foundation papers in landscape ecology* (pp. 11–18). New York: Columbia University Press.

Bolstad, P. V., Swank, W., & Vose, J. (1998). Predicting Southern Appalachian overstory vegetation with digital terrain data. *Landscape Ecology, 13*(5), 271–283.

Borcard, D., & Legendre, P. (2002). All-scale spatial analysis of ecological data by means of principal coordinates of neighbour matrices. *Ecological Modelling, 153*, 51–68.

Burnett, C., & Blaschke, T. (2003). A multi-scale segmentation/object relationship modeling methodology for landscape analysis. *Ecological Modelling, 168*, 233–249.

Cushman, S. A., & Huettmann, F. (Eds.). (2010). *Spatial complexity, informatics, and wildlife conservation*. Tokyo: Springer.

Cushman, S. A., & McGarigal, K. (2002). Hierarchical, multiscale decomposition of species-environment relationships. *Landscape Ecology, 17*, 637–646.

Dorner, B., Lertzman, K., & Fall, J. (2002). Landscape pattern in topographically complex landscapes: Issues and techniques for analysis. *Landscape Ecology, 17*, 729–743.

Drăguţ, L., Walz, U., & Blaschke, T. (2010). The third and fourth dimensions of landscape: Towards conceptual models of topographically complex landscapes. *Landscape Online, 22*, 1–10. https://doi.org/10.3097/LO.201022.

Hoechstetter, S., Walz, U., Dang, L. H., & Thinh, N. X. (2008). Effects of topography and surface roughness in analyses of landscape structure – A proposal to modify the existing set of landscape metrics. *Landscape Online, 3*, 1–14. https://doi.org/10.3097/LO.200803.

Yao, J., Peters, D., Havstad, K., Gibbens, R., & Herrick, J. (2006). Multiscale factors and long-term responses of Chihuahuan desert grasses to drought. *Landscape Ecology, 21*(8), 1217–1231.

Khoroshev, A. V. (2003). Spatial organization of landscapes as a function of the block structure of territories. *Proceedings of Moscow University, Series 5 Geography, 1*, 9–15. (in Russian).

Khoroshev, A. V. (2019). Multiscale organization of landscape structure in the middle taiga of European Russia. *Landscape Online, 66*, 1–19.

Khoroshev, A. V., & Aleshchenko, G. M. (2008). Methods to identify geosystems with a commonalty of intercomponent relationships. *Geography and Natural Resources, 29*(3), 267–272.

Khoroshev, A. V., Eremeeva, A. P., & Merekalova, K. A. (2013). Evaluation of intercomponent linkages in the steppe and taiga landscapes in relation to modifiable spatial unit. *Proceedings of Russian Geographical Society, 145*(3), 32–42. (in Russian).

Khoroshev, A. V., Merekalova, K. A., & Aleshchenko, G. M. (2007). Multiscale organization of intercomponent relations in landscape. In K. N. Dyakonov, N. S. Kasimov, A. V. Khoroshev, & A. V. Kushlin (Eds.), *Landscape analysis for sustainable development. Theory and applications of landscape science in Russia* (pp. 93–103). Moscow: Alex Publishers.

Malanson, G. P. (1996). Modelling forest response to climatic change: Issues of time and space. In S. K. Majumdar, E. W. Miller, & F. J. Brenner (Eds.), *Forests - a global perspective* (pp. 200–211). Easton: Pennsylvania Academy of Sciences.

Malanson, G. P., Zimmerman, D. L., Kinney, M., & Fagre, D. B. (2017). Relations of alpine plant communities across environmental gradients: Multilevel versus multiscale analyses. *Annals of the American Association of Geographers, 107*(1), 41–53. https://doi.org/10.1080/24694452. 2016.1218267.

McGarigal, K., Zeller, K. A., & Cushman, S. A. (2016). Multi-scale habitat selection modeling: Introduction to the special issue. *Landscape Ecology, 31*, 1157–1160.

McGarigal, K., & Cushman, S. A. (2005). The gradient concept of landscape structure. In J. Wiens & M. Moss (Eds.), *Issues and perspectives in landscape ecology* (pp. 112–119). Cambridge: Cambridge University Press.

Messier, C., & Puettmann, K. J. (2011). Forests as complex adaptive systems: Implications for forest management and modeling. *L'Italia Forestale e Montana, 66*(3), 249–258.

Miguet, P., Jackson, H. B., Jackson, N. D., Martin, A. E., & Fahrig, L. (2016). What determines the spatial extent of landscape effects on species? *Landscape Ecology, 31*, 1177–1194. https://doi.org/10.1007/s10980-015-0314-1.

Neef, E. (1967). *Die theoretischen grundlagen der landschaftslehre*. Gotha-Leipzig: Haack.

Oline, D., & Grant, M. C. (2002). Scaling patterns of biomass and soil properties: An empirical analysis. *Landscape Ecology, 17*, 13–26.

O'Neill, R. V. (1988). Hierarchy theory and global change. In T. Rosswall, R. G. Woodmansee, & P. G. Risser (Eds.), *SCOPE 35. Scales and global change: Spatial and temporal variability in biospheric and geospheric processes* (pp. 29–45). Chichester: Wiley.

Peters, D.P.C., Pielke, R.A. Sr., Bestelmeyer, B.T., Allen, C.D., Munson-McGee, S., Havstad, K.M. (2004). Cross-scale interactions, nonlinearities, and forecasting catastrophic events. Proceedings of the National Academy of Sciences USA, 101, 15130–15135.

Sebastiá, M.-T. (2004). Role of topography and soils in grassland structuring at the landscape and community scales. *Basic and Applied Ecology, 5*, 331–346.

Solnetsev, N. A. (1948). The natural geographic landscape and some of its general rules. In *Proceedings of the second all-union geographical congress, Vol. 1* (pp. 258–269). Moscow: State Publishing House for Geographic Literature. (in Russian). See also in J. A. Wiens, M. R. Moss, M. G. Turner, D. J. Mladenoff (Eds.) (2006). *Foundation papers in landscape ecology* (pp. 19–27). New York: Columbia University Press.

Solon, J. (1999). Integrating ecological and geographical (biophysical) principles in studies of landscape systems. In J. A. Wiens & M. R. Moss (Eds.), *Issues in landscape ecology. 5th IALE-world congress* (pp. 22–27). Snowmass.

Turner, M. G., & Gardner, R. H. (2015). *Landscape ecology in theory and practice. Pattern and process*. New York: Springer.

Weaver, K., & Perera, A. H. (2004). Modelling land cover transitions: A solution to the problem of spatial dependence in data. *Landscape Ecology, 19*, 273–289.

Wu, J., & David, J. L. (2002). A spatially explicit hierarchical approach to modelling complex ecological systems: Theory and applications. *Ecological Modelling, 153*, 7–26.

Part V
How Patterns Control Dynamic Events

Chapter 17
Structure and Long-Term Dynamics of Landscape as a Reflection of the Natural Processes and History of Nature Use: The Example of the Northwest of European Russia

Grigory A. Isachenko

Abstract Landscape is understood as a set of *landscape sites* (i.e., the stable part of the landscape described by the landform, upper layer of soil-forming bedrock, and moistening regime) and *long-term landscape states*. Landscape dynamics is regarded as the change of the long-term states over time. We present the main cornerstones of the landscape dynamics concept. Each type of landscape sites has its peculiar set of landscape-dynamic trajectories (series of changes of long-term landscape states). The research was performed in the northwestern European Russia, where the diversity of natural conditions is supplemented by change of state borders, ethnic groups, priorities, and systems of nature use and management for centuries. Based on the field data, we describe the main processes responsible for the long-term dynamics of taiga landscapes: dynamics of forest stands, forest regeneration after fires and windfalls, reforestation on abandoned agricultural lands, and peat bogs changes owing to drainage. The most effective way of study, representation, and simulation of landscape changes is landscape-dynamical mapping, including creation of map series based on the unified system of landscape sites. A series of large-scale maps created for the key studied area (map of landscape sites and long-term landscape states, impact map, map of present-day processes) is presented.

Keywords Landscape dynamics · Landscape site · Long-term landscape state · Northwestern European Russia · Landscape-dynamical mapping

G. A. Isachenko (✉)
Saint Petersburg State University, Saint Petersburg, Russia

© Springer Nature Switzerland AG 2020
A. V. Khoroshev, K. N. Dyakonov (eds.), *Landscape Patterns in a Range of Spatio-Temporal Scales*, Landscape Series 26,
https://doi.org/10.1007/978-3-030-31185-8_17

17.1   Introduction

The discussion about the contributions of natural and anthropogenic factors to the formation and functioning of landscape has never stopped in landscape science. Supporters of the most radical viewpoints in this discussion, although relatively rare nowadays, believe either that landscape is a completely natural object or that it is a product of human activity. At the same time various "intermediate" conceptions are numerous. The author develops his concept dealing with this issue on the basis of landscape-dynamic research, realized in the northwestern European Russia. In this region the diversity of natural conditions is supplemented by change of state borders, ethnic groups, priorities, and systems of nature use and management for centuries.

The studies, carried out in the 1990s, have allowed to formulate the concept of landscape dynamics (Isachenko 2007), substantive items of which are stated below.

17.2   Concept of Landscape Dynamics

First, in each landscape (or natural territorial complex) various components and elements have different *characteristic time scales* (time of full change of object or time of one full cycle at cyclic character of changes). For example, vegetation cover (including forest) changes much faster than landforms under the influence of denudation. Therefore, we can distinguish high-frequency (relatively fast-changing) and low-frequency (slow-changing) components and elements of landscapes. The same applies to the results of human activity in the landscape: some of them have longer duration than the others.

Second, depending on characteristic time scale, in any elementary landscape unit, the stable part, or a *landscape site*, and a dynamical part, which is described by a set of *states* of different duration, are distinguished. The site is described by the basic elements as a landform and the upper layer of soil-forming bedrock, for example, a plain on lacustrine-glacial sand. In similar climatic conditions (e.g., in the temperate-continental taiga of the northwestern European Russia) the main characteristics of natural landform (shape, inclination, exposure, etc.) and the upper layer of soil-forming bedrock (lithologic composition, etc.) unambiguously determine the character and degree of moistening regime and the mode of matter migration (eluvial, trans-eluvial elementary landscape units, etc.).

The longevity of existence of most natural landscape sites is measured by periods of $n \cdot 10^3 - n \cdot 10^4$ years. Typically, the sites maintain their essential characteristics being subjected to widespread areal anthropogenic impacts: forest felling, fires, atmospheric pollution, etc.; a radical change of site properties (or even the destruction of some sites) is generated by man on a relatively small area.

Natural landscape sites can be modified as a result of agricultural use (accompanied by changes in micro-relief, humus horizon of soil, moistening regime), drainage of peat bogs, etc. There are also completely anthropogenic (technogenic) landscape sites as road embankments, dams, weirs, quarries, etc.

Landscape boundaries are drawn on the maps first of all as site boundaries. These boundaries are not absolute but show location of strips with the greatest territorial contrasts. In other words, the characteristics of environment vary more strongly between sites than within the limits of sites. The boundaries of man-modified and especially anthropogenic sites are sharper and quite often linear.

Third, "states" of landscapes (natural territorial complexes) are classified in their relation to the annual cycle. With reliance upon this relation, it is possible to distinguish *short-*, *medium-*, and *long-term states* of landscapes (with duration less than one year, from one year up to ten years, and tens and first hundreds of years, respectively).

The changes of long-term landscape states occur at one to three orders of magnitude faster than those in landscape sites (except for catastrophic changes): this is why the network of landscape sites, classified according to the area natural peculiarities and the scale of mapping, can be seen as a "reference frame" of the territory. Long-term states of natural landscapes, as a rule, correspond to plant communities and some properties of soil profile (e.g., the thickness of humus, podzolic and peat horizons in taiga landscapes).

Fourth, "dynamics of a landscape" is understood as a sequence of all the states of different duration, and also a set of transitions between states. Transitions that have the certain duration also can be examined as states (using larger time scale). Accordingly, it is possible to consider short-term (diurnal, seasonal, etc.), medium-term, and long-term dynamics of landscapes. Long-term changes in landscapes, many of which are comparable in duration to human life, are less studied than the most contrasting seasonal changes in landscapes.

Fifth, irreversible changes of a relief and substrate (i.e., the upper layer of soil-forming parent rock) under the impact of processes with characteristic time scale, as a rule, of more than 1000 years, are examined as *evolution of landscapes.*

Sixth, landscape dynamics, as a rule, is a result of *superposition of processes* of different causality.

Processes of the first category have spontaneous character, that is, occur without any human participation, and sometimes without any opportunity of such participation.

Spontaneous processes can be caused either by exogenic (e.g., waterlogging under influence of climate changes) or endogenic factors (increase of erosion due to neotectonic uplift of the territory). Some processes are determined by self-development of certain natural objects, for example, overgrowing of the surfaces that have been exposed after retreat of a lake or sea. Most spontaneous processes are caused by simultaneous action of several external factors; for example, forest regeneration on clearings is complicated and slowed down by periodic local fires, overgrowing by herbs, and partial waterlogging.

Spontaneous processes can also unrecognizably transform man-made landscape sites after cessation of human activity. This happened, for example, with medieval cities of Angkor (Cambodia) and Machu Picchu (Peru), which had been completely covered by natural vegetation and affected by denudation and processes of soil formation.

The second category of processes is connected to human activity, that is, with various human impacts. Impact is understood as the event caused by external factors inducing relatively fast change of a state of landscape. Three basic groups of human impacts on landscapes are distinguished: *point-source*, *linear*, and *areal impacts*.

Anthropogenic influence can be short term, playing role of initial impetus with subsequent start of spontaneous processes (e.g., natural forest regeneration after clear cutting), or long term (use of the territory as an arable land for centuries).

The effects of various impacts (forest cutting, air pollution, roads construction, recreation, etc.) will be different in various landscapes, depending on landscape site and current state/states.

In developed landscapes the consequences of anthropogenic impacts are always superimposed on natural processes, including their stimulation. Many natural processes (e.g., tectonic movements) are fundamentally irreversible even in large cities with the highest degree of transformation of the natural environment. At the same time, some of the natural processes in highly urbanized landscapes can be minimized or even eliminated by creating man-made landscape sites.

Seventh, each impact can be treated as a starting point of the subsequent *dynamic trajectories of a landscape*. The trajectory represents a sequence of long-term states. The number of possible trajectories of the given landscape as a result of any impact (or during its realization) usually exceeds one.

The stronger the connections in a landscape and the more limited the number of plant and animal species that can occupy ecological niches vacated because of impact, the smaller is the number of trajectories that is possible. In taiga landscapes of northwestern European Russia, the minimal variety of dynamic trajectories is typical for extreme landscape sites, for example, crystalline rocks and oligotrophic bogs.

Eighth, every superposition of the impacts complicates a dynamic trajectory of a landscape and generates various "lateral branches." Therefore, any landscape type at any period of time is represented in space by various states of one or several dynamic trajectories (e.g., different stages of forest regeneration after different-time clearings and fires). Thus, the landscape structure of any developed and nondeveloped territory is always more or less mosaic. This diversity (fragmentation) of landscapes is well visible at space images and cause considerable difficulties for mapping.

Ninth, character and intensity of human impacts on a landscape in each historical period are determined by a set of economic, social, political, and ethnic factors realized in *regional system of landscape management* (*nature use*). To study modern states and tendencies of change in any landscape, it is necessary to analyze its changes at previous historical periods. This approach is close to the concept of *path dependence*, according to which landscapes can be understood only in their historic context (Jones 1991, etc.). According to Roberts (2002) our spatial and environmental understanding of the past must be structured in terms of four conditions: (1) antecedence, (2) change, (3) stability, and (4) contingency.

Hence, the key idea of landscape-dynamical analysis is dividing characteristics of elementary landscapes into two categories: *site* characteristics and *state* characteristics. Sites are relatively stable in time, while states are mobile and change due to spontaneous processes and numerous impacts. Both sites and states can be classified and typified.

17.3   Nature Peculiarities of Northwestern European Russia

The northwest of European Russia is a territory of more than 500,000 km² that had been subjected to the Valdai (Würm) glaciation in late Pleistocene. The glaciation has left indelible traces in a geological structure, landforms, and shape of modern landscapes. These are very complicated (chaotic) relief, wide occurrence of glacial, fluvio-glacial, and lacustrine-glacial deposits, and high abundance of lakes. The region is located at the contact of two largest tectonic structures: the Baltic crystalline shield and East-European plate. Within the limits of the Baltic shield the most ancient (Archean and early Proterozoic) crystalline rocks are exposed; to the south they are covered by a thick layer of Paleozoic and Cenozoic sedimentary rocks.

Most part of the region belongs to taiga zone with three subzones: the northern, middle, and southern ones. Absolute dominance of coniferous boreal forests and oligotrophic peat bogs are characteristic. Coniferous trees are presented by pine (*Pinus sylvestris*) and spruce (*Picea abies, Picea obovata, Picea fennica*); list of deciduous trees includes no more than 15 species.

Present-day landscapes of northwestern European Russia bear not only imprints of natural processes, but also notable legacy of long-lasting (several thousand years long) human activity in the region. The territory for many centuries was an object of geo-politic interests of Russia (and its predecessors), Sweden, and Finland. State borders here were changed many times. Accordingly, human impacts on landscapes during the different historical periods were various, including those that had opposite directions. One of the typical examples is destruction of woods and drainage of bogs for increasing agricultural lands that dominated almost over the end of the nineteenth century (and up to 1930s on the territory belonging to Finland). In the twentieth century the tendency was replaced by overgrowing and bogging of abandoned agricultural lands. Since the mid-twentieth century, the basic influences on landscapes have been substantially connected with needs of large cities, first of all St. Petersburg with five-million population.

Our investigations in the area have resulted in typology of landscape sites for taiga landscapes of northwestern European Russia, including 30 main types of sites (Isachenko 2007). Classes of landscape sites were identified due to peculiarities of the relief and soil-forming bedrocks. Thus, peat bog sites were distinguished with respect to the peat mineralization (trophic status). This typology of landscape sites was applied to landscape mapping in scales from 1:5000 up to 1:500,000.

The classification of long-term states of landscapes takes into account the following characteristics: (1) role of vegetation in formation of a vertical structure and lateral/radial structure of landscape; (2) dominance of various types of vegetation and, accordingly, a degree of manifestation of environment-forming functions (forest, herb vegetation, and so forth); (3) spontaneity of vegetation development or necessity of human control over it; and (4) development of soil formation processes (Isachenko 2007). The number of long-term states is much more than the number of types of landscape sites.

The landscape-dynamic studies in taiga landscapes of northwestern European Russia started in the 1990s with application of a complex of methods: stationary investigations, time-related interpretation of spatial combinations and series, analysis of forest stand structure, ecological-floristic analysis, and comparison of different-time maps and remote images.

In 1991, the Ladoga region Landscape-ecological Field Station was organized as a part of University of St. Petersburg. The station is located 150 km to the north of St. Petersburg, near the Ladoga lake. Observations have been performed at 25 test plots with area ranging from 60 to 2500 m². In 2006, the permanent observations on nature protected areas of St. Petersburg have started. By 2018, the landscape monitoring network includes 55 test plots with area ranging from 100 to 2500 m². Observations on the test plots are carried out with periodicity from two times a year up to once every 5 years.

17.4   Results

The results of our studies provide opportunity to describe the main processes of long-term dynamics in the taiga of northwestern European Russia. Tree species that are strong dominants can be regarded as system-forming elements of taiga landscapes. Special attention was paid to the study of long-term dynamics of forest stands.

17.4.1   Dynamics of Forest Stands

Dynamics of forest landscapes without change of dominating tree species is the first kind of dynamics. The constancy of forest stands structure is characteristic of spruce (*Picea sp.*) forests, growing for a long time without catastrophic influences (fires, cuttings, windfalls), and also for landscape sites with extreme conditions where the pine has no competitors among other tree species (tops of crystalline ridges with thin discrete eluvium, dunes, oligotrophic bogs). The observation results showed that the quantity of undergrowth in pine forests is not always enough for regeneration of dominant tree species. In stably developing spruce forests, unlike pine ones, numerous uneven-age undergrowth of dominating tree species always exists.

Long maintenance of a constancy of tree species composition in small-leaved forests as a whole is observed less often due to smaller life expectancy of dominant small-leaved trees—birch (*Betula pendula, Betula pubescens*) and aspen (*Populus tremula*). Nevertheless, at the several studied test plots small-leaved stands demonstrate no evidence of change in dominant tree species. In some cases, it is connected with specifics of landscape sites (e.g., toe slopes of granite ridges with permanent moistening by ground water), whereas in the others, with absence of seed spruce trees nearby. Such conditions often take place in course of reforestation of abandoned arable lands (see below).

Change of dominant tree species in forest stands is most often observed in the forests regenerated after clear cuttings (less often after fires) on plains on morainic deposits, limnetic clay, and loam. The most common dynamic trajectory here is the invasion of spruce under canopy of fast-growing small-leaved trees (birch and/or aspen). Spruce gradually forms a dense second upper layer and then, replacing small-leaved trees step by step, gets domination in forest stands. Introduction of spruce under the canopy of pine forests, observed almost everywhere on sandy plains, is much slower process since it is suppressed by periodic fires.

Forests at drained sites on a glacial-lacustrine sand or sandy loam and on morainic deposits in last decades are characterized by slowed down replacement of other species (pine, birch, aspen) by spruce. We observed decrease in annual increment in spruce stands, increasing share of damaged trees (including the ones affected by various diseases), and mass fallouts of spruce trees. The reason for these phenomena seems to be comprehensive.

Forest regeneration successions on fire places are characteristic for sites that are periodically exposed to summer droughts—drained plains on sand and loamy sand, tops of granite ridges, etc., predominantly covered by pine forests. During the first years after the fire event, saplings of small-leaved trees (birch, aspen, species of willows—*Salix sp.*) and pine have appeared. In 4 or 5 years after the fire, the density of seedlings and undergrowth has reached a maximum (up to 500 thousand plants per hectare!). In the next 10–15 years, the abundance of undergrowth decreased by an order of magnitude due to self-thinning; herewith the share of young pines gradually increases. In 11–12 years after a fire some specimens of tree plants constitute a forest stand, reaching a trunk diameter more than 6 cm. In 20 years after the fire the young pine stand with more or less noticeable participation of birch is formed with actively increasing biomass.

The coverage of moss-lichen, grass, and dwarf shrub layers has been restored up to prefire rates in 6–7 years after the fire. In 20–30 years after fire, explerent species (mainly *Chamerion angustifolium*) and meadow grassy mesophytes, peculiar to earlier stages of postfire successions, went extinct from plant cover.

Dynamics of forest stands after windfalls is being investigated in spruce forests at different landscape sites. It is worth noting that spruce trees with very small rooting depth (usually less than 0.5 m) are being subject to windfalls very frequently especially since 2010. Windfalls affect basically the senior generations of spruce that is replaced in gaps with new generations (so-called digression phase of gap dynamics). Our results testify that in 10 years the windfall area can increase 15 times as high as before.

On one of the test plots the *Piceetum oxalidosum* forest on morainic plain in 10–11 years has lost about 90% of forest stand and was transformed to so-called windfall complex. The most probable scenario for succession is the formation of young birch forest with participation of aspen, spruce, and broad-leaved trees (maple *Acer platanoides*, etc.). In course of 3 years, the quantity of spruce undergrowth on the area of completely destroyed previous forest stand has increased by 45%; the number of undergrowth of deciduous trees during the same time decreased by 70%.

17.4.2 Afforestation on Abandoned Agricultural Lands

Due to the socio-economic reasons, since the mid-twentieth century, the northern and central areas of European Russia have been subjected to a great change of land use system. One of the results was the termination of agricultural land use on the area comprising thousands of square kilometers (Isachenko 2001). The dynamics of abandoned arable lands, meadows, and pastures can be described by four main long-term states:

1. Meadows with rare saplings of small-leaved trees (grey alder: *Alnus incana*, birch, aspen, willow) and willow shrub, up to 15 years after cessation of use
2. Closed young deciduous groves and patches of meadows with willow shrub (up to 25–30 years after abandoning)
3. Closed stand (15–20 m in height, crown density is more than 40%) with domination of small-leaved trees and, sometimes, with undergrowth of conifers (spruce or pine); the duration of this state reaches decades or even a century
4. Coniferous-small-leaved or small-leaved-coniferous middle-aged forest with some participation of meadow species in ground cover (impossible if coniferous seed trees are lacking in nearby area)

Due to cessation of drainage network functioning, each stage of afforestation can be accompanied by rebogging directed to formation of herb-sedge moors.

According to the research results, overgrowing of abandoned meadows by trees and shrubs can be significantly slowed or even interrupted in case of forming dense tall herb cover (e.g., *Filipendula ulmaria*), which prevents the growth of trees.

If coniferous seed trees are abundant enough close to overgrowing areas in 70–100 years after the cessation of agricultural use, regeneration of small-leaved-coniferous forest with floristic composition close to initial taiga community is highly probable. The cumulative effect of overgrowing of abandoned agricultural land is manifested in increase of area of small-leaved forests in the taiga and hemiboreal zone.

17.4.3   Dynamics of Peat Bogs Under Influence of Drainage

Draining land reclamations of peatlands and boggy forests have been carried out on the area comprising thousands of square kilometers in the northwestern European Russia since the second half of the nineteenth century with the purposes to improve conditions for forest growth, to get new agricultural lands, to exploit peat resources, etc. Long-term consequences of land reclamation are rather various and depend on initial types of peatlands, water discharge from mires, peat thickness, character of topography and hydrography in surroundings, duration of the drainage period, depth and density of a drainage network, and other factors.

In case of successful deep drainage, highly productive coniferous forests are developing on the former mires (eutrophic and mesotrophic, in particular). After the termination of land reclamation works in the end of the twentieth century functioning of drainage networks was not supported, therefore in vast areas secondary bogging (rebogging) is developing. In undergrowth the share of boreal forest dwarf shrubs and grassy meso-hygrophytes decreases, while coverage of grassy hygrophytes, bog dwarf shrubs, sedge, and sphagnum mosses increases.

17.5   Landscape-Dynamical Mapping

Landscape-dynamical mapping is the most effective way of representation and simulation of the landscape changes (Isachenko 1995; Isachenko and Reznikov 1995). The principal idea of this method is creation of map series based on the unified system of natural territorial units (landscape sites). The series of maps includes: basic landscape map, map of impacts on landscapes, map of present-day processes, and landscape-dynamical scenarios maps.

The *basic landscape map* (map of landscape sites and long-term states) serves as a framework for all the maps in the series. It is created on the basis of field survey using remote sensing analysis and other sources (topographic maps, digital elevation models). The taxonomic rank order of landscape units to be mapped is defined both by map scale and features of regional landscape structure.

Landscape sites are indicated by letters of the Latin alphabet. The contours of long-term landscape states are the parts of the contours of landscape sites; every type of state is indicated by figures (Fig. 17.1).

Map of impacts on landscapes is also created based on field studies. Besides, the analysis of old maps, remote images, and historical data is used. Impact is characterized by its age, intensity, and results for different landscape components and elements. Every type of impact requires corresponding methods of cartographic display: contours for areal impacts, lines for linear ones, and symbols and diagrams for point-source ones. We have established the optimal period for mapping of

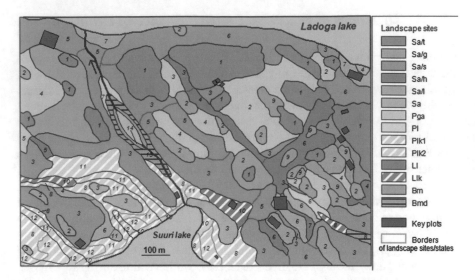

Fig. 17.1 Landscape map of the area near the Ladoga region Landscape-ecological Field Station, University of St. Petersburg (Leningrad oblast)

Landscape sites (indicated by letters of the Latin alphabet): Sa/t, rocky tops and near-top slopes of granite ridges (selga) with outcrops of granite and thin discrete eluvium; Sa/g, gently tops and near-top slopes of selga (up to 5°) with thin continuous eluvium and moraine; Sa/s, gently and medium-steep (5–20°) slopes of selga with deluvium and morainic deposits (up to 1 m); Sa/l, gently foots of selga (3–8°) with deluvium and moraine partially covered by limnetic clay and sand, with permanent moistening by ground water; Sa/h, narrow inner-selga depressions with alterations of granite outcrops and loose deposits of different composition; Sa, low gently slope nondifferentiated selga; Pga, wavy plains (or footsteps) on thin morainic loam and sandy loam, with rare granite outcrops and different drainage; Pl, drained gently inclined (2–5°) terraces on limnetic loam and clay; Plk1, slightly inclined (1–2°) wet terraces on limnetic loam and clay, cultivated in the past; Plk2, gently inclined (2–5°) terraces on limnetic loam and clay, cultivated in the past; Ll, flat boggy terraces on limnetic loam and clay, with thin peat layer; Llk, the same, cultivated in the past; Bm, mesotrophic mires of depressions; Bmd, meso-eutrophic mires of flowing depressions, drained in the past

Long-term landscape states (indicated by figures): *1*, open pine woodlands (*Pinus sylvestris*) with lichens and heath (*Calluna vulgaris*) and granite rock complexes on primitive soils; *2*, light pine forests with bilberry (*Vaccinium myrtillus*), heath, green mosses, and lichens (including postfire woods), on Entic Podzols (podbur); *3*, pine and small-leaved-pine (*Betula pendula*, *Populus tremula*) forests with bilberry and grasses, on Cambisols; *4*, spruce (*Picea abies*) and small-leaved-spruce forests with bilberry and green mosses, on Umbric Albeluvisols; *5*, spruce and small-leaved-spruce grassy-bilberry and grassy-woodsour (*Oxalis acetosella*) forests, on Umbric Albeluvisols and Cambisols; *6*, small-leaved grassy-bilberry forests on Cambisols soils; *7*, small-leaved grassy-fern forests on Gleyic Umbrisols; *8*, small-leaved grassy forests of overgrowing meadows on Gleyic Umbrisols and Gleyic Histosols; *9*, pine and birch-pine swampy dwarf shrub-sphagnous forests on Histosols; *10*, willow shrub (*Salix sp.*) with hygrophytic herbs on Gleyic Umbrisols and Gleyic Histosols; *11*, alder groves (*Alnus incana*), willow shrub, and grass-forb meadows on Gleyic Umbrisols; *12*, tall herb (*Filipendula ulmaria*, etc.) and grass-forb meadows on Gleyic Umbrisols; *13*, meso-eutrophic bogs with birch, black alder (*Alnus glutinosa*), willow shrub, hygrophytic herbs, and sedge on Histosols; *14*, open mesotrophic mires with sphagnum and sedge

impacts on taiga landscapes as 50–100 years. For example, this time is needed for regeneration of forest vegetation after clear cuttings and crown fires. Note that superposition of impacts, which is observed in most taiga landscapes, should be taken into account.

The map of impacts on landscapes of the territory near Ladoga region Field Station, University of St. Petersburg, contains information about square impacts on landscapes during the last 50 years (this term approximately corresponds to the period when land use system has been greatly changed) (Fig. 17.2). Every type of impact, as well as one case of impacts superposition, is displayed on the map in peculiar color. Most impacts were initiated by human activity or its cessation. The map also shows that about half of the territory has no significant marks of impacts over the last 50 years.

Map of present-day processes in landscapes to a great extent is a result of synthesizing two previous types of maps—the data of observations at test plots and the data obtained with use of other tools of landscape-dynamic studies (see above). On the map of study area in the northwest Ladoga region, 12 types of processes are represented (Fig. 17.3). As a rule, every long-term process manifests itself in different components (elements) of landscape according to their characteristic time scale; no landscape process can be completely described by single quantitative parameter.

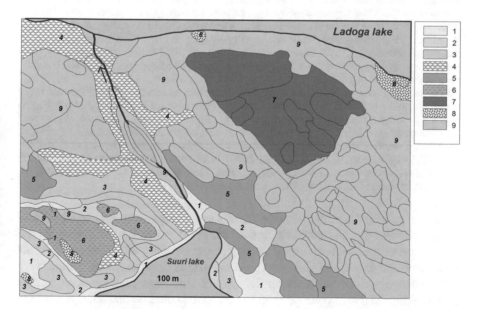

Fig. 17.2 Impacts on landscapes of the area near the Ladoga region Landscape-ecological Field Station, University of St. Petersburg (Leningrad oblast), for the period of last 50 years (indicated by figures)
Use of the territory for agriculture: 1, up to late 1940s; *2,* up to 1960s; *3,* up to 1990s; *4,* selective cuttings of different time; *5,* ground fires; *6,* selective cuttings and ground fires; *7,* crown fire in 1970s; *8,* excessive recreation; *9,* absence of impacts. The borders on the map correspond to the borders of landscape sites and long-term landscape states (see Fig. 17.1)

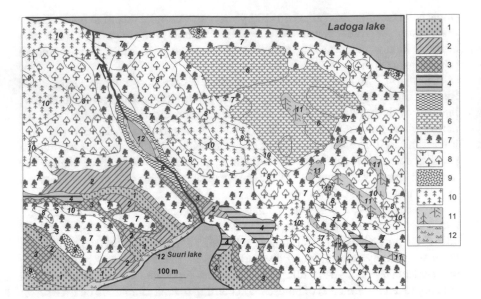

Fig. 17.3 Present-day processes in landscapes of the area near the Ladoga region Landscape-ecological Field Station, University of St. Petersburg (Leningrad oblast) (indicated by figures)
Processes on abandoned agricultural lands: *1*, transformation of grass-forb meadows to tall herb meadows, with young growth of small-leaved trees and willow shrub (first stage of overgrowing); *2*, unclose young small-leaved forests and/or willow shrub altering with meadow plots (second stage of overgrowing); *3*, small-leaved grassy forests growth (third stage of overgrowing); *4*, overgrowing by small-leaved trees and willow shrub combined with bogging-up; *5*, intensive bogging-up with the death of tree stand
Processes in landscapes, not used for agriculture: *6*, postfire successions on selga tops and upper slopes (intensive growth of pine and dwarf shrub-greenmoss cover forming); *7*, growth of pine and/or small-leaved trees combined with increasing of grassy coverage; *8*, growth of pine with stabilization of lower layers; *9*, growth of pine and small-leaved trees combined with degradation of lower layers owing to recreational press; *10*, growth of spruce and gradual replacement of small-leaved trees by spruce; *11*, growth of pine and/or birch combined with slow bogging-up; *12*, growth of sphagnous cover and overgrowing of lakes. The borders on the map correspond to the borders of landscape sites and long-term landscape states (see Fig. 17.1)

Thus, the principal method of mapping processes in landscapes is quality background realized by color and hatching. So far, as landscape for the estimated time periods can be subjected to several impacts, the present situation here can be characterized by superposition of several processes (trends). For example, overgrowing of abandoned agricultural lands on flat moistened terraces on clay may combine with bogging-up (No. 4 on the map in Fig. 17.3). It is obvious that the legend of the map is in detailed correspondence with the processes in plant cover that respond quickly to various impacts and can be easily observed. As for the study area, the significant part of observed processes (5 of 12) is caused by abandonment of agricultural lands.

17.6   Conclusions

Application of the concept of the landscape-dynamic analysis and methods of landscape-dynamical studies gives the results allowing to study more deeply natural and anthropogenic change of landscapes and to respond to the challenges of modern practical tasks of territorial development and environmental assessment. Further field research will provide revealing of new dynamical trends and long-term states of landscapes of the studied region, specifying their characteristic time scale (intensity of processes) and increasing feasibility of simulation of long-term landscape dynamics.

Acknowledgments The studies are supported by the Russian Foundation for Basic Research (project no. 19-05-01003).

References

Isachenko, G. A. (2001). Desolation processes in the landscapes of the European Russia in XX century. In *Development of European landscapes. IALE European conference proceedings* (Vol. 1, pp. 281–285). Tartu: Institute of Geography, University of Tartu.

Isachenko, G. A., & Reznikov, A. I. (1995). Landscape-dynamical scenarios simulation and mapping in geographic information systems. In *Proceedings of 17th international cartographic conference* (Vol. 1, pp. 800–804). Barcelona: Institut Cartogràfic de Catalunya.

Isachenko, G. A. (1995). Landscape mapping: new possibilities for environmental monitoring. In *Proceedings of 17th international cartographic conference, Vol. 1* (pp. 791–799). Barcelona: Institut Cartogràfic de Catalunya.

Isachenko, G. A. (2007). Long-term conditions of taiga landscapes of European Russia. In K. N. Dyakonov, N. S. Kasimov, A. V. Khoroshev, & A. V. Kushlin (Eds.), *Landscape analysis for sustainable development. Theory and applications of landscape science in Russia* (pp. 144–155). Moscow: Alex Publishers.

Jones, M. (1991). The elusive reality of landscape. Concepts and approaches in landscape research. *Norsk Geografisk Tidsskrift, 45*, 229–244.

Roberts, R. K. (2002). Space, place and time: Reading the landscape. *Journal of the Scottish Association of Geography Teachers, 31*, 39–51.

Chapter 18
Seasonal Dynamics in the Context of Polystructural Organization of Landscapes

Olga Yu. Gurevskikh and Oksana V. Yantser

Abstract Dynamic seasonal aspect in studying landscape structure involves the identification of the order of states changes in time. The specifics of the intercomponent relationships are determined by hierarchical levels of geosystems and lead to different quantitative and qualitative characteristics of seasonal processes. The most visible of them are manifested in plants that indicate changes in climatic indices. To study seasonal development, we applied quantitative tools: method of registrars term, method of integrated descriptions, and method of integrated phenological characteristics. Study of dynamic changes was performed in the Sverdlovsk region, which is located within the three physiographic countries, two geographical zones, and climatic sectors. We identified local and regional characteristics of the seasonal vegetation dynamics, controlled by polystructural organization of the landscape. The study identified indicators and criteria for dynamic seasonal changes for the representative natural systems. Seasonal changes are clearly visible in the lower level landscape units, which are the least resistant and the least durable. Regional level units have more heterogeneous conditions resulting in variability of seasonal processes; the characteristics of seasonal development within their boundaries tend to be more generalized.

Keywords Polystructural organization · Hierarchy · Geosystems · Seasonal dynamics · Quantitative methods · Phenological studies · Pheno-indicator

O. Y. Gurevskikh · O. V. Yantser (✉)
Ural State Pedagogical University, Ekaterinburg, Russia

© Springer Nature Switzerland AG 2020
A. V. Khoroshev, K. N. Dyakonov (eds.), *Landscape Patterns in a Range of Spatio-Temporal Scales*, Landscape Series 26,
https://doi.org/10.1007/978-3-030-31185-8_18

273

18.1   Introduction

The modern landscape structure of a territory is characterized by a complex combination of geosystems of different genetic type and hierarchical level, and is formed under the combined effect of natural and anthropogenic factors. The mosaic character and heterogeneity of landscape structure is generated by the heterogeneity of geographical context. The latter is expressed by complex combinations of different types of natural and anthropogenically modified sites, which are in various states belonging to one or more dynamic trajectories. The elements of landscape structure are characterized by qualitative properties and quantitative parameters that are subject to changes with various periodicity depending on the state of geosystems. From the standpoint of the landscape-dynamic approach, the state of geosystem is regarded as a spatio-temporal uniformity, distinguished by the criteria of consistent composition and ratio of the system-forming elements and processes (Isachenko 2014).

One of the essential characteristics of geosystem is dynamics in general, and seasonal dynamics, in particular. Seasonal dynamics is manifested in the rhythmic change of short-term states within a one-year cycle, lasting from several weeks to several months. Their duration varies from several weeks to several months, due to the annual changes in the thermal and water regimes. Duration is the main measure of geosystem state, the period of the Earth rotation around the Sun being the measurement unit. At a logarithmic scale, the seasonal-dynamic states are short term with a period of $<10^0$ years (Isachenko 2014). Dynamic changes generate prerequisites to the evolutionary landscape transformations. Reversible dynamic changes indicate the ability of the landscape to return to its original state; its rhythmicity is an important property of dynamics, being an integral part of the progressive development (Isachenko 1991; Mamay 1992; Schwartz 2013).

The main method for studying seasonal dynamics is the phenological observations, which allow us to track the rhythmic changes in natural components. Natural components of geosystems differ in rates of changes and the inertness of development. One of the most mobile elements is vegetation, which adapts quickly enough to changing conditions and processes. It can be characterized visually; hence, it acts as a physiognomic landscape indicator of dynamic changes (Hudson and Keatley 2010).

The cycle of seasonal changes in vegetation has a certain average trajectory with certain range of probable specific states of the system. To describe the seasonal rhythm, dynamic parameters are used such as rate, trajectory, phase shifts, and synchronism of phenomena. For these purposes a group of methods for quantitative phenological observations was well founded by V.A. Batmanov: term registrars, yield indicators, and ecometric and descriptive methods (Kupriyanova 2010).

Seasonal states of the geosystem are characterized by *pheno-indicators*—quantitative indicators that mathematically describe the course of dynamics and illustrate the sum of scores that characterize the vegetation state. As indicators of seasonal dynamics complex, we applied phenological indices of geosystems vegetation and tested them in geocomplexes of different rank order in the Sverdlovsk region (Skok et al. 2014; Skok and Yantser 2016).

18.2   Methods

The method of complex phenological indicators means that the phenological state of each plant species is determined by evaluating each accounting unit according to the generative and vegetative development standards (a sequence of phenophases). In the course of long-term research, observations are performed regularly on the same plots (phenofield) located within certain *facies*.[1] The result of field observations is a scoring of the state of each species. The abundance of species observed in a certain phenophase is translated into relative indicators, that is, the percentage of plant species that are in a certain phase on a specific date. The ratio of these indicators, which is the summarized phenological characteristic (SFC) of the vegetation in the natural complex, reflects its phenological state on the day of observations (Fig. 18.1). For each SFC, the average coefficient of its phenological state is calcu-

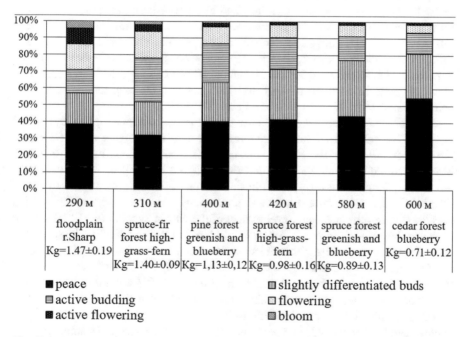

Fig. 18.1 Long-term average annual summarized phenological characteristic of generative development in facies of the taiga belt on June 8

---

[1] The smallest elementary morphological unit of a landscape in Russian landscape science. See Chap. 1 and Glossary for details.

lated ($\bar{K}_g$, generative development; $\bar{K}_v$, vegetative development), defined as the mean score of the phenological state of the site, supplemented by the standard error value ($\pm$ m).

In the case of a repeated examination of the site, an index of phenological speed can be calculated. This allows translating the scores of points per day and comparing the rate of seasonal dynamics of the vegetation in *facies*, which differ greatly in species composition. For the long-term research conducted on the same date, average long-term annual phenological coefficient $\bar{K}_{al}$ is calculated. This provides opportunity to apply this indicator for identifying and justifying the boundaries of the seasons in natural complexes. Differences in $\bar{K}_g$ or $\bar{K}_v$ among individual phenofields are referred to as *pheno-deviation*. The difference between the individual phenofield and $\bar{K}_{al}$ is called *ecological pheno-anomaly*: A ($A = \bar{K}_{al} - \bar{K}$). When studying the long-term dynamics of the facies eco-pheno-deviations in scores are calculated, for convenience of comparison being translated per day. Our study showed that the differences are statistically significant with 95% probability.

Comparison of the average score with phenophases of the standard (winter rest of buds, germination of buds, active budding, fruit blossom and ripening, coloring the leaves) allows us to analyze the state of the seasonal development of the geocomplexes. Quantitative phenological indicators of vegetation objectively characterize the stages of seasonal dynamics, and allow translating the results into time units mathematically, calculating the rate of the processes, and substantiating the differences between geosystems of different types and rank orders. At the same time, these indicators are appropriate for identifying special events in seasonal dynamics. The latter can be described qualitatively and quantitatively by textual and visual characteristics of the state of individual species, range of temperatures, and characteristics of the state of abiotic objects (Table 18.1).

Another widely applied method in phenology is the *descriptive integrated method*. The evaluation of the phenological state of an individual species, for example birch (*Bétula péndula, Betula verrucósa*), is made by counting the units that have passed the threshold value. The registration unit is a healthy individual, without any mechanical damage, which has all the characteristics of a species. As the threshold value we choose a seasonal phenomenon observed on a certain day, a specific event in the seasonal development of the unit of account (e.g., "10 percent of yellow leaves in the crown," "more than half of the leaves fallen"). The state of at least 100 accounting units within the same facies is assessed. Number of units meeting the criterion is counted, and a certain score is assigned. After that we calculate the percentage of units that have passed the threshold value as well as the observation error. The larger is the sample size, that is, the closer is the number of scanned accounting units to the number of all available items, the more accurately the percentage can be determined. The computed errors are necessary to establish the significance of differences in the percentages of the accounting units that passed the threshold value, when comparing geocomplexes. To determine the mathematical validity (reliability) of the distinction, the significance of the difference t is calculated:

Table 18.1 Indicators and criteria for the stages of seasonal development of the northern and middle taiga landscapes (within the boundaries of the Sverdlovsk Region)

Seasonal development stage Phenological criteria indicators	Factors	Taiga belt						Subalpine belt		Tundra belt
		Flood-plain of the Sharp river (H = 290 m)	Piceeto-Abietum magnoherboso-filicosum forest (H = 300 – 310 m)	Pinetum silvativa myrtilloso-hylocomiosum forest (H = 400–420 m)	Piceetum magnoherboso-filicosum forest (H = 420 – 430 m)	Piceetum myrtilloso-hylocomiosum forest (H = 580 m)	Pinetum sibirica myrtillosum forest (H = 600 m)	Betuletum uligini-vacciniosum crook-stem forest (H = 760 m)	Betuleto-Pinetum sibirica fruticulosum forest (H = 790 m)	Plot of moist herb tundra (H = 850 m)
The beginning of vegetational season	Temperature (air) of vegetation start, °C	+2.7	+2.6	+3.7	+3.5	+2.8	+2.7	+1.3	+1.1	+1.0
The brown aspect of the landscape, the first grass seedlings, the appearance of primroses, the swelling of the buds of trees and shrubs	Soil temperatures at a depth of 10 cm, t °C	+1.2	+1.5	+2.0	+1.9	+1.4	+1.1	+0.2	–	+0.1
	Air humidity	89	79	77	86	83	72	92	–	94
	Eco-deviation of the threshold	−2.7± 0.11	−2.7± 0.12	–	−1.7 ± 0.10	–	−0.6 ± 0.13	+0.1 ± 0.17	–	
	strong green haze birch, days	−2.5 ± 0.15						+0.1 ± 0.17		

(continued)

Table 18.1 (continued)

Seasonal development stage Phenological criteria indicators	Taiga belt						Subalpine belt		Tundra belt
Factors	Flood-plain of the Sharp river (H = 290 m)	*Piceeto-Abietum magnoherboso-filicosum* forest (H = 300–310 m)	*Pinetum silvativa myrtilloso-hylocomiosum* forest (H = 400–420 m)	*Piceetum magnoherboso-filicosum* forest (H = 420–430 m)	*Piceetum myrtilloso-hylocomiosum* forest (H = 580 m)	*Pinetum sibirica myrtillosum* forest (H = 600 m)	*Betuletum uligini-vacciniosum* crook-stem forest (H = 760 m)	*Betuleto-Pinetum sibirica fruticulosum* forest (H = 790 m)	Plot of moist herb tundra (H = 850 m)
Phenological indicators of vegetation development Kg	0.6 ± 0.12								
Kv	2.2 ± 0.10								
Early autumn Mass ripening of fruit and hips of the *Rosa acicularis* Lindl., *Sorbus aucuparia* L., *Vaccinium vitis-idaea* L., *Rubus saxatilis* L., and *Padus avium* Mill. — Temperature (air) of vegetation start °C	+9	+9.9	+10	+9.7	+9.5	+9.3	+8.1	+7.6	+7.2
Eco-deviation of threshold 10% yellow leaves in the crown of the birch, days	−2.5	−2.7	–	+0.1	–	−0.3	−2.2 ± 0.3	−2.4± 0.3	–
The beginning dieback of leaves of the *Dryopteris filix-mas* (L.) Schott and *Vaccinium myrtillus* L.	−1.0 ± 0.4						−2.3 ± 0.3		
Phenological indicators of vegetation development Kg	6.5 ± 0.14								
Kv	7.0 ± 0.10								

$$t = \frac{M_1 - M_2}{\sqrt{m_1^2 + m_2^2}}$$

where M_1 is the percentage of accountable item that has crossed the threshold value in one section, M_2 in another section; m_1 and m_2 are, respectively, their error values.

The significance of the difference t is relative value. If t is greater than or equal to 1.96, the difference is mathematically proven at the threshold of acceptance probability $p = 0.95$. If t is less than 1.67, the difference between the sections is within the tolerance limits. To make comparison easier, the obtained percentages M are recalculated into *eco-anomaly*. The latter index, calculated at any segment of the observation route and expressed in days, corresponds to deviations of the phenological index from the conventional phenological zero that is established for the entire territory (Yantser and Terent'eva 2013).

The phenological zero is established as the median value, assuming that the values under consideration are close to the normal distribution. For each variant of threshold, which is used to compare the seasonal dynamics of areas, we calculated the standard deviation (σ), which characterizes the duration of the seasonal phenomenon in time: the larger the σ, the slower the process. To calculate σ, observations are repeated every 2–3 days at each plot.

18.3   Study Area

Seasonal dynamics in each geographic region manifests itself in inherent calendar terms and the alternation of the seasonal phenomena onset. The northern Sverdlovsk region within the medium-altitude mountains of the Northern Urals was selected as a representative area, to illustrate the research method of phenological studies (Gurevskih et al. 2016). This area is characterized by a complex mosaic landscape structure, due to a combination of abiotic conditions (geological, geomorphic, zonal, sectoral, barrier, altitudinal belt, and solar orientation) that manifest themselves at various levels of the geosystems hierarchy.

Low-level geosystems *facies* belong to the same class of middle-taiga type of altitudinal zonation that is characteristic of medium-altitude Northern Urals in the East-Russian subsector. The *facies* under consideration belong to either taiga or subalpine or tundra altitudinal belt.

The *taiga belt* occupies the slopes of massifs and ridges up to an altitude of 750–760 m and is characterized by overmoistening. The dominant vegetation is northern-taiga forest types, the formation of which is associated with overmoistening and temperature inversions in intermountain hollows. The *middle-taiga belt* occupies well-drained convex ridgelines of medium-altitude massifs. *Subalpine belt* is influenced by the severe hydrothermal conditions, strong winds, and thick snow cover; it occupies the slopes in the interval of elevations 760–950 m. Open forests, crook-stem forests, shrubs, tundra, and meadows are typical for this belt. Lichen and moss-berries communities cover outcrops of intrusive rocks. The *mountain-tundra* landscape belt occupies the slopes at a height of more than 950–970 m. Lichen communities cover steep slopes. Gentle slopes and altiplanation terraces are occupied by permafrost tundra, dominated by *empetroso-uligini-vacciniosum* or *arctouso-cladinosum* communities.

The territory of representative area is part of the Shegul'tano-Vyjsky middle-mountain range and the Uls-Vil'vensky middle-ridge areas. Shegul'tano-Vyjsky area in the relief corresponds to flat-topped massif Denezhkin kamen' (1492 m). The Sharpinsko-Bystrinskij ridge, the Zheltaya sopka, and the Kulakovskij, Veresovyj, and Pihtovyj ridges stretch in various directions from the Denezhkin Kamen'. Moderately humid spruce forests prevail, most of them *Piceetum parviherboso-hylocomiosum* or *Piceetum vaccinioso-hylocomiosum*. Uls-Vil'vensky area is composed of more strongly dislocated metamorphic rocks. In the relief, it corresponds to the Main Ural Range, with characteristic peaks—the Mountains of Pallas (1337 m), Humboldt (1410 m), and Lepekhin (1339 m). The largest area is occupied by spruce forests, mainly *Piceetum magnoherboso-filicosum*.

18.4   Results

The model area is located in the Sol'vinsk intermontane depression on the border of landscape areas. After establishing the etalon of the threshold value, we performed calculation of the deviation of the pheno-indicator (eco-deviation) for each *facies* (in days). The error of eco-deviation was calculated by the formula:

$$m = \pm\sqrt{\frac{\sum\left(M - \bar{M}\right)^2}{n(n-1)}}$$

where $\sum\left(M - \bar{M}\right)^2$ is the sum of squares for the difference of the eco-deviation at a given point and the average eco-deviation in the area and n is the number of sites in which observations were conducted. Average errors are necessary for revealing mathematically grounded reliability of differences between sites and areas.

We conducted the phenological studies at three levels of geosystem organization: the facies of altitudinal belts, altitudinal belts, and landscape areas. Field studies of seasonal dynamics were carried out at the level of *facies*, since the seasonal changes are clearly manifested due to a homogeneous structure. The results of studying the facies of the taiga belt made it possible to identify the characteristic indicators and criteria for the boundaries of the seasons.

In spring, the range of phenomena occurrence among individual facies is maximal. The phenological differences between facies increase from the early spring to summer, but decrease from the early autumn to late autumn (Yantser and Terent'eva 2013). The first half of spring is under the significant influence of the winter season and its consequences. The dynamic state of the early spring is characterized by a continuous snow cover, a stable transition of mean daily air temperatures through 0 °C, a gray-white phenological aspect of the landscape, and $\overline{K}_g$ =0.0; $\overline{K}_v$ =0.1. The final melting of snow and the awakening of plants occur in different facies not simultaneously. Snow cover in the phenofields located in the lower part of the taiga belt degrades faster than anywhere in the study area. The second half of spring is distinct by increasing radiation, rapid warming of the soil and the near-surface air, as well as the beginning of vegetation. In spring, the correlation coefficient (0.8–0.9) between the onset of phenological phenomena and the thermal regime is the highest. In summer and autumn, correlation is less pronounced. In the first half of spring, the generative development of plants in the facies of the lower parts of the slopes is much faster than in the higher ones. The beginning of the vegetative season is characterized by a 100% clearing of the soil from the snow cover and complete melting of the ice. The greatest advance in development is observed in river valleys. A stable lag in comparison with the average $\overline{K}_g$ was detected in the *Pinetum sibirica myrtillosum* forest, located at elevation 310 m above the river valley. By the end of the first decade of June, the magnitude of the deviation decreases in all facies, with the exception of the middle and upper parts of the slope where the lag in the development of vegetation increases and reaches a maximum. Vegetation in facies of the taiga belt, which has various absolute heights and ecological conditions, proceeds in different ways. In the facies of the lower slopes, the maximum of advance was revealed at the beginning of spring and accounts for 8.0 days. A stable lag is observed in the facies of the upper part of the belt. In the *Piceetum myrtilloso-hylocomiosum* forests ($H = 580.0$ m), the delay decreases by the end of spring, and in *Pinetum sibirica myrtillosum* forest ($H = 600.0$ m), on the contrary, it increases.

The results of observations of birch greening, obtained with the descriptive integrated method, confirm the delay in the development of this species, by the "strong green haze" criterion, with increasing absolute altitude. The differences between the facies of the lower and upper parts of the belt account for 2.1 days.

The summer period is characterized by the transition of average daily air temperatures (t) above +12 °C, the transition of the minimum t above +10 °C, and the maximum t above +18 °C. At this time, soil temperatures at different depths become similar. Indicators of the season are the transitions of $\overline{K}_g$ through 1.8, $\overline{K}_v = 3.5$; blossoming of *Prunus padus*, flowering of *Sorbus aucuparia, Sorbus sibirica*, and *Rosa canina*. In summer, differences in the seasonal development of the facies belonging to the same altitudinal belt are within the limits of random deviations, which is due to stabilization of the atmospheric circulation, general warming of air and soil, and homeostatic vegetation reactions.

The end of vegetation in the autumn is characterized by ground frosts, the formation of rime and the freezing of puddles in the mornings, the transition $\overline{K}_g$ through 8.8; $\overline{K}_v$ =7.8; by the end of staining of foliage in all trees, bushes, and grassy species; and by a massive leaf fall. In autumn, the facies of the upper and lower parts of the slopes complete development earlier than the middle part by 0.8–2.0 days. This is due to temperature inversions in the lower parts of the slopes and to the influence of the altitudinal belt conditions.

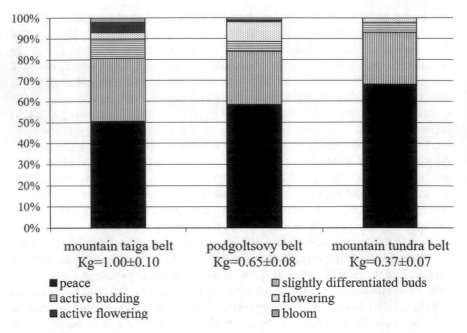

Fig. 18.2 Long-term average annual summarized phenological characteristic of generative development in altitudinal belt on June 8

A steady transition of daily average and maximum t below 0 ° C and the establishment of a permanent snow cover indicate the dynamic state of the onset of winter. There are no signs of vegetation: $\bar{K}_g$ =8.9; $\bar{K}_v$ =7.9.

In the seasonal dynamics of vegetation at the altitudinal belt level, we detected more significant differences (Fig. 18.2). In spring, the advance in the development of the phytocomponent of the taiga belt in comparison with the subalpine belt is 8.0–10.0 days, while that with tundra belt is 11.0–15.0 days. The delay in the development of processes in the subalpine belt is explained by moisture excess due to a thicker snow cover in winter that affects its late degradation in the spring, and, accordingly, the later drying and warming of the soil. In the taiga belt, the water amount resulting from snowmelt is less than in the subalpine belt by 2.2 t/ha, and more than in the tundra belt by 15.1 t/ha. In the tundra and subalpine belts, the snow density is higher, which is associated not only with lower air temperatures causing snow recrystallization but also with the effect of snow compaction by winds from the south and south-west. In the tundra belt in early spring, more than 50% of the area is covered by snow. The delay in the development of vegetation in the spring in the tundra belt as compared with those located below is associated with the peculiarities of the hydrothermal regime: slow drying out, and warming of the soil and the near-surface layer of air due to lower temperatures. The results of observing birch greening with the descriptive integrated method, by the "strong green haze" criterion, were confirmed by the delay in the development of the species in the sub-alpine belt in comparison with the taiga belt. The differences between belts account for 2.4 days (Table 18.1).

For the early summer we revealed a tendency to advanced development of veg-etation in the taiga in comparison with the subalpine belt. In the generative development of the tundra belt the delay in comparison with the taiga belt is 9.0 days, and that with the subalpine belt is 6.0 days. This period is characterized by high rates of development with maximum values (0.17 points per day) in the facies of the tundra belt. At the end of July, the rate of development in the tundra belt con-tinues to increase. This results in earlier seasonal events of the tundra belt compared to the taiga belt (2.2 days) and to the subalpine belt (2.9 days). By early August, this trend becomes mathematically significant: the vegetation development in the tundra belt leads that in the subalpine belt by 7.5 days. Differences in the development of the phytocomponent of the taiga and subalpine belts in this period account for 5.0 days, and between taiga and tundra belts 2.5 days.

Observations of the autumn coloring of birch leaves, according to the "10 per-cent yellow leaves in the crown" criterion (Table 18.1), confirmed an earlier occur-rence of the phenomenon in the upper parts of the taiga belt and in the subalpine belt. This is explained by a decrease in air and soil temperatures with an increase in absolute altitude. In the lower part of the taiga belt, an earlier coloring of birch leaves compared to the standard was observed, which is explained by temperature inversions in river valleys. The differences between the taiga and the subalpine belts are 1.3 days.

By the beginning of September, the advance in the development of the subalpine belt vegetation is 7.5 days and that of tundra is 8.3 days as compared with the taiga belt. The delay in the development of the subalpine belt in comparison with the tundra belt is 3.3 days. At the end of vegetative period, the development of vegeta-tion in the subalpine and in the tundra belts is completed faster than in the taiga belt by 5.0 days. In general, the vegetation period in the subalpine belt is reduced by 12.0 days, and in the tundra zone by 32.0 days in comparison with the taiga belt.

In the landscape areas of the medium-altitude Northern Urals, seasonal dynamics becomes more complex and proceeds at different rates, depending on the combina-tion of system-forming processes. In spring, immediately after the snowmelt, the vegetation development of the Shegul'tano-Vyjsky mid-mountain area occurs faster by 2.0–6.5 days. The advance in spring development is explained here by the lesser height of the snow cover (on average, by 25 cm, the amount of water resulting from snowmelt is twice as low and accounts for 14.6 t/ha), by the faster drying and warm-ing of soil and air, and also by prolonged insolation.

In early summer, when the temperature conditions of the landscape areas are relatively uniform, the humidity factor plays a more important role. The develop-ment of the generative organs of the plants in the Uls-Vil'vensky mid-ridge area is faster by 7.4 days, compared with the Shegul'tano-Vyjsky mid-mountain range. Vegetation is faster by 2.0 days in the Shegul'tano-Vyjsky area. In autumn, unlike spring, the development of vegetation proceeds and ends more quickly (by 3.0–4.0 days) in Uls-Vil'vensky area, which is located further to the west. The Shegul'tano-Vyjsky area is characterized by higher air and soil temperatures, due to longer illumination during the day, which determines longer duration of the processes.

A representative area for phenological studies was selected within the territory of the Denezhkin kamen' nature reserve, where the number of anthropogenic modifications is relatively small. This allowed us to analyze the "pure" natural processes that determine the patterns of seasonal dynamics in more details. Anthropogenic impacts on the landscape change not only its landscape structure, but also its natural dynamic trajectory.

18.5   Conclusion

The results of phenological studies allow detecting the processes in geosystems of different hierarchical levels. The study of the triad—*facies*, *altitudinal belts*, and *landscape area*—allows establishing general and particular trends of dynamic development, determined by different-scale contributions. Seasonal dynamics in geosystems of different levels depends on the framework conditions and local processes. Indicators of the general state of vegetation—complex phenological indicators—allow us to identify differences in the rate of the processes and also the boundaries of the seasons and the criteria for their subdivisions. Short-term dynamics at higher levels is generalized and includes the characteristic parameters of local geosystems.

In the conditions of modern climatic changes, systematic long-term research of seasonal dynamics makes it possible to reveal general trends in the timing of occurrence of phenomena in the organic and inorganic nature. Application of the presented indicators and criteria for the seasonal development of geosystems of different rank provides the opportunity to track the patterns of phenological changes in time and space.

References

Gurevskih, O. Yu., Kapustin, V. G., Skok, N. V., & Yantser, O. V. (2016). *Physical-geographical regionalization and landscapes of Sverdlovsk region*. Ekaterinburg: Ural State Pedagogical University. (in Russian).

Hudson, I. L., & Keatley, M. R. (2010). *Phenological research: Methods for environmental and climate change analysis*. Dordrecht: Springer.

Isachenko, A. G. (1991). *Landscape science and physical-geographical regionalization*. Moscow: Vysshaya shkola. (in Russian).

Isachenko, G. A. (2014). The concept of long-term landscape dynamics and modern challenges. In K. N. Dyakonov, V. M. Kotlyakov, & T. I. Kharitonova (Eds.), *Issues in geography. Vol. 138. Horizons of landscape studies* (pp. 215–233). Moscow: Kodeks. (in Russian).

Kupriyanova, M. K. (2010). V.A.Batmanov – founder of novel direction in phenology. In *State-of-art in modern phenology and development perspectives* (pp. 42–56). Ekaterinburg. (in Russian).

Mamay, I. I. (1992). *Landscape dynamics. Research methods.* Moscow: MSU Publishing House. (in Russian).

Schwartz, M. D. (Ed.). (2013). *Phenology: An integrative environmental science.* Dordrecht: Springer.

Skok, N. V., Ivanova, Y. R., & Yantser, O. V. (2014). Application of quantitative phenological methods to characterization of mountainous belt in the Middle Urals. *Proceedings of Tomsk University, Series Natural and Technical Sciences, 19*(5), 1569–1572. (in Russian).

Skok, N. V., & Yantser, O. V. (2016). Phenological research methods in study of landscape dynamics: Review. *Proceedings of Bashkirsky University, 21*(1), 91–100. (in Russian).

Yantser, O. V., & Terent'eva, E. Yu. (2013). *General phenology and methods of phenological research.* Ekaterinburg: Ural State Pedagogical University. (in Russian).

Part VI
How Patterns Respond to Climatic and Anthropogenic Changes

Chapter 19
Dendrochronological Indication of Landscape Spatiotemporal Organization in the Northern Taiga of West Siberian Plain and Elbrus Region: Astrophysical and Geophysical Drivers of Bioproductivity

Kirill N. Dyakonov and Yury N. Bochkarev

Abstract We hypothesized that temporal organization is related to intralandscape differentiation in dominant and subdominant units and is controlled by both regional and local scale landscape patterns. Regression methods were used to relate dendrochronological data from the northern West Siberia and the Caucasus to astrophysical predictors (distance between the center of the Sun and the Solar System barycenter, and monthly values of Wolf numbers characterizing solar activity) and geophysical predictors (geomagnetic *Aa* index and temperature and precipitation values for various periods). Dendrochronologies were processed in order to extract fluctuations for each temporal hierarchical level. Fluctuations and various frequencies of external astrophysical and geophysical factors determine hierarchy in chronoorganization of landscape space, which may be synchronous or asynchronous depending on local conditions. Spatiotemporal organization of landscape functioning is better pronounced in the severe conditions of the northern West Siberia than in mild conditions of the Caucasus. Variability in solar and geomagnetic activity, position of the barycenter of the Solar System, meteorological regime act as the important controls over intra-century and inter-century dynamics of bioproductivity both in mountainous regions and at the northern border of the taiga. At the northern border of the forest zone, these factors are more powerful, with dominance of climatic ones.

Keywords Dendrochronology · Spatiotemporal organization · Climate · Fluctuations · Astrophysical factors · Geophysical factors

K. N. Dyakonov (✉) · Y. N. Bochkarev
Lomonosov Moscow State University, Moscow, Russia

© Springer Nature Switzerland AG 2020
A. V. Khoroshev, K. N. Dyakonov (eds.), *Landscape Patterns in a Range of Spatio-Temporal Scales*, Landscape Series 26,
https://doi.org/10.1007/978-3-030-31185-8_19

19.1 Introduction

Morphological structure of a landscape develops under a strong control of abiotic environment with great contribution of lithogenesis. This results in multiplicity of patterns and complex genesis. Landscape cover is shaped by numerous matter and energy flows, part of which is system-forming; hence, we face systems multiplicity in landscape space (Raman 1972; Reteyum 1988; Solnetsev 1981; Kolomyts 1998; Khoroshev 2016).

Physical-geographical hierarchies are identified based on various approaches such as structural-genetic, landscape-geochemical, basin ones. Spatial organization of landscape cover is commonly described by hierarchy of homogeneous units, which proved its relevance for practical purposes (e.g., landscape planning). However, this approach is insufficient for studying functional and dynamic issues and forecast of future landscape states.

In the last one-third of the twentieth century, stationary physical-geographical studies in the former USSR provided deep insights into both spatial and, in particular, temporal organization of landscapes, or geosystems (D.L. Armand, V.B. Sochava, N.L. Beruchashvili, I.I. Mamay, Yu.G. Puzachenko, K.N. Dyakonov, V.A. Snytko and others). Sochava (1978), Beruchashvili (1986), and Mamay (2005), who gained great experience in long-term research, introduced the notion of "geosystem state" and proposed appropriate classifications.

Our conception of geosystem spatiotemporal organization (or physical-geographical organization) involves stable orderliness and structural arrangement in space and time which are manifested on the Earth's surface as a heterogeneous mosaic of individual geosystems within a range of scale levels as well as in regular alternation of hours-long, daily, seasonal, annual, intra- and inter-century-long states, and functioning regimes (Dyakonov 1988).

The need to investigate spatiotemporal organization of a landscape was encouraged by intrinsic logics in development of landscape science, more precisely, by implementation of system approach. Social requirement for landscape-ecological studies was a good stimulus in 1960–1980s. For example, great experience in landscape analysis was gained during elaboration of the projects for redistribution of runoff in the basins of the Pechora, the Vychegda, the Kama, and the Volga Rivers aimed at stabilizing the Caspian Sea level, in the Ob and the Irtysh basins aimed at development of irrigation, and at solution of the Aral Sea problem. These huge projects involved creation of large hydrotechnical system, including water reservoirs and channels. Method of geographical analogies was applied to make forecast of consequences. Detected regularities of spatiotemporal organization of landscapes in the impact zones of existing reservoirs were extrapolated for the newly projected ones (Reteyum 1970; Dyakonov 1975; Vendrov and Dyakonov 1976).

To get insight into spatiotemporal organization of landscapes, a researcher faces the need to characterize how general and specific regularities of hierarchically organized cycles in geosystem functioning are manifested at various scale levels. It is well known that primary productivity can be considered as the main integral indicator of functioning. Landscape dendrochronology is believed to be one of the most important tools to study inter- and intra-century-long cycles (Dyakonov et al. 2003). The experience of landscape-dendrochronological investigation was gained in various subzones of taiga, in mixed-forest and forest-steppe zones, as well as in the Caucasus (the Elbrus district). This chapter presents the results obtained in the northern West Siberia and the Elbrus district.

Our research focused on the following issues.

- Establishment of temporal organization of landscape functioning at the northern border of the taiga zone in the West Siberian plain and at the uppermost border in the Elbrus district. We hypothesized that temporal organization is related to intralandscape differentiation in dominant and subdominant units and is controlled by both regional and local scale landscape patterns.
- Assessment of contributions of geophysical, astrophysical, climatic, and coenotic drivers as well as of joint effects of solar (Wolf numbers), geomagnetic (*Aa* index), and climatic cycles into dynamics of increments of coniferous trees at various time scales, depending on local intralandscape conditions.
- Testing the hypothesis that contribution of solar activity as a factor of bioproduction dynamics is higher at the uppermost border of forest landscapes in mountains than at the northernmost border in plains since the thinner atmosphere in mountains absorbs and reflects less radiation. A.L. Chizhevsky (1973), in his pioneer works, emphasized the critical contribution of electromagnetic activity of the Sun to dynamics of natural processes in landscapes. For the review of the problem, see in Raspopov et al. (2005).

19.2 Study Area

The research was performed in two regions. At the northern border of the taiga zone in the West Siberian plain (the Nadym district, 65°34′ N, 72°32′ E), we studied the landscape of flat or slightly inclined, relatively drained lacustrine-alluvial plains with permafrost spots, ice mounds (up to 5 m high), and shallow thermokarst lakes. The second study area is located in the central Caucasus, the Elbrus district (43°20′ N, 42°30′ E), in the range of elevations 1450–2500 m a.s.l. The landscapes are dominated by *Pinetun sosnovskyi herbosum* forests on Cambisols, *Pinetum sosnovskyi hylocomioso-herbosum* forests on Protic Regosols (proluvial soils), and *Pinetun sosnovskyi herbosum* forests of Fluvisols.

19.3   Materials and Methods

We collected data on annual radial tree increment in the Nadym district (five sample plots) and in the Elbrus district (14 sample plots). To derive increment time series, in the Nadym district we sampled 142 trees of three species – *Pinus sibirica, Pinus sylvestris, and Larix sibirica* and in the Elbrus district we sampled 120 trees *(Pinus sosnovskyi)*. Pressler corer was used to collect samples from the two opposite sides of a stem of 12–20 trees at each plot in line with integrated description of the landscape unit. In the Nadym district, the data set involved all main types of geosystems as follows (Table 19.1): (1) poorly drained locality *(mestnost*[1]) with ice mounds, (2) floodplain locality with permafrost spots, and (3) well-drained locality without permafrost, in which we sampled landscape units *(urochisches)* differing in moisture supply. In the Elbrus district, samples were collected in the following geosystems: (1) in the vicinity of the tree line, (2) in the middle section of forested north-facing and south-facing slopes, (3) at the mudflow cones of various age (the Garabashi river), and (4) in the steppified open forests on the north-facing slope of the Baksan river valley.

As astrophysical predictors, we considered distance between the center of the Sun and the Solar System barycenter, and monthly values of Wolf numbers characterizing solar activity.

As potential geophysical and meteorological explanatory variables, we used geomagnetic *Aa* index and temperature and precipitation values for various periods. The choice of geomagnetic *Aa* index is explained by close relationships between, on the one hand, the Earth's magnetic activity and motion of the Sun (Fig. 19.1), and, on the other hand, between the Earth's magnetic activity and thermic regime of the troposphere (Fig. 19.2). Coordinates of the Solar System barycenter were estimated in EROS software, elaborated in the Pulkovo State Astronomic Observatory. The contribution of atmospheric moisture was not taken into consideration in this study due to the lack of available data.

Tree rings width was measured using Lintab 5 followed by cross-dating. To measure tree rings using scanned images of cores we applied Cybis Coorecorder (version 2.0.14) module in ITRDB (International Dendrochronological Library) software elaborated for the statistical processing of dendrochronological data. To process annual increments, we used standardization technique, that is transferred absolute values to relative ones (Cook 1985; Fritts 1976). Published data show that series of annual radial increments are influenced by biological age factor. The latter

---

[1] *Urochishche* is the term for hierarchical level of landscape morphological units (natural territorial complexes, NTC) higher than *facies* and lower than *mestnost* used in Russian landscape science. See Glossary and Chap. 1 for details.

Table 19.1   Characteristics of sample plots used for sampling tree cores

Index, tree species sampled	Landform and location	Elevation a.s.l., m	Forest type, canopy closure	Average tree height, m	Age, years
Elbrus district, the Caucasus					
C3, pine	SW slope (15° – 20°), Terskol	2300–2400	*Pinetum herbosum* steppified forest, 0.05–0.3	20–25	150–230
C4, pine	SSE-facing debris flow fan (5°), with channels covered by herbaceous vegetation, Garabashi	2280–2320	*Pinetum hylocomioso-herbosum* forest, 0.3–0.5	20–25	100–200
C5, pine	Mudflow channels within SSE-facing debris flow fan (5°), Garabashi	2280–2320	*Pinetum herbosum* forest or pine forest without herb cover at the channel margins, 0.2–0.3	20–25	100–220
C6, pine	Marginal sector of the S-facing debris flow fan (7°–10°), Garabashi	2300–2320	*Pinetum hylocomioso-herbosum* forest, 0.3–0.5	25	200–300
C7, pine	Ridges between mudflow channels, SSE-facing debris flow fan (5°), Garabashi	2280–2320	*Pinetum herbosum* forest or pine forest without herb cover, 0.3–0.5	15–20	40–50
C8, pine	The Baksan river valley bottom (0–3°) subject to avalanches, Terskol	2150	*Pinetum herbosum* forest, 0.3–0.4	20	80–100
C9, pine	S-facing foot slope (20°–30°), Terskol	2300–2350	*Pinetum herbosum* steppified forest, 0.05–0.2	20	150–200
C10, pine	Tree line at N-facing slope (15°–20°), Terskol, Cheget	2450–2500	*Pinetum herbosum* forest, 0.3–0.4	20–25	150–200
C11, pine	N-facing slope (20–30°), Terskol, Cheget	2200–2300	*Pinetum herbosum* forest, 0.5–0.7	25–30	100–170
C13, pine	The Baksan river valley bottom (0–3°) with ridges and runnels, Terskol	2120–2140	Pine forest without herb cover, locally *Pinetum hylocomioso-herboso-vacciniosum* forest, 0.6–0.7	25–30	150–260
C14, pine	NW-facing slope (25–35°), Tyrnyauz	1450–1500	Sparse *Pinetum herbosum* steppified forest, <0.05	8–12	

(continued)

Table 19.1 (continued)

Index, tree species sampled	Landform and location	Elevation a.s.l., m	Forest type, canopy closure	Average tree height, m	Age, years
Nadym district, West Siberia					
K1, cedar	Lacustrine-alluvial drained sandy plain (0–3°) without permafrost. Peat thickness 10 cm	20	*Betuleto-Lariceto-Pinetum* or *Pineto(sibirica)-Lariceto-Pinetum fruticuloso-hylocomioso-cladinosum* forests, 0.3	12–15	120–160
L1, larch					150–200
C1, pine					120–150
C14, pine	Flat well-drained sandy plains without permafrost and peat		*Lariceto-Pinetum cladinosum* forests and low-density stands, 0.01–0.15	8–10	120–160
K13, cedar	Swamped lacustrine-alluvial hummocky plain (0–3°) without permafrost in the vicinity of floodplain. Peat thickness 40–50 cm		*Pinetum(sibirica) fruticuloso-caricoso-polytrichoso-sphagnosum* sparse forest with larch, spruce and birch, 0.05–0.15	10–15	120–270
K10, cedar	Hummocky floodplain of the Long'yugan river with permafrost in the vicinity of oxbows. Peat thickness > 40 cm and seasonally thawed layer 40 cm		*Betuleto-piceeto-pinetum(sibirica) vaccinioso-calamagrostidoso-chamaemorosum* forest, 0.4–0.5	17	100–150
K9, cedar	Ice mound up to 5 m high. Peat thickness > 100 cm thickness and seasonally thawed layer 40–60 cm		*Pinetum(sibirica) nanobetuloso-vaccinioso-ledosum* or *chamaemoroso-caricoso-cladinosum* sparse forest, <0.05	10–15	250–430

is commonly described by so-called "large growth curve" and is excluded from consideration when one studies external influences on increment (Bitvinskas 1974).

By so doing, we calculated second-order indices – individual chronologies. They were averaged to derive generalized chronology of increment indices for the sample plot. ARSTAN software was used to standardize samples for indexing and excluding age trend. Raw increment data and derived dendrochronologies were processed in order to extract fluctuations for each temporal hierarchical level. Increment fluctuations were ranked by means of periodograms from Fourier analysis of generalized chronologies.

We performed stepwise regression, in which generalized chronologies of annual radial increment for sample plots were used as dependent variables. Wolf numbers, geomagnetic *Aa* index, meteorological, and climatic indices were taken as explanatory variables. All the regression models were built separately for the low-frequency, mid-frequency, high-frequency increment fluctuations, and for interannual fluctua-

Fig. 19.1 Dependence of the Earth's magnetic activity on the position of the Sun in relation to the barycenter of the Solar System (1868–2011). 1, positive anomaly; 2, negative anomaly. Calculated from data British Geological Survey

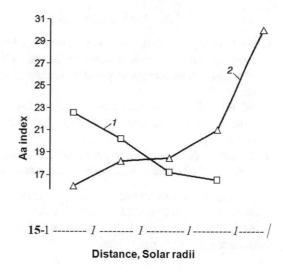

Distance, Solar radii

Fig. 19.2 Magnetic activity of the Earth during positive and negative global anomalies of air temperature (1880–2011): 1, positive anomaly; 2, negative anomaly. Difference is significant by Mann–Whitney U-test. Calculation from data: British Geological Survey и National Climatic Data Center, National Oceanic and Atmospheric Administration

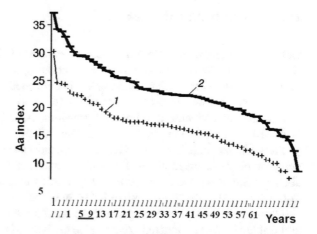

tions. We compared multiple correlation coefficients and F-values to evaluate general dependence of various increment cycles on joint effect of external factors.

To assess external effects on annual radial increment, we analyzed synchronism coefficients (SC) for increment fluctuations. We modified SC (Huber 1943) by calculating at each plot maximum number of records with changed sign of increment (positive or negative) in comparison with the previous year and dividing it by total number of series in a group. By so doing, we obtained SC for two certain years in a range between 0.5 and 1. These SC values were averaged for the whole growth period of the trees or for certain growth period. This allowed judging about degree of external influences on increment. Degree of increment synchronism among elementary geosystems (*facies*) and within *facies* is believed to be a very simple and highly informative index.

We tested the following basic hypotheses.

1. If dynamics of tree increments is determined by intrinsic properties of a geosystem, then average synchronism among trees within *facies* should be higher than synchronism among generalized dendrochronologies of various geosystems in a region.
2. SC for radial increments among various facies is directly related to degree of external influences and, hence, can be treated as a measure of geosystem sensitivity.
3. Change in level of synchronism indicates change in factors that influence increment, given that abiotic and biocenotic conditions are stable.

High synchronism among various geosystems is believed to indicate dominant contribution of one or several external factors. If synchronism among various *facies* is lower than that among individual trees within one *facies*, then contrast between neighboring *facies* is significant.

19.4   Results and Discussion

Generalized dendrochronologies based on Fourier analysis allowed establishing wavy character of biological production operating at various frequencies in the northern West Siberia (Fig. 19.3). We distinguished monotonous trends (large growth curves), low-frequency fluctuations (with 50-year-long period and more), mid-frequency fluctuations (about 30-year-long period), high-frequency fluctuations (10–15-year-long period), and interannual fluctuations (5-year-long-period and less). Wavy multifrequency character of landscape processes is determined by wavy character of external and Earth phenomena ranging from Earth's revolution about the Sun and rotation on its axis to cycles of solar activity, long-period fluctuations connected to the Solar System barycenter, and cycles of orogenesis.

Now we shall consider the dynamics of annual radial increment at various hierarchical levels. We conducted comparative analysis of synchronism in fluctuations of generalized chronologies among sample plots and of mean synchronism among individual (within-*facies*) chronologies. It turned out that the longer the fluctuation period, the less the synchronism in fluctuations among generalized chronologies of various geosystems. However, this is not true for interannual fluctuations in the Nadym district since their synchronism is less than that of high-frequency fluctuations (Table 19.2). Mean synchronism of individual chronologies among sample plots depending on fluctuation period decreases insignificantly. In the Elbrus district, for mid-frequency, high-frequency, and interannual fluctuations increment synchronism among sample plots is much higher than that within plots (Table 19.2). Low-frequency fluctuations follow the opposite rule. In the Nadym district, for low-frequency, mid-frequency, and high-frequency fluctuations synchronism among sample plots is lower than mean synchronism for all the sample plots. This difference is the greatest for low-frequency fluctuations.

These findings enable us to conclude that geosystem-specific differences in dynamics of annual radial increment in the Elbrus district are significant for low-frequency fluctuations (with 50-year-long and more period) only. Climate-driven low-, mid-, and high-frequency fluctuations in the upper Baksan basin are manifested in a similar way. On the contrary, in the Nadym district, dynamics of particular geosystems differ at the levels of low-, mid-, and high-frequency fluctuations by response to cyclic climatic changes.

It is important to know whether the abovementioned rule holds true for the whole growth period of trees since the early nineteenth century. Figure 19.4 shows evidence that in the Elbrus district synchronism among geosystems was higher than within-geosystem synchronism for mid- and high-frequency and interannual incre-

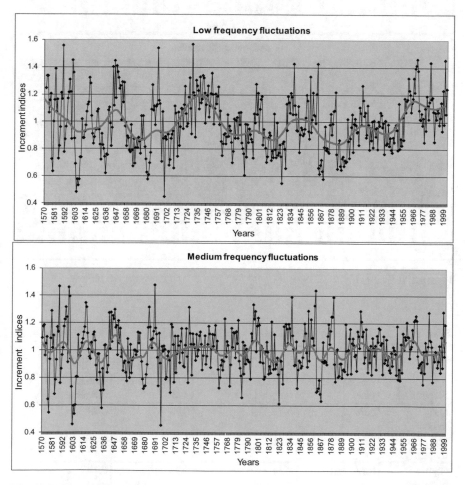

Fig. 19.3 An example of various-frequency fluctuations in radial increment of *Pinus sibirica* in the northern West Siberia (1641–2001). Top-down: low frequency, medium frequency, high frequency, interannual fluctuations

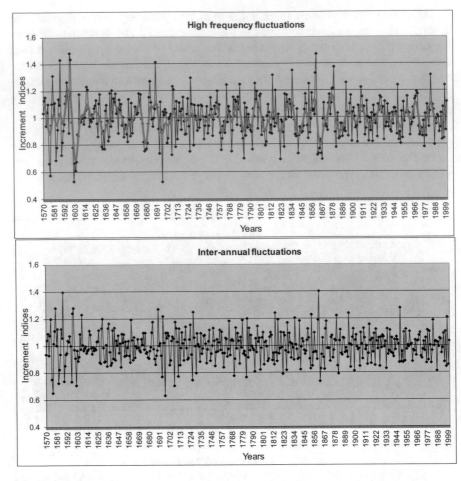

Fig. 19.3 (continued)

Table 19.2 Synchronism in fluctuations of increments among *facies* and within *facies* in the Caucasus and northern West Siberia

	Elbrus district (Caucasus)		Nadym district (West Siberia)	
Fluctuations	Among sample plots	Average from sample plots	Among sample plots	Average from sample plots
Low-frequency	0.68	0.72	0.73	0.78
Mid-frequency	0.73	0.71	0.76	0.77
High-frequency	0.78	0.74	0.79	0.79
Interannual	0.79	0.73	0.77	0.75

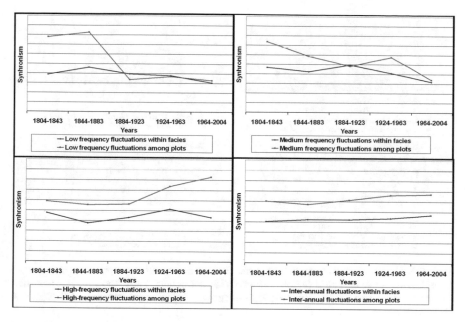

Fig. 19.4 Averaged coefficients of synchronism for increments at various frequencies among sample plots (purple lines) and within facies (blue lines). The Elbrus district

ment fluctuations. As for low-frequency fluctuations, in the first half of the nineteenth century, synchronism among geosystems was much lower than within-geosystem one and decreased by the late nineteenth and early twentieth centuries. We found no evidence of such trend for within-geosystem dynamics. Hence, in the early and middle nineteenth century, low-frequency increment fluctuation in all geosystems were determined by low temperatures of the Little Ice Age (Fernau stage). Differences in low-frequency increment fluctuations among geosystems were minimum as well. The same holds true for fluctuations at the other time scale levels. Since the late nineteenth century, low-frequency fluctuations have controlled the specifics of increment dynamics in various geosystems. Thus, the Little Ice Age masked differences among geosystems. In contrast, synchronism in high-frequency fluctuations among geosystems increased in the second half of the twentieth century as compared to the Little Ice Age (Fig. 19.4). Most likely, this kind of fluctuation is driven by the other climatic factors.

In the Nadym district, during the whole of the twentieth century, increment synchronism among trees within sample plots was higher than that among sample plots for low- and mid-frequency fluctuations (Fig. 19.5). Evidently, differences in increment dynamics among geosystems were significant in the twentieth century. However, during the Little Ice Age in the early and middle nineteenth century, synchronism in mid-frequency fluctuations among sample plots was higher than that within plots, while for low-frequency fluctuations these values were almost similar (Fig. 19.4). This finding is in full agreement with that for the Elbrus district: differences in increment dynamics among geosystems were insignificant.

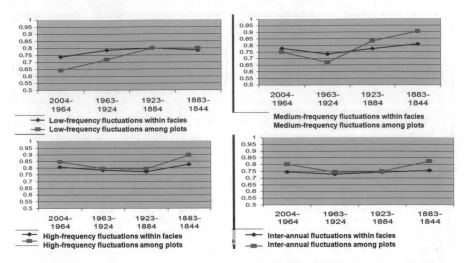

Fig. 19.5 Changes in synchronism coefficients for increments within sample plots (facies) (blue lines) and among sample plots (purple lines) for various chronointervals. Top left: low-frequency. Top-right: mid-frequency. Bottom left: high-frequency. Bottom right: interannual fluctuations. West Siberia

Both in the Elbrus and the Nadym districts for high-frequency and interannual fluctuations, we found evidence that at any moment increment synchronism *among* geosystems is higher than *within* geosystems. In both regions, we detected certain increase in synchronism by the present time (Figs. 19.4 and 19.5). The geosystems in the Nadym district are more highly specific in comparison with the Elbrus district. This is confirmed by higher synchronism within sample plots as compared to that among plots at any frequencies. We explain this by more severe growth conditions in the northern West Siberia. In the Nadym district, trees, in compliance with Liebig's law of limiting factors, are more sensitive to external conditions, mostly thermic ones, which are highly in contrast among geosystems. In the Elbrus district, factors substitution is more well-pronounced in compliance with Shelford's law of tolerance. However, during the Little Ice Age, these differences were insignificant.

It is worth noting that the tree line in the Elbrus district has an important peculiarity. The mean periodical increment of pines accounts for 1.28 mm per year, long-term variability is 18.3%, mean tree height is 15–20 m. In northern West Siberia, these values account for 0.36 mm per year, 22.2%, and 6–10 m, respectively. Earlier publications showed that the maximum variability in radial increment, accounting for 26–29%, occurred at the northernmost and southernmost borders of the forest zone (Dyakonov 1981). The values derived for the Elbrus region are close to that for the middle taiga landscapes of the East European Plain and the West Siberian Plain. Thus, we can conclude that *the mountainous forests in the Caucasus occupy the elevations lower than their thermic limit which in plains is close to the mean July temperature of approximately + 11°C.*

Table 19.3 Multiple coefficients of correlation averaged by sample plots at various hierarchical levels of fluctuations

Hierarchical level of chronoorganization	Climatic indices		Solar and geomagnetic activity		All factors	
	Elbrus	Nadym	Elbrus	Nadym	Elbrus	Nadym
Mid-frequency fluctuations	0.36	0.56	0.40	0.24	0.56	0.64
High-frequency fluctuations	0.40	0.55	0.43	0.25	0.58	0.61
Interannual fluctuations	0.44	0.58	0.31	0.17	0.58	0.62

Thus, we revealed various synchronism of increment fluctuations at different time scales among geosystems and within geosystems. It means that the factors affecting the dynamics of bioproductivity at intra-century and inter-century time scales differ much. Manifestation of differences in synchronism depends on intra-facies properties. In this connection, we face the need to establish the contributions of astrophysical and geophysical factors to increment dynamics. The answer for this question will enable us to get insight into biogeophysical essence of spatiotemporal organization.

In the Elbrus district, multiple coefficients of correlation are higher in models that relate increments to indices of solar and geomagnetic activity and climatic parameters than in those models that consider climatic parameters only (Table 19.3). This holds true for all the geosystems except for the Tyrnyauz city district in the middle section of the Baksan basin, where steppified pine open forests occur at elevations of 1450–1500 m. Solar activity in July enters almost all the models as explanatory variable being negatively related to increments. This is not a surprise, since annual ring is forming in this month to a large extent. F-value is higher in the models that relate high-frequency fluctuations to solar and geomagnetic activity (11-year-long cycles providing the largest contribution) than in models with climatic independent variables only. This result confirms the hypothesis that the solar activity directly affects dynamics of bioproductivity at the tree line, which, in this case, is not determined by climatic factors (Bochkarev and Dyakonov 2009).

All the models demonstrate positive correlation between increments and solar and geomagnetic activity in October of the previous year, which probably affects emergence of the next-year ring or influences indirectly via some meteorological processes. For three geosystems (north-facing slope of the Cheget massif, Garabashi mudflow cone, and bottom of the Baksan valley), we detected that increments are negatively influenced by solar and geomagnetic activity in May as well as by average winter and annual geomagnetic activity.

In the Elbrus district, we found smaller local and regional contrasts in dependence of increments on meteorological factors in comparison with that on solar and geomagnetic activity. The range of multiple correlation coefficients accounts for 0.12–0.20, depending on time scale for the former and 0.18–0.35 for the latter (Table 19.4).

In the northern West Siberian plain, positive effect of solar activity in July on increment was detected only for *Pinetum cladinosum* forest located within the

Table 19.4 Range of averaged multiple coefficients of correlation at various hierarchical levels

Hierarchical level of chronoorganization	Climatic indices		Solar and geomagnetic activity		All factors		Difference between the Elbrus and the Nadym districts		
	Elbrus	Nadym	Elbrus	Nadym	Elbrus	Nadym	Climatic indices	Solar and geomagnetic activity	All factors
Mid-frequency fluctuations	0.12	0.15	0.35	0.08	0.20	0.24	−0.03	+0.27	−0.04
High-frequency fluctuations	0.16	0.24	0.31	0.20	0.25	0.21	−0.08	+0.11	+0.04
Interannual fluctuations	0.20	0.24	0.18	0.18	0.14	0.28	−0.04	0.00	−0.14

lacustrine-alluvial sandy plain without permafrost. Probably, this type of pine forest is less sensitive to variability of external factors due to insufficient water supply in summer. Small thickness of organic soil horizon may contribute to the low sensitivity as well.

In the other plots, we found no evidence that solar and geomagnetic activity and meteorological parameters produce joint effect on inter-century and intra-century dynamics of bioproductivity. Geomagnetic activity seemed to be significant, but we failed to find explanations for spatial contrasts in its influence on local level geosystems. Climatic (thermic) signal turned out to be a more powerful driver for tree increments, as well as ubiquitous occurrence of Histic horizon in soils with thickness ranging from 10 to 100 cm.

Table 19.4 shows ranges of average multiple correlation coefficients between the increments and the climatic and geophysical factors for various time scales. In West Siberia the values are larger than in the Elbrus district by 0.03–0.08 for the meteorological parameters. In contrast, the range of correlation coefficients between increments and solar and geomagnetic activity is smaller in West Siberia by 0.27 for mid-frequency fluctuations and by 0.11 for high-frequency ones. Hence, astrophysical factors are less significant for intra-century and inter-century variability of increments. At the time scale of interannual high-frequency fluctuations, no difference was detected.

Similar conclusions were obtained at submeridional transect in Finland for the contribution of solar activity and climatic factors to tree ring chronology with weak nonlinear relations (Ogurtsov et al. 2008). Climatic signal masks solar signal almost completely. Nevertheless, the cited authors established significant linkage between long-term cycles of the solar activity (80–200 years and more) and climatic cycles both in boreal plain regions (Ogurtsov et al. 2001, 2002) and at the tree line in mountains (Raspopov et al. 2005).

Now we consider in detail correlation linkages between annual radial increments and astrophysical and geophysical factors, including position of the Sun in relation to the barycenter of the Solar System (BCSS), for particular plots. The role of BCSS in biospheric processes is still poorly-studied. This direction of research in Russia was initiated by A.Yu. Reteyum, who develops the ideas by Landscheidt (1999). For multiple correlations between increments and corresponding variables for two regions, see Table 19.5. We revealed inter-*facies* contrasts in this relationship. Solar activity was significant in three *facies*, and air temperature in five *facies* from the seven investigated ones. Position of the Sun in relation to BCSS was significant in all the *facies*. The least significant dependence of *Pinus sibirica* increments on astrophysical and geophysical factors (r = 0.33) was observed in the geochemically heteronomous floodplain facies of the Long'yugan river, with permafrost under peat deposits. In contrast, the strongest correlation was observed in the zonal geochemically autonomous facies of flat, relatively well-drained, lacustrine-alluvial plain, with thin peat layer without permafrost. The same holds true for the slightly inclined swamped lacustrine-alluvial plain without permafrost. In these facies, averaged multiple coefficient of correlation accounted for 0.75–0.79. Air and soil tempera-

Table 19.5 Multiple coefficients of correlation between radial increments and geophysical and astrophysical factors in particular facies

	Factors						All factors
	Geophysical			Astrophysical			
Sample plot, tree species	Temperature	Aa index	Temperature and Aa index	Wolf numbers	Position of the Sun in relation to the BCSS	Wolf numbers and position of the Sun in relation to the BCSS	
Siberia 1 *Pinus sibirica*	0.39	0.47	0.57	–	0.33	0.33	0.61
Siberia 1 *Larix sibirica*	0.57	–	0.57	–	0.27	–	0.79
Siberia 1 *Pinus sylvestris*	0.43	0.35	0.53	–	0.32	0.13	0.75
Siberia 14 *Pinus sylvestris*	0.46	0.41	0.67	0.23	0.34	0.40	0.71
Siberia 13 *Pinus sibirica*	–	0.36	0.38	0.20	0.24	0.20	0.75
Siberia 10 *Pinus sibirica*	0.21	0.23	0.21	0.24	0.33	0.33	0.33
Siberia 9 *Pinus sibirica*	–	0.38	0.35	–	0.12	0.18	0.69
Caucasus 3 *Pinus sosnovskyi*	0.31	0.19	0.47	0.20	0.33	0.40	0.66
Caucasus 4 *Pinus sosnovskyi*	0.28	–	0.45	–	0.42	0.42	0.72
Caucasus 5 *Pinus sosnovskyi*	0.37	–	0.37	–	0.26	0.27	0.60
Caucasus 6 *Pinus sosnovskyi*	–	–	0.49	0.26	0.31	0.40	0.59
Caucasus 8 *Pinus sosnovskyi*	0.29	0.20	0.42	0.50	0.37	0.57	0.64
Caucasus 9 *Pinus sosnovskyi*	0.25	0.27	0.45	–	0.23	0.23	0.46
Caucasus 10 *Pinus sosnovskyi*	0.31	–	0.46	0.36	0.28	0.46	0.67
Caucasus 11 *Pinus sosnovskyi*	0.31	0.25	0.31	0.26	0.35	0.63	0.56
Caucasus 13 *Pinus sosnovskyi*	0.37	0.23	0.53	0.29	0.35	0.49	0.71
Caucasus 14 *Pinus sosnovskyi*	0.29	0.60	0.80	0.40	0.28	0.27	0.67

tures were the binding geophysical factor in the drained *facies*, and geomagnetic *Aa* index – in the swamped *facies*. The Sun position is the most important astrophysical factor of the two considered.

Table 19.6 shows the averaged multiple coefficient of correlation between increments and geophysical and astrophysical factors. In the northern West Siberia, the averaged coefficient accounts for 0.66. The factors can be ranked by significance as follows (Table 19.4): geomagnetic *Aa* index (r = 0.31), distance between the Sun and the BCSS (r = 0.28), solar activity (r = 0.10).

Inter-century increments dynamics is related to fluctuations of temperature and precipitation in permafrost geosystems. Slopes of ice mounds devoid of peat mantle turned out to be the most sensitive, while ice mounds with peat deposits are more stable with self-regulating capacity due to better isolation of ice from heat flow. This finding confirms that in the northern West Siberia, permafrost and peat are able to mask external astrophysical signals.

In the Elbrus district, data from ten facies show that averaged multiple coefficient of correlation is close to that for the northern West Siberia (r = 0.63). In high-altitude mountains, the contribution of the Sun position and, in particular, the solar activity (r = 0.32 for each factor, r = 0.41 for their joint effect) are higher. The effect of geomagnetic *Aa* index is lower than in West Siberia, while that of temperature is higher (r = 0.28). The joint effect of geophysical factors is similar in both regions.

Inter-*facies* contrast in *r* is less pronounced in the Caucasus in comparison with West Siberia. Range of correlation coefficients accounts for 0.26 in the Caucasus and for 0.46 in West Siberia. Solar activity in the Caucasus is more important since it was significant in seven facies out of ten. The effect of the Sun position was more significant as well (r = 0.32) and detected in all facies. The temperature factor was significant in nine facies out of ten. On the contrary, geomagnetic *Aa* index is less significant than in West Siberia (r = 0.17 and r = 0.31, respectively). In general, the contribution of geophysical factors is approximately equal.

In the Elbrus district, slope gradient is an important control over increments dynamics. Increments are strongly correlated with astrophysical and geophysical indices on flat surfaces and gentle slopes (3–5°), as well as on the north-facing slope at elevations of 2450–2500 m. Weak correlations were detected on steep slopes (20–30°) and in azonal *facies* on the south-facing mudflow cones.

Thus, both in the northern West Siberia and in the Elbrus district, local landscape conditions (slope gradient and aspect, mudflow cones, etc.) are able to correct external signals.

19.5   Conclusion

1. Fluctuations and various frequencies of external astrophysical and geophysical factors determine hierarchy in chronoorganization of landscape space, which may be synchronous or asynchronous depending on local conditions.

Table 19.6 Averaged multiple coefficients of correlation between radial increments and geophysical and astrophysical factors

| | Factors | | | | | |
| | Geophysical | | Astrophysical | | | |
Region	Temperature	Aa index	Temperature and Aa index	Wolf numbers	Position of the Sun in relation to the BCSS	Wolf numbers and position of the Sun in relation to the BCSS	All factors
Siberia	0.25	0.31	0.47	0.10	0.28	0.22	0.66
Caucasus	0.28	0.17	0.48	0.32	0.32	0.41	0.63

2. In the northern West Siberia, we revealed fluctuations in increments at four hierarchical levels and corresponding features of spatiotemporal hierarchical organization of geosystems. At the high (*mestnost*-related) spatial scale, we detected low-frequency chronoorganization with fluctuation period more than 50 years. Mid- and high-frequency fluctuations in increments are inherent for the lower levels, such as *urochishches* and *facies*.

3. In the Elbrus district, such kind of spatiotemporal hierarchy is less pronounced. Dynamics of increments has specific features in middle and high altitudes, the latter being less differentiated by chronoorganization of increments due to relatively favorable growth conditions and large contribution of local level factors.

4. Analysis of synchronism in tree radial increments within *facies*, among *facies*, and among physical-geographical provinces provides the opportunity to assess how the signal from external geophysical factors is being transformed by local landscape and biological factors. By taking into consideration various chronointervals and fluctuation frequency, it is possible to compare contributions of external and internal factor of production process. High synchronism among *facies* and regions allows extrapolating the results of stationary research to adjacent or remote territories. This is the methodical essence of the method of geographical analogies.

5. Temporal changes in increment synchronism are characteristic of mid- and low-frequency fluctuations in relatively favorable growth conditions. Synchronism increases during unfavorable climatic periods (such as the Little Ice Age) as well as a result of changes of long-term states, such as mudflows in the Caucasus or progressive bogging in West Siberia.

6. Spatiotemporal organization of landscape functioning is better pronounced in the severe conditions of the northern West Siberia than in mild conditions of the Caucasus. This is believed to be a manifestation of Liebig's law of minimum and Shelford's law of tolerance. In the northern West Siberian plain, dynamics of increments and corresponding annual productivity depend on annual fluctuations of climatic parameters, temperature being the most important of them. Increment is controlled by air temperature indirectly via heating of soils and parent rocks. In plots with peat deposits thicker than 10–15 cm, soil warming is low, and, hence bioproductivity responds with 2–5 -year-long delay.

7. Variability in solar and geomagnetic activity position of the barycenter of the Solar System, meteorological regime act as the important controls over intracentury and inter-century dynamics of bioproductivity both in mountainous regions and at the northern border of the taiga. At the northern border of the forest zone, these factors are more powerful with dominance of climatic ones. Consideration for the position of the Sun in relation to the barycenter of the Solar System increases the density of linkage between increments and geophysical and astrophysical factors. The contribution of the latter is better pronounced in the Elbrus district.

8. The contribution of the solar and geomagnetic activity is higher in the Caucasus than in West Siberia. At the time scale of interannual fluctuations, this contribution is almost insignificant.
9. Permafrost and bogging in the northern West Siberia neutralize the influence of astrophysical signals via contrasts in peat thickness and seasonally thawed layer. In the Elbrus district, external astro- and geophysical signals can be neutralized on steep slopes and mudflow and avalanche cones.

Acknowledgments The research was financially supported by Russian Foundation for Basic Research (project 19-05-00786).

References

Beruchashvili, N. L. (1986). *Four dimensions of a landscape*. Moscow: Mysl. (in Russian).
Bitvinskas, T. T. (1974). *Dendroclimatic studies*. Leningrad: Gydrometeoizdat. (in Russian).
Bochkarev, Y. N., & Dyakonov, K. N. (2009). Dendrochronological indication of landscape functioning at the northernmost and uppermost borders of forest. *Proceedings of Moscow University, Series 5 Geography, 2*, 37–50. (in Russian).
Chizhevsky, A. L. (1973). *The Earth echo of solar storms*. Moscow: Mysl. (in Russian).
Cook, E. R. (1985). *A time series analysis approach to tree-ring standardization*. Ph.D. thesis. Tucson: Arisona University Press.
Dyakonov, K. N. (1975). *Impact of large plain water reservoirs of the riparian forests*. Leningrad: Gydrometeoizdat. (in Russian).
Dyakonov, K. N. (1981). Temporal variability of geosystem properties in West Siberia. *Proceedings of Academy of Sciences of the USSR, Geographical Series, 6*, 91–101. (in Russian).
Dyakonov, K. N. (1988). *Landscape geophysics. Method of balances*. Moscow: MSU Publishing House. (in Russian).
Dyakonov, K.N., Belyakov, A.I., & Bochkarev, Y. N. (2003). Landscape dendrochronology as an actual research direction. In *Dendrochronology: Achievements and perspectives* (p. 36). Krasnoyarsk. (in Russian).
Fritts, H. C. (1976). *Tree-rings and climate*. London/San Francisco.
Huber, B. (1943). Über die Sicherheit Jahresringchronologischer Datirung. *Holz als Roh- und Werkstoff, 6*(10/12), 263–268.
Khoroshev, A. V. (2016). *Polyscale organization of a geographical landscape*. Moscow: KMK. (in Russian).
Kolomyts, E. G. (1998). *Polymorphism of landscape-zonal systems*. Pushchino: ONTI PIC RAN. (in Russian).
Landscheidt, T. (1999). Extrema in sunspot cycle linked linked tusn' motion. *Solar Physics, 189*(2), 415–426.
Mamay, I. I. (2005). *Landscape dynamics. Research methods*. Moscow: MSU Publishing House. (in Russian).
Ogurtsov, M. G., Kocharov, G. E., & Lindholm, M. (2001). Solar activity and regional climate. *Radiocarbon, 43*(2), 439–447.
Ogurtsov, M. G., Nagovitsyn, Y. A., Kocharov, G. E., & Jungner, H. (2002). Long-period cycles of the sun's activity recorded in direct solar data and proxies. *Solar Physics, 211*, 371–394.
Ogurtsov, M. G., Raspopov, O. M., & Helama, S. (2008). Climatic variability along a North-South transect of Finland over the last 500 years: Signature of solar influence or internal climate oscillations? *Geografiska Annaler, 90A*(2), 141–150.

Raman, K. G. (1972). *Spatial polystructurality of topological geocomplexes and experience of its identification in the conditions of the Latvian SSR*. Riga: Latvian State University Publishing House. (in Russian).

Raspopov, O., Dergachev, V., Kozyreva, O., & Kolstrom, T. (2005). Climate response to de Vries solar cycles: Evidence of Juniperus turkestanica tree rings in Central Asia. *Mem. S.A.It., 76*, 760.

Reteyum, A. Y. (1970). Changes in natural conditions in the impact zone of Rybinskoye water reservoir. In *Impact of water reservoirs in forest zone on adjacent territories* (pp. 23–24). Moscow: Nauka. (in Russian).

Reteyum, A. Y. (1988). *The terrestrial worlds (on holistic studying of geosystems)*. Moscow: Mysl. (in Russian).

Sochava, V. B. (1978). *Introduction to the theory of geosystems*. Novosibirsk: Nauka. (in Russian).

Solnetsev, V. N. (1981). *System organization of landscapes*. Moscow: Mysl. (in Russian).

Vendrov, S. L., & Dyakonov, K. N. (1976). *Water reservoirs and natural environment*. Moscow: Mysl. (in Russian).

Chapter 20
Carbon Balance in Forest Ecosystems and Biotic Regulation of Carbon Cycle Under Global Climate Changes

Erland G. Kolomyts, Larisa S. Sharaya, and Natalya A. Surova

Abstract The strategy of prognostic landscape-ecological studies on climate-induced changes in the biological cycle and carbon balance in forest ecosystems as leading factors of the biotic regulation of the environment is presented. We describe methods for constructing analytical and cartographic empirical statistical models that make it possible to reveal the local mechanisms of biotic regulation and identify the zonal/regional types of forest formations capable of stabilizing the continental biosphere in the changing climate. The prognosis of changes in biotic regulation of the carbon cycle according to the scenarios of forthcoming greenhouse warming and cooling by global HadCM3 and E GISS models accordingly are described.

Keywords Ecosystem · Biogeocoenosis · Climatic changes · Empirical statistical modeling · Mapping · Ecological prediction

20.1 General Features of Landscape Ecological Modeling

Landscape ecology describes the behavior of geo(eco)systems as integrated natural/territorial formations. To this end, the following alternative approaches could be relevant, organism-discrete and functional-continual (Puzachenko 1984), which reveal different aspects of geo(eco)system structure and function. Both approaches have been used in this project.

The landscape ecological analysis of spatial organization of natural complexes and their anthropogenic dynamics and carbon balance is based on the construction of *discrete empirical statistical models* of their structure and function, according to Rozenberg (1984). These models can be classified as self-organizing models.

E. G. Kolomyts (✉) · L. S. Sharaya · N. A. Surova
Institute of Ecology of the Volga River Basin, Russian Academy of Sciences, Togliatti, Russia

© Springer Nature Switzerland AG 2020
A. V. Khoroshev, K. N. Dyakonov (eds.), *Landscape Patterns in a Range of Spatio-Temporal Scales*, Landscape Series 26,
https://doi.org/10.1007/978-3-030-31185-8_20

They are substantially different from the known simulation models (Printice et al. 1992; Smith et al. 1992; Holten et al. 1993; Kudeyarov 2007; Zavarzin 2007 etc.), which apply more or less a priori approach. Empirical statistical models use a posteriori approach, i.e., the results of field studies are used not for verification of some calculation data but as a factual basis of modeling per se. It yields the empirically substantiated results of much higher spatial resolution, rather than extensive geographic generalizations. In contrast to simulation models, the *method of local discrete modeling of natural ecosystems*, which we proposed and implemented, can be treated much more reasonably as a biogeocoenotic method – in the spirit of V.N. Sukachev's theory (1972).

It is important to emphasize that such models allow predictive analysis with wide use of the results of traditional field landscape ecological surveys, as well as various cartographic materials.

At the same time, it should be noted that simulation and empirical statistical models cannot be considered as interchangeable. Each of these classes of models has its pros and contras similar to any type of modeling. As it has been shown by comparative analysis of the models of plant responses to climatic changes (Lischke et al. 1998, 2006), simulation (dynamic) modeling provides a detailed "time scan" of the structural or functional ecosystem parameter under consideration but have no opportunity of reasonable spatial extrapolation of the results. In the meantime, empirical statistical models, being static, can describe the territorial diversity of predicted situations but suffer from uncertainly over time. In this context, we believe that landscape ecological prediction for a century-long period should focus not so much on the structural evolution of ecosystems but on the *directed change of their function*, i.e., on the shifts in the small biological cycle (phytomass production and decomposition, as well as humification). The latter takes the first few years in the taiga zone and occurs during the year in the subzone of broadleaved forests (Kolomyts 2003). These characteristic time scales of functional relaxation are proportional to exogenous changes in the carbon content of forest phytomasses (living and dead) and mobile humus of the soil (Dobrovolysky and Nikitin 2006; Osipov 2004). This approach is in complete agreement with the subject of scientific research.

From the standpoint of geographical and functional ecology (Odum 1983; Gerasimov 1985), the mechanisms of biotic regulation of the carbon cycle under global climatic changes can be described by the dynamics of internal turnover and balance of plant matter in ecosystems. We have performed the empirical statistical modeling of functional parameters of topogeosystems on the basis of *biochorological principle* proposed by N.V. Timofeev-Ressovsky (Timopheev-Resovsky and Tyuryukanov 1966; Timofeev-Resovsky 1970): living matter spatially organize the matter and energy cycle in biogeocoenoses as discrete elementary structural units of the biosphere.

Further development of the concept of biotic regulation is seen as a study of local and regional mechanisms of the biological cycle, which ensures stability of natural environment in accordance with the Le Chatelier's principle (Gorshkov 1994). The sphere of *biogeocoenoses* (*landscape facies*[1]) as the elementary biochorological units, i.e., ecosystems of the local (topological) level, is the most complex and active part of natural environment, its functional "core" (Sochava 1974). This elementary unit accepted in geographical ecology in Russia corresponds to categories "Site," "Ecoelement," and "Landtype Phase" in classifications developed in Australia-Britain, Canada, and the USA, respectively (Klijn and de Haes 1994). Spatial diversity of the biological cycles is generated mainly by structural heterogeneity of biogeocoenoses. As far as we know, the concept of biotic regulation of carbon cycle in the biosphere in such an aspect of the problem statement has not been developed so far.

According to the known theoretical works (Krapivin et al. 1982; Gorshkov 1994; Kondratyev et al. 2003), each ecoregion is treated as a statistical ensemble of biogeocoenoses (landscape *facies*): elementary biochorological units interact rather weakly with each other but are highly integrated internally on the basis of stabilizing selection. Each *facies*-level ecosystem acquires particular functional properties associated with the pattern of the local matter and energy flows. This approach makes it possible to reveal the behavioral patterns of biotic communities under different geomorphic and edaphic conditions on the basis of huge empirical information collected in sample plots and processed statistically.

20.2   Conceptual Foundations of Landscape-Ecological Predictions

The previous works by the authors integrating results of field and desktop studies provided opportunity to develop a substantiated and internally consistent *ideology of local and regional landscape-ecological predictions*. This concept takes into account basic statements by predecessors and can be summarized briefly as follows:

1. The stable functioning of geo(eco)systems is ensured first of all by the soil-biotic mechanisms of their monosystem organization and self-regulation (Sochava 1974, 1978). These mechanisms are based on *intercomponent interactions* – the vertical (radial) landscape connections promoting the stabilizing selection, which maintains the modal phenotype of this system, according to Pianka (1978), and providing it with certain *geotropic areal stability* (Kolomyts et al. 2015).
2. The impetus to geo(eco)system transformation at a new climatic signal is given by the intensification of *intercomplex interactions* (Zlotin 1987; Tishkov 1988), i.e., horizontal (lateral) connections and relations *in the system of landscape couplings* (Glazovskaya 2007), or catenas. This effect is determined by the multivalued transformation of background signal in the elements of micro- and

---

[1] *Facies* is the elementary morphological unit of a landscape in Russian terminology. See Glossary and Chap. 1 for details.

meso-catenas under the influence of local geomorphological and hydro-edaphic factors (Vysotsky 1960).

3. The developing disruptive selection promotes the formation of two and more modal phenotypes (Pianka 1978). There is inevitable *spatial diversity of local and regional* functional and structural *rearrangements* of ecosystems at the same background signal (Kolomyts 2003, 2008). This system-forming process leads to a new *evolutionary climax* of phytobiota (Krishtofovich 1946) under the given zonal-regional conditions.

4. In case of disturbed catena organization of ecosystems, *phytobiota* as the most dynamic geocomponent *plays the key transforming role and determines the "fate" of the entire ecoregion* (Semenov-Tyan-Shansky 1928; Simonov 1982). The main driving force of ecosystem rearrangements is *competitive relationships between the ecological niches of plant communities* (Svirezhev and Logofet 1978) in the space of new hydrothermic fields. Competition results in ecological diversification, i.e., the division of niches that favors the development of more complex biotic formations (Pianka 1978). Evolution is implemented via migration processes in the plant cover (Vasilyev 1946), and, therefore, ecosystems shift to the *new type of areal stability, migration*, which is accompanied by the enhanced efficiency of utilization of matter and energy resources, always more or less limited.

The proposed ideological concept is a working tool for applying the landscape approach to ecological prediction. On this basis, we have developed specific ways of constructing *predictive models in the framework of geographical ecology.*

20.3   Empirical Material for Modeling

The forecast analysis was based on the materials of large-scale landscape-ecological surveys carried out by the authors (1987–1998) by a specially elaborated method at six experimental test sites of the Middle and Upper Volga Region. These test sites embrace a wide range of zonal plant formations – from south forest-steppe to mixed forests (Fig. 20.1). Each of these test sites is representative for a particular *regional ecosystem*, with the corresponding conventional name: Zhiguli, By-Sura, Green Town, Shchelokovsky Farmstead, Vyksa, Kerzhenets, and By-Oka-terrace reserve (BOTR).

Six *facies* groups (Table 20.1) which comprise the major diversity of forest types and geotop types have been identified based on 45–50 sample plots in each of the test sites. *Facies* groups arranged in the vector system of local micro-catenas in the direction of edaphic moistening increase, from eluvial (E) and trans-eluvial (TE) geotops through slope transit (T) and transaccumulative (TA) toward riverain accumulative (A) and bed super-aqual (Saq), according to classification by M.A. Glazovskaya (1964). The middle-site types of regional level have the same system of landscape linkages (Polynov 1956). The eluvial-accumulative (EA) sites in interfluve depressions have the excessive edaphic moistening as well. We unite

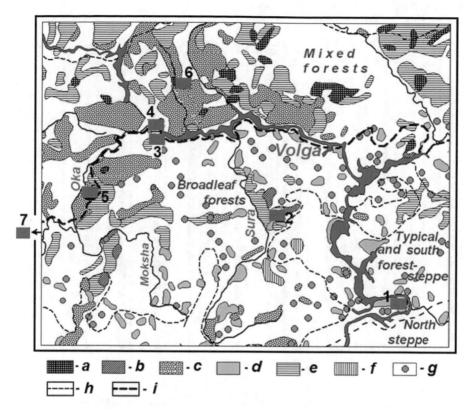

Fig. 20.1 Zonal-regional hydrothermic conditions and forest formations in the Upper Volga River basin and location of experimental test sites

Forests, according to Isachenko and Lavrenko (1974): *a*, fir, silver fir, and black alder; *b*, broadleaved-pine, sometimes with fir; *c*, pine and fir-pine; *d*, broadleaved; *e*, birth-aspen and birch-lime; *f*, birch and pine-birch; *g*, extra-scale areas of broadleaved forests; *h*, boundaries between the natural zones/subzones; *i*, Main Landscape Border of the Russian Plain, by Milkov (1981)

Experiment test sites: 1, Zhiguli low-mountain range; 2, By-Volga high plain, high right bank region of By-Sura river (By-Sura); 3, high-plain *urochishche* Green Town; 4, high-plain farewell Shchelokovsky Farmstead; 5, Oka-Tesha forested lowland in the low By-Oka region (vicinity of the town Vyksa); 6, Kerzhensky reserve (trans-Volga outwash forested lowland); 7, By-Oka-terrace reserve (lowland middle left-bank region of By-Oka river)

these sites together with super-aqual ones. Group of flat interfluve (*plakor*) *facies* was distinguished among them, according to Vysotsky (1960). They occupy usually the eluvial sites and reflect the zonal-regional background of the given territory. The facial groups of the other geotops have been considered as extrazonal categories, i.e., as representatives of other zonal types of geographic environment, not necessarily neighboring but also rather distant.

To calculate carbon balance, we used the phytomass parameters (tons/ha) as follows: (1) mass of stems, branches, and bark of trees and shrub, *BS*; (2) root mass, *BR*; (3) total mass of leaves, *BV*; (4) forest litter mass, *ML*; (5) debris, dead mass of stems and branches (brushwood and dead-wood), *WD*; and (6) humus mass in organic-mineral layers of the soil, *HU*.

Table 20.1 Groups of biogeocoenoses (landscape *facies*) of different Middle Volga ecoregions represented by corresponding experimental sites

Ecoregions (experimental test sites)

Zhiguli low-mountain range (Zhiguli)

1. Light pine forests with occurrence of steppe species on the steep well-insolated slopes
2. Pine forests on sand-loamy soils (*subors*) and pine-broadleaved forests on the steep neutral and poorly insolated slopes
3. Maple-lime oak and aspen forests with nut grove on the poorly insolated slopes of middle steepness
4. Elm-maple-lime forests on the gently sloping interfluves
5. Shadowy broadleaved forests on the lower parts of steep slopes
6. Lime, maple, and aspen forests of the bottoms of small deep valleys

High right bank region of By-Volga River (Green Town)

1. Pine forests on sand-loamy soils (*subors*) on the hilly interfluves
2. Spruce forests and spruce-lime oak forests on the flat sandy interfluves
3. Spruce-lime oak forests and birch forests on the flat sandy-loamy interfluves
4. Spruce-pine forests on the gentle sandy slopes

High right bank By-Sura region (By-Sura)

1. Pine and spruce-pine forests on the upper parts of slopes of sandy ancient-alluvial plain
2. Pine-broadleaved and aspen forests on the flat interfluve of secondary moraine plain
3. Oak-lime and aspen forest with nut grove on the loamy flat interfluves
4. Spruce-pine forests on the lower parts of slopes of sandy-loamy ancient-alluvial plain and high floodplain
5. Spruce-lime-oak and pine-birch-aspen forests on the gently concave slopes of high sandy-loamy interfluves
6. Swamped spruce-pine forests on the interfluve depressions of moraine and outwash sandy-loamy plain

High right bank region of By-Volga River (Shchelokovsky Farmstead)

7. Maple-lime oak forests on the near-watershed well-insolated loamy slopes
8. Lime and aspen forests on the poorly insolated and neutral loamy slopes
9. Maple-lime oak forests on the gently sloping loamy interfluves
10. Maple-lime oak forests on the transit well-insolated loamy slopes

Description	No.	Description	No.
Pine and spruce forests on the depressions of sandy interfluves and on terraces	5	Oak, lime, and aspen forests on the transit poorly insolated and neutral loamy slopes	c/11
Swamped floodplain small-leaved forests in the valleys of small rivers	6	Oak and lime forests on the lower parts and bottoms of loamy slopes	12
Oka-Tesha forested lowland (Vyksa)		*Trans-Volga outwash forested lowland (Kerzhenets)*	
Pinetum or *Piceeto-Pinetum cladinoso-hylocomiosum* forests on the top of sandy ridges	1	Pine forests with spruce and birch on sandy ridges	1
Piceeto-Pinetum vacciniosum forests on the loamy-sandy gently-rolling interfluves and slopes	2	Pine forests with birch and spruce on slightly undulating interfluves of sandy ancient-alluvial plain	2
Piceetum and *Piceeto-Pinetum oxalidosum* forests on the sandy flat interfluves and gentle slopes (with small depth of loam occurrence)	3	Spruce and spruce-birch forests on high interfluves of sandy-loamy moraine-outwash plain	3
Oak-pine-spruce forests with nemoral herbs on the well-drained sandy-loamy slopes and their foots	4	Spruce-oak-lime and small-leaved forests on high floodplains and terraces	4
Pinetum and *Piceeto-Pinetum hylocomioso-myrtillosum* forests on the sandy-loamy gently concave slopes	5	Coniferous and small-leaved forests on the depressions of moraine-outwash plain	5
Pinetum and *Alnetum glutinosa polytrichoso-sphagnosum* forests on the loamy small stream valleys and floodplain depressions	6	Swamped mixed forests on the floodplain depressions and small stream valleys	6
By-Oka-terrace reserve (BOTR)			
Pine-birch forests, with aspen and lime, on gently bulging sandy interfluves and upper slopes	1	Lime-aspen-birch forests on high and middle parts of sandy-loamy slopes (with small depth of carbonate eluvium)	4

(continued)

Table 20.1 (continued)

Ecoregions (experimental test sites)		
Pine-spruce and spruce-pine forests on the gentle sandy-loamy interfluves		2
Pine-lime-oak and pine-lime forests on sandy-loamy interfluves (with small depth of carbonate eluvium)		3
Spruce and spruce-pine forests on middle and lower parts of sandy-loamy slopes		5
Swamped coniferous and small-leaved forests on the floodplain depressions and small stream valleys		6

Convention meanings to symbols of biogeocoenosis groups (for all tables and figures). Dominant forest stand: *a* pine (*Pinus sylvestris*), *b* spruce (*Picea excelsa*), *c* oak (*Quercus robur*), *d* lime (*Tilia cordata*), elm (*Ulmus glabra*), *e* broadleaved species without the division, *f* birch (*Betula pendula*), aspen (*Populus tremula*), *g* black alder (*Alnus glutinosa*). Grass cover: *h* forb-meadow steppe. Soil-forming rocks: *i* sand, *j* sandy loam, *k* middle and heavy loam, *l* sandy-loamy moraine with small boulders, *m* carbonate bedrock

The hydrothermic trends for the period up to 2150 have been taken from two global coupled atmosphere ocean general circulation models (AOGCMs): (1) E GISS (Goddard Institute of Space Search, USA), more moderate (Hansen et al. 2007), and (2) HadCM3, Version A2 (Hadley Centre of Climate Research, UK), more extreme (Gordon et al. 2000; Pope et al. 2000). The first model predicts the relatively temperate global warming. It is one of the most preferable general circulation models, according to the results of testing statistical significance of the data obtained and simulation of the state of contemporary climate, including its seasonal characteristics (Santer 1985). It has been successfully used in a number of prognostic-ecological researches (Izrael 1992; Holten et al. 1993; Smith and Shugart 1993; etc.). Ecological prognosis based on the second model permits to evaluate the forest ecosystem reaction on such global climatic changes which may exceed the level of tree endurance and cause the disintegration of both nemoral and boreal forests on the large spaces.

20.4   Modeling Methods

Our purpose was to establish the local mechanisms of biotic regulation of the carbon cycle and the regional patterns of this regulation on the basis of predicted changes in discrete parameters of the small biological cycle in forest biogeocoenoses at the specified variants of climate prediction for a particular period of time. The potential of biotic regulation of the carbon cycle was assessed by the hydrothermic ordination of discrete metabolic parameters of forest ecosystems under different zonal-regional and local conditions of the Volga River basin.

The thermo- and hydro-edaphic ordination of metabolic characteristics of topo-ecosystems was performed using two geophysical parameters: the temperature of soil at the 50 cm depth (t_{50}) and summer productive moisture reserves in the 0–50 cm soil layer $(W\ 50)$. This parameter is most closely connected with atmospheric humidification. The functional characteristics of forest ecosystems also show the highest correlation with these parameters. As one can see (Table 20.2), correlations are not always rather high though quite significant (Pearson's test of significance, $P < < 0.05$). The weak connections can be interpreted only as a certain general tendency of the given metabolic parameter response to the influence of geophysical trend on the background of significant "noise" effect of the other local-scale factors.

To estimate changes in the carbon contents in individual biotic components and forest biogeocoenoses in general, we used the traditional tools of forest management (Kobak 1988; Tselniker 2006) based on the estimation of the dynamics of living phytomass and dead organic matter (including humus mass), which ensure the best results in CO_2 balance calculations over long periods of time.

Table 20.2 Examples of the models describing the response of the biological cycle parameters in biogeocoenoses of different regional ecosystems (the Middle Volga Region) to soil temperature and soil moisture content during vegetation period

Model equation	Regional ecosystems (see Fig. 20.1)	Parameters (see in the text)	Coefficients			Statistical characteristics		
			b_0	b_1	b_2	R	P	S_y
$y = b_0 + b_1x_1 + b_2x_2$	By-Sura	BV	21.93	−0.7415	**−0.7135**	0.664	0	3.09
		HU	5.99	0	**55.75**	0.696	0	20.29
	Green	BR	8.712	5.838	**−0.0678**	0.582	$0.1 \cdot 10^{-3}$	17.92
	Town	HU	16.34	−0.2402	**0.07426**	0.713	0	
	Shchelokovsky farmstead	ML	27.18	−2.333	**0.04896**	0.717	0	6.51
$y = b_0 + b_1x_1 + {}_2x_1^2$	Shchelokovsky farmstead	WD	642.1	−114.7	5.142	0.936	0	2.21
	Zhiguli	ML	44.08	−0.3611	0.00091	0.699	0	17.75
$y = b_0 + b_1x_2 + {}_2x_2^2$	Shchelokovsky farmstead	BS	−34.93	0.9472	0.00403	0.769	0	90.49
		HU	12.5	0.1712	0	0.649	0	9.31
	Kerzhenets	WD	22.85	−0.2588	0.00149	0.72	0	12.78
		HU	8.84	0.0091	0.0006	0.851	0	8.03
	By-Oka-terrace reserve	BS	244.6	0.3956	−0.00312	0.557	$0.7 \cdot 10^{-3}$	71.4
		HU	25.31	0.0159	0.00042	0.577	$0.4 \cdot 10^{-3}$	16.32
$y = \exp(b_0 + b_1/x_1)$	Zhiguli	BS	3.706	15.42		0.611	0	39.3
		BR	1.768	26.26		0.614	0	12.25
	Green town	ML	0.0845	35.08		0.506	0	16.0
		WD	−4.389	64.63		0.642	0	7.0
	Vyksa	ML	0.3274	46.81		0.798	0	0.34

Footnote. In equations: x_1 soil temperature on depth of 50 cm (t_{50}), x_2 stored soil moisture in the 0–50 cm layer (*W-50*). Statistical characteristics: *R* coefficient of correlation, *P* significance level (Pearson's criterion), S_y standard deviation. Statistically significant equation parameters are shown in bold type

The change $\Delta C(Fa)$ of carbon flow in the soil-plant-atmosphere system was calculated as follows:

$$\Delta!\left(Fa\right) = \Delta!\left(Rm\right) - \Delta!\left(NPP\right), \tag{20.1}$$

where Fa is CO_2 flow above the plant cover; Rm is CO_2 emission as a result of vital activity of soil and aboveground saprotrophs (mainly bacteria and fungi) which decompose humus, forest litter, and skeletal dead mass; and NPP is pure primary production of biogeocoenoses (aboveground + root). Using the considered discrete parameters of minor biological cycle (see Chap. 3, Sect. 3.3.4), the above equation can be written in expanded form:

$$\Delta!\left(Fa\right) = \Delta C\left(WD\right) + \Delta C\left(ML\right) + \Delta C\left(HU\right) - \Delta C\left(BS\right) - \Delta C\left(BV\right) - \Delta C\left(BR\right). \tag{20.2}$$

This balance equation was used to calculate possible changes of carbon flows between soil-plant cover and atmosphere in different periods of prediction (by GISS and HadCM3) for each facies group in all six regional ecosystems (see Table 20.1). The carbon contents in different kinds of phytomass and in labile humus of soil were calculated using conversion coefficients, based on the known literature sources (Kobak 1988; Isaev and Korovin 1999; Osipov 2004, etc.). The values of some of these coefficients (e.g., by BV, ML, and HU) were site-specific and differed between *facies* groups on a test site depending on the zonal and local conditions of habitat.

Each parameter in the right part of the Eq. (2) may have either positive or negative values. With positive values, the first three items indicate an increase of CO_2 emission from soil-plant cover into atmosphere and the remaining three items correspond to a decrease of this flow. In the latter case, the dead mass pool acts as an additional carbon source, while the living phytomass plays the role of its sink (deposit). With negative values of the abovementioned parameters, the pattern is quite the opposite: in the reduced branch of biological turnover, carbon dioxide released into the atmosphere decreases, while the autotrophic biogenesis becomes less intensive and consumes less CO_2, by this compensating the resulting deficiency of carbon dioxide in the atmosphere.

The resulting total balance of changes in carbon exchange between biogeocoenoses and atmosphere $\Delta C(Fa)$ is expected to show whether this group of forest biogeocoenoses consumes additional amount of CO_2 from atmosphere due to the shifts in biological turnover induced by the global warming or, on the contrary, becomes a source of its additional emissions. In the former case, there is a negative feedback realizing the Le Chatelier's principle for stabilization or even weakening of the primary thermo-arid climatic signal; in the latter case, there is a positive feedback, which leads to intensification of the greenhouse effect of atmosphere and, consequently, the warming itself.

Predicted *carbon balance* in local ecosystems for five periods is presented in Fig. 20.2. The zero balance shows the maintenance of the base level of influence, for which the direct and reverse carbon flows between biogeocoenoses and atmosphere

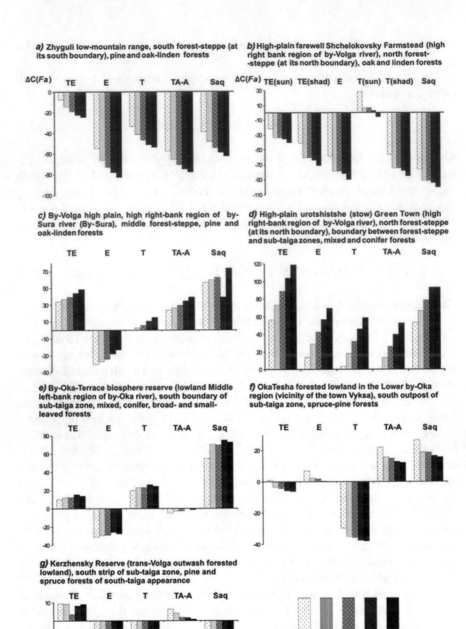

Fig. 20.2 Landscape couplings of predicted C balance dynamics in forest biogeocoenoses under different zonal/regional conditions of the Middle and Upper By-Volga regions, according to the scenarios of HadCM3 model

$\Delta C(Fa)$, carbon balance (t/ha); TE, E, T, ..., local sites (see in the text); solar exposed slopes – sun, high light level; shad, low and neutral light level

are balanced. The negative and positive carbon balances show that the biotic regulation of carbon cycle is directed at additional destabilization of biosphere in the first case and at maintenance of its stability under changing climate in the second case.

The novel methods of geomorphometry were used in the cartographic part of predictive analysis of carbon balance (Shary 2001; Shary et al. 2002). At the local level, the large-scale mapping (1:50,000–1:100,000) was based on predetermined, statistically significant linkages between the structural and functional characteristics of local ecosystems with morphometry of the relief (Zalihanov et al. 2010). Morphometric parameters were derived from the NASA SRTM elevation data (Shuttle Radar Topography Mission, the project of radar survey of relief by Shuttle, 2000) as matrices with 50×50 m cell size recalculated into the Gauss-Kruger projection for the ninth $6°$ zone. For instance, the following multiple regression equations were obtained for the ecoregion of the Zhiguli low-mountain range:

(a) Total C sources for the base period:

$$C_{base} = 222.6 - 55.6 \times GA - 25.6 \times Q - 12.6 \times H; \quad r_S = 0.87, \ P < 10^{-6};, \quad (20.3)$$

(b) The change in total C stocks by 2075:

$$\Delta C_{2075} = -99.8. + 47.1 \times GA + 25.3 \times Q + 18.0 \times H; \ r_S = 0.88, \ P < 10^{-6}., \quad (20.4)$$

where r_S is the Spearman correlation coefficient and P significance level.
Equations for the plain ecoregion of By-Oka-terrace reserve are as follows:

(c) Total C sources for the base period:

$$C_{base} = 243.69 - 92.82 \times MCA - 41.62 \times kh + 22.31 \times Z; \quad r_S = 0.48, \ P < 10^{-2}; \quad (20.5)$$

(d) The change in total C stocks by 2075:

$$\Delta C_{2075} = -5.25 + 10.18 \times MCA + 8.07 \times Q - 3.21 \times kA; \quad r_S = 0.50, \ P < 10^{-3}, \quad (20.6)$$

where Z, elevation a.s.l.; Q, slope illuminance with position of the Sun at the southern azimuth (instant intensity of direct solar irradiation); GA, steepness of slopes; H, mean landform curvature; kh, horizontal curvature showing flux convergence regions with the growth of carbon deposits; MCA, maximal catchment area; and kA, total accumulation curvature separating zones of relative accumulation and removal from areas of relative transit ($\Delta C(Fa)$ is maximum in the transition zone).

Regional maps were composed using the NASA SRTM elevation data. The Oka river basin was taken as a model territory of the regional level. The matrix for this basin was transformed into the Kavraysky VII INT projection for the European Russia with 900 m resolution. Beforehand, we applied "GIS Eco" software (Shary 2001) to perform the statistical analysis of spatial differentiation of changes in the total C content for predictive periods for the plant formations and their combina-

Table 20.3 Equations of the relations between carbon content change (ΔC(*Fa*) in the forest formations of the Oka river basin by 2075 and the most important environment factors, according to HadCM3 model

Groups of plant formations	Regression equations
a) Spruce and broadleaved-spruce forests	$\Delta\mathrm{C}(Fa) = 76.7 \bullet Z + 46.7 \bullet T + 24.0 \bullet Q - 63.8$
b) Pine and broadleaved-pine forests	$\Delta\mathrm{C}(Fa) = -24.0 \bullet Z + 21.0 \bullet MCA - 7.1 \bullet Q + 12.1$
c) Broadleaved forests	$\Delta\mathrm{C}(Fa) = -76.2 \bullet T + 42.9 \bullet MCA + 12.1 \bullet GA + 5.1$
d) Second-growth forests substituting for pine and broadleaved-pine forests	$\Delta\mathrm{C}(Fa) = 83.1 \bullet k_{max} + 76.6 \bullet MCA + 9.7 \bullet T - 61.6$
e) Second-growth forests substituting for spruce, pine, and broadleaved forests (all second-growth forests in the basin)	$\Delta\mathrm{C}(Fa) = 37.5 \bullet MCA - 18.4 \bullet GA - 10.9 \bullet T + 12.7$
f) Primary and second-growth forests of the basin (all forests in the basin)	$\Delta\mathrm{C}(Fa) = 28.8 \bullet MCA + 23.9 \bullet k_{max} - 7.2 \bullet T - 7.8$

Footnote. Predictors in regression equations are as follows: Z absolute elevation, MCA maximal catchment area, GA slope gradient (degrees), k_{max} ridge landforms (maximal curvature), Q insolation (instant intensity of direct solar radiation), T average July temperature

tions. For multiple regression equations used to compose the predictive maps, see Table 20.3.

Transition from the local to regional level of prediction with mapping at 1:2,500,000 scale was carried out using original method of *inductive-hierarchical extrapolation*. We developed this method on the basis of the empirically established phenomenon of polyzonal nature of local ecosystems as a response to global climate changes (Kolomyts 2008). The patterns of the modification of the zonal/regional climatic background by local geomorphic and hydro-edaphic factors serve as the basis for analyzing *regional systems of local zonal structure*. They consist of vector series of upland biogeocoenoses reflecting the zonal/regional background of the area and extrazonal topological level ecosystems as representatives of another, often remote, zonal types of the geographic environment.

Each type or subtype of the plant formation shown in a coarse-scale geobotanical map was identified by a specific group of biogeocoenoses from the upland-extrazonal series (Table 20.4). These series characterize the regional spatially ordered (microcatenary) system of local zonal structure that is adequate to the vector of the predicted climate changes and, hence, may serve as a model of the main trends of ecosystem rearrangements.

In this case, such parameters were the total carbon base stocks in the groups of forest formations, the total carbon balance for the predicted periods. Finally, the weighted average values $\Sigma\mathrm{C}(bas)$, $\Delta\mathrm{C}(Fa)$ were calculated according to the area share in the given areal range of particular groups of formations on the basis of information analysis of each group of plant formations of the basin.

Table 20.4 The predicted changes in total carbon content (Δ C) according to the EGISS climate model on the test areas (points) representing plant formation group 84 (broadleaved-pine and *Pinetum composita, herboso-fruticulosum*)

Test site	Sample plot number	Slope exposure	Site type	Basal C content	ΔC for predicted periods	
					2050 (cooling)	2200 (warming)
Kerzhenets	42	N	E	153.38	−44.12	−46.06
Vyksa	26	Flat surface	E	197.73	27.19	26.08
	51	NE	T	261.7	−69.74	−52.29
	57	E	TE	116.07	74.05	68.99
	58	SE	T	223.85	−47.57	−49.58
	64	Flat surface	E	200.32	−39.79	−42.2
Green Town	11	SW	T	143.22	37.12	33.68
	20	SE	TE	194.32	−3.15	21.52
	26	Flat surface	TA	255.75	−84.61	−59.83
	48	Flat surface	T	154.7	42.09	65.12

The landscape map at the level of species of landscapes and types of *mestnost*[2] (map of *meso-sites*) was composed in parallel (see Kolomyts 2005, p. 21). In this case, the original map of the landscapes groups (sensu V.A. Nikolaev 1978) was used.

Then, each range of this formation was presented as a *multi-vector set of meso-catenas* and was subdivided into regional types of localities, or *mesogeotopes*, from the eluvial and trans-eluvial types to the accumulative and super-aqual in compliance with geochemical classification (Polynov 1956; Glazovskaya 1964). For this purpose, new methods of geomorphometric statistics were applied with the appropriate correction of the results by means of the landscape and soil maps. An intermediate map of the types of *meso-localities* was composed for the entire forested part of the region. Assuming mesocatena to be a homomorphic image of microcatena, we assigned biogeocoenoses of each previously identified group to mesogeotypes of the corresponding geobotanical range. The resultant vegetation-catena mosaic was then related to the basic or prognostic metabolic parameters of biogeocoenoses, which were used as local representatives of different zonal/regional types or subtypes of the geographic environment. In this case, the list of parameters included the total carbon base stocks in the groups of forest formations, the total carbon balance for the predicted periods. Finally, the weighted average values $\Sigma C(bas)$, $\Delta C(Fa)$ were calculated according to the area share in the given areal range of particular groups of formations on the basis of information analysis of each group of plant formations of the basin.

The methodology described here is an innovative approach to the regional mapping of the basic carbon content and predicted carbon balance in forest ecosystems based on the results of field landscape-ecological surveys.

---

[2] *Mestnost* is the morphological unit of a landscape at the level higher than *facies* and *urochishche* in Russian terminology. See Glossary and Chap. 1, for details.

20.5 The Carbon Balance of Forest Ecosystems Predicted for Different Climatic Trends

20.5.1 Analytical Predictive Modeling

The local mechanisms of biotic regulation of the carbon cycle were investigated on the basis of predicted changes in the biological cycle of forest biogeocoenoses under scenarios of climate prediction up to 2150–2200, according to the global models HadCM3 and E GISS. The second model shows a more complex picture: the cold arid trend until the middle of the twenty-first century followed by the weak climate warming and humidification (Table 20.5).

Our calculations have demonstrated that the sensitivity of parameters of the biological cycle to external impact is not adequate to the scope of changes, which occur in the course of this impact. The main results of empirical-statistical prediction are given below:

1. The decomposition part of biological cycle is the most sensitive to climatic impacts. Nevertheless, during sufficiently long periods of time, *the maximal (by absolute values) changes occur not in branch but in autotrophic biogenesis indicated by general net production of forest communities.* Moreover, skeletal tree-shrub phytomass (*BS*), mainly wood gain, is the great part of the shifts of productivity and carbon content (Fig. 20.3).

   Such regularity is outlined in both climate prediction models and is typical for all the forest ecosystems under consideration (Tables 20.6 and 20.7). In 2100, the changes in total carbon content in its extreme values will range from −78÷100) t/ha at the zonal forest-steppe ecotones (test grounds Zhiguli and Shchelokovsky

Table 20.5 The predicted climatic parameters for different periods by three types of hydrothermal trends, according to HadCM3 (version A2) and EGISS models

HadCM3 model									
Test ground	Basic C_{hum}	2075 (thermo-arid trend)				2150 (thermo-arid trend)			
		Δt_{Jan}	Δt_{Jul}	Δr_{year}	C_{hum}	Δt_{Jan}	Δt_{Jul}	Δr_{year}	C_{hum}
Zhiguli	0,95	2,7	3,2	−38	0,71	5,9	7,1	−50	0,36
By-Sura	1,07	0,8	3,1	−16	0,76	7,2	7,0	−34	0,45
BOTR	1,52	3,2	4,0	−96	1,12	6,5	5,1	−31	0,74
E GISS model									
		2050 (cold-humid trend)				2200 (thermo-humid trend)			
		Δt_{Jan}	Δt_{Jul}	Δr_{year}	C_{hum}	Δt_{Jan}	Δt_{Jul}	Δr_{year}	C_{hum}
Zhiguli	0,95	−1,0	−0,5	−19	0,99	1,9	1,2	13	0,80
By-Sura	1,07	−1,0	−1,0	15	1,25	1,7	1,3	16	0,92
BOTR	1,52	−1,3	−1,9	61	2,02	2,6	0,9	58	1,47

Note: The base period falls within the interval of weather observations of 1885–1985. Climatic parameters: Δt_{Jan}, Δt_{Jul}, and Δr_{yr}, the deviations of the mean January and mean July temperatures, annual precipitation, respectively, from their base values; C_{hum}, the Vysotsky-Ivanov annual atmospheric humidity factor

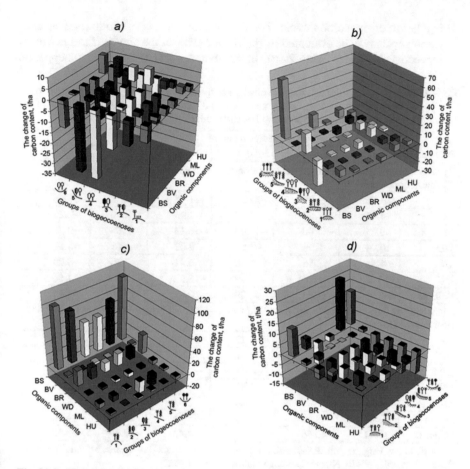

Fig. 20.3 The changes of carbon content (t/ha) predicted for year of 2150 in separate biogeocoenotic pools for different zonal-regional conditions of the Volga River basin, according to HadCM3 model

Test grounds: a, Zhiguli; b, By-Oka-terrace reserve; c, Green Town; d, Kerzhenets. For conventional meanings of biogeocoenoses groups, see Table 20.1

Farmstead) to +115÷120) t/ha in coniferous/broadleaved forests near the northern boundaries of the nemoral forest subzone (Green Town). The changes of carbon content in *BS* will be, respectively, −25÷40 t/ha and + 85÷105 t/ha. It means that in the coming century, the content of carbon conserved in skeletal phytomass of forest ecosystems of the Middle Volga River may change for ±25÷50% and more of the base value on the average. This fact points to quite a significant regulating role of the productivity of mixed and nemoral forests in their carbon exchange with atmosphere.

2. The global warming will induce in some cases an abrupt drop and in other cases significant increase of forest productivity. These effects are expected to have direct response in the carbon balance and to be the key control over the biotic

regulation of the carbon cycle. Zonal-regional contrasts of biotic regulation of carbon cycle may be evaluated by the value $\Delta(Fa)$ for the eluvial (flat-interfluve) topoecosystems (see Table 20.6). In the southern and middle forest-steppe, an

Table 20.6 Predicted partial and balance changes in the carbon content (tons/ha) relative to the base period in flat-interfluve forest biogeocoenoses in different zonal-regional conditions of the Middle and Upper Volga Region, according HadCM3 model

Parameters	Basic values	Changes in predicted periods				
		2025	2050	2075	2100	2150
Zonal ecotone of forest and steppe (Zhiguli low-mountain range – National Nature Park "Samarskaya Luka")						
BS	101.75	−23.05	−27.32	−31.55	−35.78	−39.66
BV	4.16	− 1.60	−1.77	−1.93	−2.10	−2.26
BR	30.25	−11.90	−14.08	−16.13	−18.06	−19.72
WD	10.24	−6.98	−7.53	−8.02	−8.45	−8.80
ML	5.35	3.56	4.90	6.34	7.78	8.99
HU	113.30	−10.54	−15.29	−18.62	−20.68	−21.66
Balance	**265.05**	**− 50.51**	**−61.09**	**−69.91**	**−77.29**	**−83.11**
Right bank of Sura river, mixed and broadleaf forest – National Nature Park "Chavash Forest" (test ground By-Sura)						
BS	123.42	−18.99	−14.77	−10.84	−3.31	2.90
BV	5.17	−0.60	−0.61	−0.84	−0.93	−1.13
BR	40.40	−8.88	−8.04	−7.82	−6.46	−5.64
WD	4.90	−1.06	−1.23	−1.46	−1.71	−1.91
ML	5.88	−1.28	−1.60	−2.03	−2.45	−2.79
HU	31.64	−9.76	−11.09	−11.45	−13.53	−14.76
Balance	**211.41**	**−40.56**	**−37.34**	**−34.42**	**−28.39**	**−23.33**
High-plain By-Volga region, at the northern boundary of nemoral forest zone. Mixed forests in the south vicinity of city Nizhny Novgorod (stow "Green Town")						
BS	134.14	13.73	26.61	38.16	49.82	60.19
BV	6.59	1.90	2.32	2.65	3.01	3.30
BR	39.48	4.44	7.99	11.09	14.28	17.06
WD	4.06	−3.46	−3.63	−3.73	−3.80	−3.84
ML	5.48	−0.13	−1.00	−1.60	−2.05	−2.38
HU	11.16	−1.71	−2.28	−2.65	−3.11	−3.43
Balance	**201.36**	**14.77**	**30.0**	**43.91**	**58.15**	**70.9**
Southern strip of mixed forest zone. Left bank of middle Oka river. Moraine-erosion plain. Pine-lime-oak forests (By-Oka-terrace reserve)						
BS	310.18	−27.14	−27.85	−27.76	−27.99	−27.91
BV	11.54	−1.98	−1.88	−1.89	−1.87	−1.88
BR	84.81	−4.11	−4.30	−4.28	−4.34	−4.32
WD	31.96	−7.44	−7.19	−7.21	−7.15	−7.17
ML	14.96	6.58	9.44	9.58	12.69	11.14
HU	22.05	2.84	2.52	2.55	2.48	2.51
Balance	**475.5**	**−31.24**	**−29.25**	**−29.0**	**−26.17**	**−27.61**

(continued)

Table 20.6 (continued)

Parameters	Basic values	Changes in predicted periods				
		2025	2050	2075	2100	2150
Southern strip of subtaiga zone on the over-Volga Region, moraine-outwash forested lowland. Fir-pine forests – Biosphere reserve "Kerzhenets"						
BS	84.7	−13.00	−12.64	−11.86	−11.85	−1.15
BV	4.41	1.21	1.19	0.82	0.86	0.60
BR	17.08	−9.44	−9.81	−11.68	−11.60	−12.59
WD	7.85	0.40	−0.26	−0.31	−0.40	−0.64
ML	10.15	−3.91	−4.16	−4.71	−4.73	−3.45
HU	13.69	−2.45	−3.41	−3.50	−3.66	−4.09
Balance	**137.88**	**−27.19**	**−29.09**	**−33.24**	**−31.38**	**−31.32**

Footnote. *BS* living skeletal phytomass, *BV* green mass, *BR* root mass, *WD* dead skeletal phytomass, *ML* forest litter mass, *HU* humus mass

unambiguous and quite clear decline of the production potential of forest communities is being observed. Carbon deposit in skeletal phytomass will decrease most quickly in flat-interfluve mesophilic elm-lime and oak forests of erosion-loamy uplands (see Fig. 20.2, *a–c*, and 20.3, *a*, *b*). Their role as a sink of atmospheric carbon will be reduced.

3. The productivity of marginal forest communities at the zonal forest-steppe ecotone will decrease to a greatest extent. In the skeletal tree-shrub pool of flat-interfluve broadleaved forests of the Zhiguli mountain range, the loss of carbon content in 2050 will vary from 2–6 to 26 t/ha and from 9–14 to 34 t/ha by E GISS and HadCM3, respectively. By the end of the twenty-first century, this deficiency will reach 32 and 39 t/ha, respectively. It is obvious that the climate prediction models do not differ too much. Simultaneously, the rate of decay of skeletal dead mass and humus will significantly increase, though the forest litter mass will grow due to the increasing deficit of precipitation (see Table 20.6, *a*). As a whole, the detritus branch here will contribute to further accumulation of CO_2 in atmosphere and, in combination with the production branch, will induce *the progressing disturbance of the Le Chatelier's principle at the southern boundary of the forest belt, resulting in intensification of the global warming.*

In the fluvioglacial plains of the nemoral forest subzone, the moderate hydromorphic birch-spruce-pine forests on the transit sites are able to markedly reduce the rates of carbon conservation in skeletal phytomass as well (see Fig. 20.2, *f*). Rather small changes in productivity are expected in xeromorphic biogeocoenoses, particularly forest-meadow-steppe communities of steep south-facing slopes and pine forests with steppe species at well-drained watersheds. Hence, their contribution to carbon deposition will be insignificant.

4. On the contrary, a significant increase of productivity is anticipated in high-plain mixed and broadleaved forests of the nemoral forest subzone and in the southern subtaiga zone. Mixed and dark coniferous forest biogeocoenoses will significantly increase both their primary productivity and reserves of living organic matter. Accordingly, conservation of atmospheric carbon in perennial skeletal

phytomass of these topoecosystems will increase and, as a consequence, the significance of this phytocoenotic pool as a carbon sink will increase as well (see Figs. 20.2, *d* and 20.3, *c*). Positive biotic regulation of the carbon cycle will be widely spread here. This process will be particularly significant in TE xeromorphic and super-aqual (Saq) hydromorphic boreal forests, which will play the role of the major carbon sink. In 2050–2075, the value of additional carbon deposit in wood will be here from 9–11 to 38–57 t/ha according to GISS and from 36–53 to 77–83 t/ha according to HadCM3; in the end of the twenty-first century, its values will be 38–87 and 75–105 t/ha, respectively. Simultaneously, increased rate of decomposition of forest litter and humus is expected. The effect of forest productivity growth, which is positive for the carbon balance, will be reduced to a certain extent by resulting CO_2 emission into the atmosphere.

5. In subtaiga outwash forested lowlands, the Le Chatelier's principle will be realized only due to the functional shifts of semi-swamped coniferous forests at the lower elements of catena, where the improvement of hydrothermic conditions in soil will induce a significant increase of productivity (see Fig. 20.2, *f*). At the same time, in flat-interfluve and neighboring transit spruce-pine and spruce-oak-lime communities, the autotrophic biogenesis will be weakening against significant activation of decomposition of aboveground dead mass and humus, resulting in the general tendency to negative biotic regulation of the carbon cycle.

Far more unambiguous and intensive changes may occur in the detritus branch of biological cycle (ΔML) on outwash forested lowlands of more northern subtaiga region (Fig. 20.3, *d*). Soil temperature increase and desertification abruptly activate the degradation of dead organic matter, with higher degree of its utilization and additional release of CO_2 to atmosphere.

At a moderate cold-humid signal, the negative values of $\Delta C(Fa)$, up to –(45–60) tons/ha, will be prevalent in upland pine/lime/oak forests of the Middle By-Oka region (the By-Oka-terrace reserve) and eluvial lime forests of loamy sediments of By-Sura region. The thermo-humid signal will cause a rather wide range of the carbon balance: from +(65–90) tons/ha in TE spruce/pine and pine/broadleaf forests of the Prisurye outwash-moraine plain up to −55÷65 t/ha in flat-interfluve and TA broadleaved forests of the Zhiguli mountain range (see Table 20.7).

20.5.2 Area-Weighted Average Rates of Carbon Balances

The resultant effect of carbon cycle regulation by forest cover depends not only on the specific values of the carbon balance $\Delta C(Fa)$ for particular biogeocoenotic groups but also on the ratio of their areas in a given ecoregion. *The catenary organization* of biogeocoenoses expressed in a certain *geomorphological spectrum of the respective types of local sites* requires calculating the carbon balance as an area-weighted average value for these types. We have compared the biotic regulation of

Table 20.7 The changes in the total carbon content (tons/ha) in different forest biogeocoenoses and in their organic components in three natural protected areas in the Oka-Volga River basin for two prognostic periods, 2050 (cooling) and 2200 (warming), according to the E GISS climate model

Characteristics		National Nature Park "Samarskaya Luka" (southern forest-steppe)		National Nature Park "Chavash Forest" (typical forest-steppe)		By-Oka-terrace reserve (southern subtaiga)	
		2050	2200	2050	2200	2050	2200
Groups of biogeocoenoses (types of sites)	1 (TE)	9,55	−12,23	77,33	90,29	20,39	29,46
		[−]	[−]	[−]	[+]	[−]	[+]
	2 (TE, E)	−16,20	−38,72	46,83	63,60	−13,13	−0,44
		[−]	[−]	[+]	[+]	[+]	[−]
	3 (E, t)	3,34	−27,76	**−61,03**	**−32,51**	**−44,80**	**−27,82**
		[−]	[−]	[+]	[−]	[+]	[−]
	4 (E–TA)	**−15,20**	**−58,88**	12,34	31,82	13,85	29,01
		[+]	[+]	[−]	[+]	[−]	[+]
	5 (TA, A)	−36,04	−65,03	7,05	27,24	7,26	11,35
		[+]	[−]	[−]	[+]	[−]	[+]
	6 (Saq, EA)	1,22	−35,79	35,72	63,65	−26,02	11,82
		[−]	[−]	[−]	[+]	[+]	[+]
	Average	−8,89	−39,74	19,71	40,68	−7,07	8,90
		[+]	[−]	[−]	[+]	[+]	[+]
Organic components	BS	−12,65	−22,93	3,56	21,69	−6,38	6,17
		[+]	[+]	[−]	[+]	[+]	[+]
	BV	−0,57	−1,14	− 0,14	0,12	−0,63	−0,08
		[+]	[−]	[+]	[+]	[+]	[−]
	BR	1,98	−5,11	2,92	5,23	−1,81	2,19
		[−]	[−]	[−]	[+]	[+]	[+]
	ML	−1,54	1,38	−1,54	−3,12	−5,02	0,29
		[+]	[+]	[+]	[−]	[+]	[+]
	WD	−2,83	−5,36	24,72	28,62	−0.68	0,74
		[+]	[−]	[−]	[+]	[+]	[+]
	HU	0,42	−6,52	−9,81	−11,85	7,45	−0,42
		[−]	[−]	[+]	[−]	[−]	[−]
	Sum	−8,89	−39,74	19,71	40,68	−7,07	8,90
		[+]	[−]	[+]	[+]	[+]	[+]

Note: Types of sites, according to Glazovskaya (1964) and Ramensky (1971): *E* eluvial, *TE* trans-eluvial, *T* transit, *TA* trans-accumulative, *A* accumulative, *Saq* super-aqual, *EA* eluvial-accumulative. Flat-interfluve biogeocoenoses are in bold. [+] and [−] denote the attenuation and intensification of climatic trend, respectively. For other symbols, see Table 20.6

the carbon cycle by upland forest biogeocoenoses, which are known from literature (Vysotsky 1960) to represent a certain zonal-regional (background) bioclimatic standard of this territory, on the one hand, and by its entire biogeocoenotic diversity (as the weighted average of $\Delta C(Fa)$ as a parameter of biotic regulation), on the other

Table 20.8 The upland (A) and area-weighted average (B) base level of organic carbon in the soil-plant cover (t/ha) and its changes over the nature protected areas in the Middle and Upper Volga Region according to the predictive climate models

		HadCM3 model		E GISS model	
Versions of the average	Basic period	2075 warming	2150 warming	2050 cooling	2200 warming
a) Zhiguli. National Nature Park "Samarskaya Luka" (southern forest-steppe)					
A	216,36	−90.81 [−]	−100.47 [−]	−15.20 [+]	−55.88[−]
		(42,0)	(46,4)	(7,0)	(22,8)
B	193,7	−73.87 [−]	−82.60 [−]	−10.81 [+]	−44,23 [−]
		(38.1)	(42.6)	(5.6)	(22.8)
b) By-Sura. National Nature Park "Chavash Forest" (typical forest-steppe)					
A	221,9	−40.75 [−]	−32.08 [−]	−61.03 [+]	−32.51 [−]
		(18.4)	(14.5)	(27,5)	(14,7)
B	167.4	24.3 [+]	34.75 [+]	16.14 [−]	37.41 [+]
		(14.5)	(20.8)	(9.6)	(22.3)
c) BOTR. By-Oka-terrace reserve (southern subtaiga)					
A	243,43	−32,84 [−]	−31,45 [−]	−44,88 [+]	−27,82 [−]
		(13.5)	(12.9)	(18.4)	(11.4)
B	200.49	9.02 [+]	10.65 [+]	−9,52 [+]	6,16 [+]
		(4.5)	(5.3)	(4.7)	(3.1)

Note. Deviations from the base value (%) are given in parenthesis. [+] and [−] denote positive and negative regulation of the carbon cycle at a given climatic trend, respectively

hand. The calculations were made for the three nature protected areas (NPA) of the Volga Region (Table 20.8).

The results showed evidence that the carbon balance values in ecosystems of upland forest communities are dramatically different from the area-weighted average values of this ecoregion not only in magnitude but also in sign. In the By-Sura and BOTR ecoregions, the upland nemoral and mixed forests must provide rather substantial negative regulation of the carbon cycle under global warming. However, positive though slightly weaker regulation is generally predicted for each of these regions. The opposite regulation contrasts are peculiar to the ecoregion of typical forest-steppe under global cooling. Extrazonal (Sprygin 1986) forest communities comprising an entire spectrum of local deviations from the zonal-regional standard and occupying up to 70–75% and more of the region area yield the results exceeding greatly the carbon balance of upland forests as the "cores of typicality" (Armand 1975), of the given zonal-regional system.

The typical example is the topography-driven diversity of positive regulation of the carbon cycle for the cold humid scenario for 2050 according to EGISS (see Tables 20.7 and 20.8). In the zonal range of landscapes in the Volga River basin under consideration (from the southern forest steppe to the southern subtaiga), such regulation is expected, first of all, from local representatives of the bioclimatic background – eluvial upland forests; the communities of accumulative sites will play a less significant role ($\Delta C(Fa) \approx$ from −15 to −45÷60 t/ha). These will be the Zhiguli and By-Sura ecoregions, where the maximum decrease in the carbon balance of ecosystems will occur due to a decrease in the skeletal mass of forest stand. In the areas with domination of upland and accumulative facies, one can expect the overall mitigating effect of forest cover on the global cooling ($\Delta C(Fa) = -9.5 \div 10.8$ t/ha; see Table 20.8).

However, in the By-Sura region of typical forest-steppe, the two extreme "poles" of local catenas, namely, TE and Saq, dominate, where forest biogeocoenoses will reduce CO_2 emission to the atmosphere (due to a decrease in the rates of debris decomposition and, to a lesser extent, an increase in timber and root mass), in spite of increased mineralization of humus. As a result, the forest cover of this ecoregion will perform the general negative regulation of the carbon cycle ($\Delta C(Fa) = 19.71$ t/ha) end by this exacerbate global cooling.

Thus, there is direct evidence of the incorrect extrapolation of model scenarios of the biological cycle and carbon balance derived from simulation modeling of the "typical biomes" (Forman 1995; Aber et al. 2001; Kudeyarov 2007; Zavarzin 2007) on the entire territory of a natural zone/subzone or a large region. For each structural subdivision of a particular zonal-regional unit, the carbon balance has to be calculated by the *geomorphological types of biogeocoenoses* in the framework of local patterns of their *catenary organization*, followed by obtaining the taxonomic (area-weighted average) standard value of the carbon balance. It yields entirely different results of balance calculations compared to the calculations by the typical (background) biomes (Tishkov 2005) as we have demonstrated by the experiment of empirical statistical modeling.

20.5.3 Carbon Balance of Forest Formations Under Conditions of Cooling and Warming

The empirical data from large-scale landscape ecological surveys made it possible to calculate the carbon balance ($\Delta C(Fa)$ of forest formations of the Oka-Volga River basin for different scenarios of global climate changes and thereby to assess the magnitude of positive or negative biotic regulation of the carbon cycle of these zonal/regional phytocoenological units at a territorial scale (Tables 20.8 and 20.9).

According to the extreme scenario of global warming (HadCM3), the area-weighted average carbon balance in forests of the major water catchment area of the Volga River basin by 2050 will have probably been within a range of $-21 \div 27\%$ to $+11 \div 17\%$ of basic carbon content, with an overall increase in $\Delta C(Fa)$ from south-east to northwest. These values are quite comparable with the data of other predictive regional estimates. For example, the calculations according to IPSL-CM2 (the SRES-A2 scenario) for the East European subcontinent showed an increase in the net primary production (NPP) and the respective carbon deposition by 20% on average by the middle of the twentieth century (Mokhov et al. 2005). It corresponds to the maximum possible increment of carbon in upland mixed forest stands of the

Table 20.9 The area-weighted average and generalized carbon balances in different types (subtypes) of forest vegetation in the territory of the Oka-Volga River basin for the cold humid (2050) and thermo-humid (2200) scenarios, according to E GISS predictive climate model

Type (subtype) of vegetation	Area m²	%	Weighted average balance, tons/ha 2050	2200	Total balance, million tons 2050	2200
I. Spruce southern taiga and broadleaved-spruce subtaiga forests	65,961	0.31	−1.67	−2.85	−11.024689 [+]	−18.827987 [−]
II. Pine middle and southern taiga forests	9745	0.05	−20.56	−19.89	−20.032470 [+]	−19.383642 [−]
III. By-Kama mixed dark-coniferous forests	48,506	0.23	14.43	14.32	69.973657 [−]	69.458707 [+]
IV. Broadleaved-pine (subtaiga) forests	37,841	0.18	8.24	15.99	31.178295 [−]	60.512105 [+]
V. East European broadleaved forests	25,894	0.12	14.40	−2.23	37.280251 [−]	−5.770327 [−]
VI. Swamped forests	7088	0.03	14.32	17.52	10.606824 [−]	12.977064 [+]
VII. Nemoral floodplains	16,267	0.08	0.15	7.91	0.244005 [−]	12.867197 [+]
Total	211,302	1.00	29.30	30.77	118.225873 [−]	111.833117 [+]

Volga River basin (see Table 20.6). For coniferous, mixed, and other forests of the United States, the changes in net primary production and carbon content were calculated according to predictive biogeochemical models CEN (CENTERE) and TER (Terrestrial Ecosystem Model), according to Aber et al. (2001). According to the scenario of UKMO-1987 climate model (the precursor of the Hadley Centre Climate Model), it was shown that the doubling of atmospheric CO_2 leads to the changes in carbon content in the range from −1.5÷1.8 to +7.8÷12.5%.

Even in accordance with the moderate predictive climate model E GISS, the specific carbon balance of the groups of forest formations (GFF) will have a considerable amplitude: within ± (5–10)÷(45–58) tons/ha, i.e., from 0.5–1.0% to 25–38% of the basic carbon content in forest ecosystems.

Under the cold humid climatic trend (the prediction for 2050 according to E GISS), the overwhelming majority of forest stands will increase the productivity and, accordingly, the total carbon stocks. This fact confirms our hypothesis that forest communities at the boreal ecotone of the East-European Plain, from the southern subtaiga and further to the south, are under conditions of insufficient atmospheric moistening (Kolomyts 2003, 2005). The effect of increased precipitation but not decreased temperature will be dominant in the cold humid trend (see Table 20.6), which will result in additional carbon sink from the atmosphere to forest ecosystems and, accordingly, exacerbate the cooling.

The number of the groups of forest formations with positive $\Delta C(Fa)$ values and, hence, negative regulation of the carbon cycle exceeds twice the number of groups with the negative balance that mitigates cooling (see Table 20.9). The carbon sink will also reach the maximum values (up to 32–35% of its base stocks) compared to its maximum emission (no more than 12–17%). The most substantial increase in cooling will occur on the part of broadleaved forests over the entire forest-steppe zone (with the exception of mesophytic lime oak forests and oak forests of By-Ural region), subtaiga *Pinetum fruticuloso-sphagnosum* forests, and, to a lesser extent, swamped forests in all zones. The cooling will be mitigated only by nemoral-grass spruce- and pine-broadleaved forests, which will provide additional CO_2 emission.

The prediction associated with the thermo-humid trend according to E GISS for 2200 shows a different pattern. Both factors (the increase in temperature, both in summer and winter, and the increase in annual precipitation) will interfere with the same sign, causing the prevalent increase in forest productivity and total carbon stocks in these forests (up to 35–43 tons/ha, i.e., more than 20–25% of its base stocks). The latter will prevent the increase in near-surface atmospheric temperatures and thereby will have a positive effect on biotic regulation of the carbon cycle. The process will be especially prevalent among the southern taiga and subtaiga pine and coniferous-deciduous formations (see Table 20.9). At the same time, the increase in atmospheric moistening will play the key role in nemoral-grass and *Pinetum cladinosum* forests, while the temperature increase will have priority in subtaiga *Pinetum fruticuloso-sphagnosum* forests and in forest bogs.

In the forest-steppe zone, the thermo-humid climatic signal will cause, on the contrary, a decrease in productivity of broadleaved and pine/small-leaved forests and, accordingly, additional CO_2 emission to the atmosphere (from 5–15 to 30 t/ha and more, i.e., from 3–5% to 35–30% of the base carbon stocks). Here, the negative effect of higher temperatures (summer temperatures, in particular) will be stronger than the positive effect of increased precipitation. As a result, the lime, lime-oak forests, birch and aspen forests, as well as broadleaf-pine forests with steppe species will play a negative role in biotic regulation of the carbon cycle, intensifying the initial global warming. The simultaneous increase in precipitation will level this process but insignificantly.

Thus, the regulation of the carbon cycle by boreal and nemoral forests of the Russian Plain must be variegated enough not only by sign but also by absolute magnitude. Such variegation among the groups of forest formations will manifest itself within each type (subtype) of vegetation; therefore, the final effect of regulation of the carbon cycle for the entire Volga River basin territory under consideration will substantially depend on the ratio of GFF areas.

In the meantime, the high specific weighted average values of the carbon balance in a particular GFF often do not correspond to its area. For example, the pine middle and southern taiga forests with the maximum negative values of $\Delta C(Fa)$, according to both climatic scenarios, occupy small areas and their total contribution to the regional carbon emission will be rather low (see Table 20.8). All in all, they will have a weak influence on both cooling and warming in the Volga region. The (±)-regulation of the carbon cycle by the most widespread (occupying more than 30% of the region area) dark-coniferous southern taiga and mixed forests will also be insignificant due to the low negative values of $\Delta C(Fa)$. On the contrary, much less common broadleaved-pine subtaiga forests will be the objects of the strongest CO_2 sink, which will predetermine their dominant role (together with the Kama basin mixed forests) in the regional biotic regulation of the carbon cycle: negative or positive under conditions of global cooling or warming, respectively. The East European broadleaved forests, in spite of their small areas, will be able to cause a significant negative regulation of the carbon cycle in the region under the cold humid trend, due to a considerable increase in productivity and the respective carbon sink. The thermo-humid trend will cause an insignificant positive regulation on their part.

In general, the forest cover of the Oka-Volga River basin must implement the commensurate regulation of the carbon cycle, both negative and positive under cooling and warming conditions, respectively (see Table 20.8, the lower line), with climate humidification in both cases. Approximately the same scales of the regulatory effect of forests can be also expected for extreme thermo-aridization. Our calculations have shown (Rozenberg et al. 2011) that the total carbon balance of forest cover of the territory under consideration for the period till 2075, according to HadCM3, must be about 110.298 million tons.

It would be interesting to compare our findings with the published data. Over the 65-year period (1985–2050), carbon accumulation in the forests of Russia must be about 23.1 Gt (Zalihanov et al. 2006). It will make up 577.5 million ha per 2.5% of area, as is commensurate with the area of the Oka-Volga River basin under consideration, i.e., more than five times higher than is expected here by 2075 according to HadCM3 and by 2200 according to E GISS.

The total carbon sink in the forests of Russia by 2010 had estimated as 150–200 Mt./year (Zamolodchikov et al. 2011). Taking this deposition to be 3.75–5.00 Mt./year for the 2.5% area, we evaluate that in the next 65 years the total carbon sink will probably amount to 245–325 Mt., also exceeding our predictive values two or three times. As we can see, though the Hadley model is considered extreme, it gives even underestimated values of changes in the carbon content of forest cover under conditions of global warming.

20.6   Mapping Predictive Scenarios of the Carbon Balance

The base and predictive matrix maps of forest ecosystem characteristics derived from of such models (Figs. 20.4 and 20.5) are a helpful tool not only for better understanding of elementary and complex intra-system interactions but also for decision-making in forestry. In the Zhiguli low-mountain range, the maximum CO_2 emission under global warming will be characteristic for gentle near-watershed poorly insolated slopes, with the minimum catchment area. On steep slopes, carbon emissions will be negligible. In the By-Oka-terrace reserve, the maximum total carbon stocks in the base period are characteristic for low-inclined near-watershed regions with the minimum catchment area. In the predicted period, these sites, as well as the relatively steep slopes, especially south-facing ones, will be characterized by the maximum CO_2 emission. The forest stands in ravines and near-talweg locations will be the pools of the maximum carbon deposition.

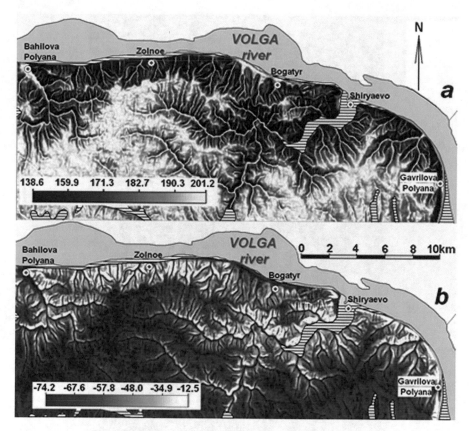

Fig. 20.4 Zhiguli low-mountain ecoregion – National Nature Park "Samarskaya Luka". Base total C content in forest ecosystems (*a*) and changes in total C for the predicted period of 2075 (*b*), in t/ha, according to HadCM3 model
Main predictors for the calculation of maps – insolation – coming to the slopes (illumination), steepness of slope, maximal catchment area. Step of lattice – 50 m. Mean change, calculated by matrix: for 2075, −65.5 t/ha; for 2150, −74.0 t/ha

Let us now consider plant formations, considering that their carbon characteristics involve not only vegetation but soils as well. The maximum basic total carbon stocks (above 200 t/ha) were observed in the southern taiga and subtaiga pine forests and the birch forests that replace them during succession. The broadleaved-pine subtaiga forests, as well as the oak and lime forests of the nemoral forest subzone and their derivatives, contain much less carbon: 140–180 t/ha. The minimum total carbon stocks (100–130 tons/ha) account for the middle and southern taiga spruce and broadleaved-spruce subtaiga forests (Fig. 20.6). The empirical data are generally in agreement with the review materials on forest phytomass with regard to individual species and their carbon stocks in the respective regions of European Russia (Usol'tsev 2003) but make them much more precise as they characterize the real forest formations but not idealized pure plantations. In the works of the cited author, it has been noted that the total carbon stocks in the territory of the Upper and

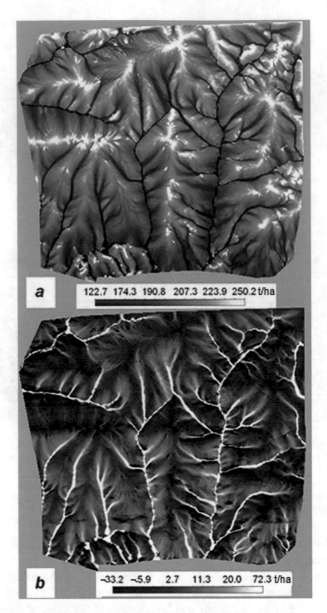

Fig. 20.5 Lowland Middle left-bank region By-Oka river – By-Oka-terrace reserve. Changes of sum carbon content (carbon deposit, t/ha) to predictive dates 2075 (*a*) and 2150 (*b*), in t/ha, according to HadCM3 model

Main predictors – illumination, maximal catchment area and horizontal curvature showing flux convergence regions, with the growth of carbon deposits. Step of lattice – 20 m. Mean change, calculated by matrix: for 2075, +0.35 t/ha; for 2150, +0.67 t/ha

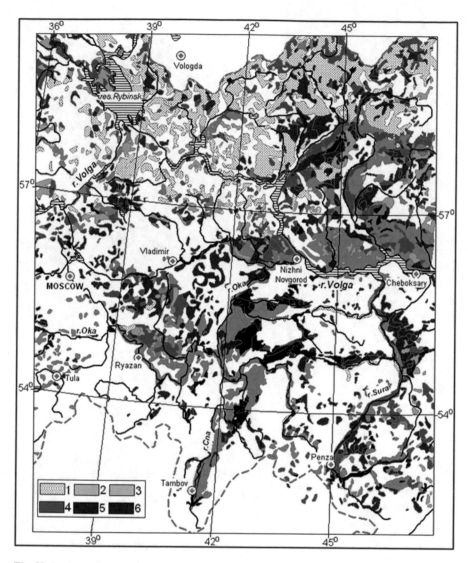

Fig. 20.6 Map of basic total carbon content in forest ecosystems of Oka-Volga River basin
Total C content, in t/ha: 1–104–116; 2–116–130; 3–130–150; 4–150–170; 5–170–206;
6–206–280

Middle Volga Region account for the following: (1) in pine and spruce forests,
100–200 t/ha (up to 350 t/ha in the pine forests of the southeasternmost regions); (2)
in oak and lime forests, 100–200 t/ha and 200–350 t/ha in the northwestern part of
the region and in its southern and southeastern part, respectively; (3) in birch forests, from 100–200 t/ha in southern taiga and subtaiga to 25–100 t/ha in the
forest-steppe zone; and (4) in aspen forests, from 10–200 t/ha in the western sector
to 200–350 t/ha in By-Ural region.

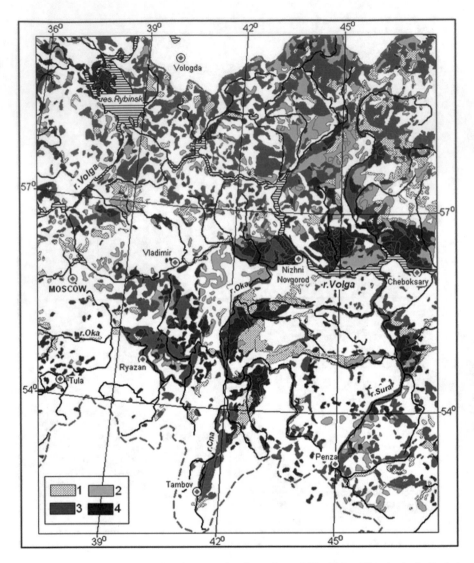

Fig. 20.7 Map of C balance (t/ha) in vegetation formations of Oka-Volga River basin for period 2050 by prognostic climate E GISS model (cool-humid trend)
The changes of total C content, in t/ha: 1 – (−76)–(−14); 2 – (−14)–0; 3–0–20; 4–20–162

The maps show evidence that substantial differences are inherent to spruce forests only. Our values for these formations approximate to the data presented in Bobkova (1987) on total carbon stocks in the phytomass of middle taiga upland spruce forests in the North of the European Russia: about 80–100 t/ha.

For the territory of the Oka-Volga River basin, the maps of the carbon balance of forest formations have been composed for the cold humid (2050) and thermo-humid (2200) climatic scenarios according to E GISS (Figs. 20.7 and 20.8). According to the former scenario, the most effective positive regulation of the carbon cycle that

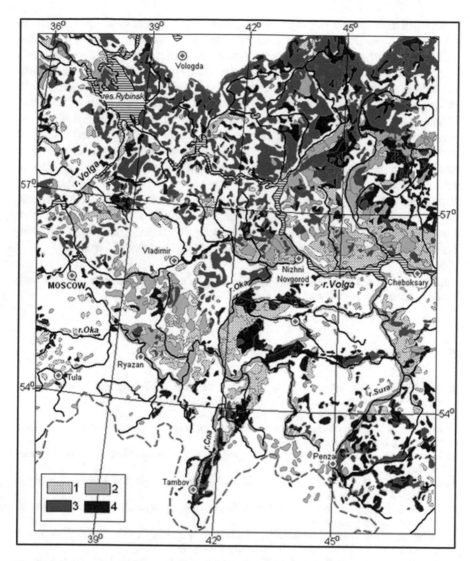

Fig. 20.8 Map of C balance (t/ha) in vegetation formations of Oka-Volga River basin for period 2200 by prognostic climate E GISS model (thermo-humid trend)
The changes of total C content, in t/ha: 1 – (−66)–(−10); 2 – (−10)–0; 3–0–30; 4–30–89

would mitigate the process of cooling will be ensured by small-leaved forests that have replaced the primary middle and south taiga and subtaiga pine forests in the Unzha and Vetluga river basins and along the Oka-Moksha right bank, where their productivity will be maximally reduced ($\Delta C(Fa) = -40 \div 75$ tons/ha).

At the same time, the abovementioned indigenous communities, as well as the broadleaved-pine subtaiga forests, will increase their productivity ($\Delta C(Fa)$ in the range from 25–40 to 100–150 t/ha), thereby exerting a negative effect on the carbon balance. The small sparse stands of broadleaved forests in the Sura and Moksha river basins will play a very weak positive regulatory role ($\Delta C(Fa) = -5 \div 15$ t/ha). In general, this scenario is characterized by the small-scale mosaic of values $\pm \Delta C(Fa)$.

The thermo-humid scenario presents a much more distinct pattern (see Fig. 20.8). The middle and southern taiga spruce forests and the substituting birch forests over the entire northern and northeastern part of the Oka-Volga River basin will reduce the productivity (due to increased atmospheric moistening) and thereby must contribute to global warming ($\Delta C(Fa)$ will account for $-15 \div 30$ to $-45 \div 65$ t/ha). The weak negative regulation of the carbon cycle ($\Delta C(Fa) = -10 \div 35$ t/ha) should be expected also from oak and lime forests of the forest-steppe zone. As we have already seen, this regulation must be much more pronounced under extreme thermo-aridization according to HadCM3. At the same time, the primary deciduous-pine forests and their derivatives (birch and aspen forests) in the broad belt encompassing the subtaiga zone and the nemoral forest subzone will show a sufficiently pronounced positive regulation of the carbon cycle as they will increase their productivity ($\Delta C(Fa) = 25$–45 tons/ha, here and there up to 60–85 tons/ha).

Acknowledgments This research was financially supported by Russian Foundation for Basic Research (grant 18-05-00024-a).

References

Aber, J., Nelson, R. P., Mcnulty, S., et al. (2001). Forest processes and global environmental change: Predicting the effects of individual and multiple stressors. *Bioscience, 51*(9), 735–751.

Armand, D. L. (1975). *Science of landscape*. Moscow: Mysl. (in Russian).

Bobkova, K. S. (1987). *Biological productivity of coniferous forests of European North-East*. Leningrad: Nauka. (in Russian).

Dobrovolysky, G. V., & Nikitin, E. D. (2006). *Soil ecology*. Moscow: Moscow University Press. (in Russian).

Forman, R. T. T. (1995). *Land mosaics: The ecology of landscapes and regions*. Cambridge: Cambridge University Press.

Gerasimov, I. P. (1985). *Ecological problems in the past, present-day and future world geography*. Moscow: Nauka. (in Russian).

Glazovskaya, M. A. (1964). *Geochemical foundations for typology and methods of research of natural landscapes*. Moscow: MSU Publishing House. (in Russian).

Glazovskaya, M. A. (2007). *Geochemistry of natural and technogenic landscapes*. Moscow: Moscow State University Press. (in Russian).

Gordon, C., Cooper, C., Senior, C. A., et al. (2000). The simulation of SST, sea ice extents and ocean heat transport in a version of the Hadley Centre coupled model without flux adjustments. *Climate Dynamics, 16*, 147–168.

Gorshkov, V. G. (1994). *Physical and biological basis of life stability. Man, Biota, environment*. New York: Springer Verlag.

Hansen, J., Sato, M., Ruedy, R., et al. (2007). Climate simulations for 1880–2003 with GISS model E. *Climate Dynamics, 29*, 661–696.

Holten, J. I., Paulsen, G., & Oechel, W. C. (Eds.). (1993). *Impacts of climatic change on natural ecosystems (with emphasis on boreal and arctic/alpine areas)*. Trondheim: NINA & DN.

Isachenko, T. I., & Lavrenko, E. M. (Eds.). (1974). *Map of vegetation of the European part of the USSR. Scale 1: 2,500,000*. Moscow: GUGK. (in Russian).

Isaev, A. S., & Korovin, G. N. (1999). Carbon in forests of the Northern Eurasia. In *Carbon cycle in the territory of Russia* (pp. 63–95). Moscow: Ministry of science and technology. (in Russian).

Izrael, Y. A. (Ed.). (1992). *Evaluation of ecological and socio-economic consequences of climate changes. Report of the II MGEIK working group*. St. Petersburg: Gidrometeoizdat. (in Russian).

Klijn, F., & de Haes, H. A. U. (1994). A hierarchical approach to ecosystems and its applications for ecological land classification. *Landscape Ecology, 9*(2), 89–104.

Kobak, K. I. (1988). *Biotic components of the carbon cycle*. Leningrad: Gidrometeoizdat. (in Russian).

Kolomyts, E. G. (2003). *Regional model of global environmental changes*. Moscow: Nauka. (in Russian).

Kolomyts, E. G. (2005). *Boreal ecotone and geographical zonality. Atlas-monograph*. Moscow: Nauka. (in Russian).

Kolomyts, E. G. (2008). *Local mechanisms of global changes in natural ecosystems*. Moscow: Nauka. (in Russian).

Kolomyts, E. G., Kerzhentsev, A. S., & Sharaya, L. S. (2015). Analytic and cartographic models of the functional stability of forest ecosystems. *Biology Bulletin Reviews, 5*(4), 311–330.

Kondratyev, K. Y., Losev, K. S., Ananicheva, M. D., & Chesnokova, I. V. (2003). *Scientific foundations of life sustainability*. Moscow: VINITI, Institute of Geography RAS. (in Russian).

Krapivin, V. F., Svirezhev, Y. M., & Tarko, A. M. (1982). *Mathematical modeling of global biospheric processes*. Moscow: Nauka. (in Russian).

Krishtofovich, A. N. (1946). Evolution of plant cover in the geological past and its main factors. In V. L. Komarov (Ed.), *Materials on history of flora and vegetation of the USSR* (Vol. II, pp. 21–87). Moscow-Leningrad: Academy of science USSR Press. (in Russian).

Kudeyarov, V. N. (Ed.). (2007). *Modeling organic matter dynamics in forest ecosystems*. Moscow: Nauka. (in Russian).

Lischke, H., Guisan, A., Fischlin, A., & Bugmann, H. (1998). Vegetation response to climate change in the Alps: Modeling studies. In *Views from the Alps: Regional perspectives on climate change* (pp. 309–350). Cambridge: MIT Press.

Lischke, H., Zimmermann, N. E., Bolliger, J., et al. (2006). TreeMig: A forest-landscape model for simulating spatio-temporal patterns from stand to landscape scale. *Ecological Modelling, 199*(4), 409–420.

Milkov, F. N. (1981). *Physical geography: Present-day state, regularities, problems*. Voronezh: Voronezh University Press. (in Russian).

Mokhov, I. I., Dufrcnsc, J.-L., Trent, II. L., ct al. (2005). Changes in drought and bioproductivity regimes in land ecosystems in regions of northern Eurasia based on calculations using a global climatic model with carbon cycle. *Doklady Earth Sciences, 405A*(9), 1414–1418.

Nikolaev, V. A. (1978). *Classification and small-scale mapping of landscapes*. Moscow: Moscow University Press. (in Russian).

Odum, E. P. (1983). *Basic ecology*. Philadelphia: CBS College Publishing.

Osipov, V. V. (Ed.). (2004). *Ecosystems of Tellerman forest*. Moscow: Nauka. (in Russian).

Pianka, E. R. (1978). *Evolutionary ecology* (2nd ed.). New York: Harper & Row Publishers.

Polynov, B. B. (1956). *Selected works*. Moscow: Academy of science USSR Press. (in Russian).

Pope, V. D., Gallani, M. L., Rowntree, P. R., & Stratton, R. A. (2000). The impact of new physical parametrizations in Hadley Centre climate model – HadCM3. *Climate Dynamics, 16*, 123–146.

Printice, I. C., Cramer, W., Harrison, S. P., et al. (1992). A global biome model based on plant physiology and dominance, soil properties, and climate. *Journal of Biogeography, 19*, 117–134.

Puzachenko, Y. G. (1984). Directions in geographical ecology. In R. I. Zlotin et al. (Eds.), *Modern problems of ecosystem geography* (pp. 15–19). Moscow: Institute of Geography AS USSR. (in Russian).

Ramensky, L. G. (1971). *Problems and methods for studying vegetation cover. Selected works.* Leningrad: Nauka. (in Russian).

Rozenberg, G. S. (1984). *Models in phytocenology.* Moscow: Nauka. (in Russian).

Rozenberg, G. S., Kolomyts, E. G., & Sharaya, L. S. (2011). Changes in carbon balance in forest ecosystems under global warming. *Proceedings of Russian Academy of Sciences, Geographical Series, 3*, 33–44. (in Russian).

Santer, B. (1985). The use general circulation models in climate impact analysis – A preliminary stay of the impacts of a CO_2 – Indicated climatic change on West European agriculture. *Climatic Ghanges, 7*(1), 71–93.

Semenov-Tyan-Shansky, V. P. (1928). *District and country.* Moscow-Leningrad: State Press. (in Russian).

Shary, P. A. (2001). Analitical GIS Eko. http://www.giseco.info. (in Russian).

Shary, P. A., Sharaya, L. S., & Mitusov, A. V. (2002). Fundamental quantitative methods of land surface analysis. *Geoderma, 107*(1–2), 1–32.

Simonov, Y. G. (1982). Principal properties of objects of geographical forecast and methods of their formal description. In *Problems of regional geographical forecast* (pp. 112–193). Moscow: Nauka. (in Russian).

Smith, T. M., & Shugart, H. H. (1993). The transient response of terrestrial carbon to perturbed climate. *Nature, 361*(6412), 523–526.

Smith, T. M., Leemance, R., & Shugart, H. H. (1992). Sensitivity of terrestrial carbon storage to CO_2-induced climate change: Comparison of four scenarios based on general circulation models. *Climatic Change, 21*, 367–384.

Sochava, V. B. (1974). Geotopology as a division of the theory of geosystems. In V. B. Sochava (Ed.), *The topological aspects of the theory of geosystems* (pp. 3–86). Novosibirsk: Nauka. (in Russian).

Sochava, V. B. (1978). *Introduction to the theory of geosystems.* Novosibirsk: Nauka. (in Russian).

Sprygin, I. I. (1986) (first published in 1926). *Materials for studying vegetation of the Middle Volga region.* Scientific heritage. Vol. 11. Moscow: Nauka. (in Russian).

Sukachev, V. N. (1972). *Selected works. Vol.1. Foundations of forest typology and biogeocenology.* Leningrad: Nauka. (in Russian).

Svirezhev, Y. M., & Logofet, L. O. (1978). *Stability of biological communities.* Moscow: Nauka. (in Russian).

Timofeev-Resovsky, N. V. (1970). Structural levels of biological systems. In *System studies. Ezhegodnik* (pp. 80–114). Moscow: Nauka. (in Russian).

Timopheev-Resovsky, N. V., & Tyuryukanov, A. N. (1966). On elementary biochorological units of biosphere. *Bulletin MOIP, Section Biology, LXXI*(1), 123–132. (in Russian).

Tishkov, A. A. (1988). Approaches to studies of biota dynamics as an object of geographical forecast. In V. S. Preobrazhensky (Ed.), *Geographical forecast and nature protection problems* (pp. 49–60). Moscow: Institute of Geography AS USSR. (in Russian).

Tishkov, A. A. (2005). *Biospherical functions of natural ecosystems of Russia.* Moscow: Nauka. (in Russian).

Tselniker, Y. L. (2006). Gas cycle of CO_2 in forest biogeocenoses. In S. E. Vompersky (Ed.), *Ideas of biocenology in forest science and forestry* (pp. 213–229). Moscow: Nauka. (in Russian).

Usol'tsev, V. A. (2003). *Phytomass of the forests of the North Eurasia. Limiting productivity and geography.* Ekaterinburg: Ural State Forest-technical University. (in Russian).

Vasilyev, V. N. (1946). Conformity to natural laws of process of plain changes. In V. L. Komarov (Ed.), *Materials on history of flora and vegetation of the USSR* (Vol. II, pp. 365–403). Moscow-Leningrad: Academy of science USSR Press. (in Russian).

Vysotsky, G. N. (1960). *Selected works.* Moscow: Sel'khozgiz. (in Russian).

Zalihanov, M. C., Losev, K. S., & Shelekhov, A. M. (2006). Matural ecosystems – The most important resource of the mankind. *Proceedings of Russian Academy of Sciences, 76*(7), 612–634. (in Russian).

Zalihanov, M. C., Kolomyts, E. G., Sharaya, L. S., et al. (2010). *High mountain geoecology in models*. Moscow: Nauka. (in Russian).
Zamolodchikov, D. G., Korovin, G. N., Grabovsky, V. I., & Korzukhin, M. D. (2011). Bimodal retrospective analysis of carbon balance in forests in 1988–2008. In A. S. Komarov (Ed.), *Matematical modeling in ecology. Eko-MatMod-2011* (pp. 100–101). Pushchino. (in Russian). http://www.ecomodelling.ru/doc/conferences/proceedings_EcoMatMod_2011.pdf. Accessed 04 Aug 2018.
Zavarzin, G. A. (Ed.). (2007). *Carbon pools and fluxes in terrestrial ecosystems of Russia*. Moscow: Nauka. (in Russian).
Zlotin, R. I. (1987). Ecological problems of biota and geosystems stability. *Proceedings of Academy of sciences of the USSR, geographical series, 6*, 45–51. (in Russian).

Chapter 21
Actual Changes of Mountainous Landscapes in Inner Asia as a Result of Anthropogenic Effects

Kirill V. Chistyakov, Svetlana A. Gavrilkina, Elena S. Zelepukina, Galina N. Shastina, and Mikhail I. Amosov

Abstract Mountain landscapes' reactions on global changes are of particular interest since they reproduce long-term trends contributing to generation of forecasts and scenarios for sustainable regional development. The series of observations in mountainous Inner Asia over the climate, the glacier balance, river runoff lasting for about half a century allowed us to clarify the spatial gradients of the geosystem structure and functioning characteristics, and, therefore, resulting in parameterizing their spatial and temporal variability. Digital elevation models (DEM) with the grid points determining the qualitative and quantitative attributes of geosystems, enabled us to analyze the modern landscape structure, to establish probabilistic relationships between the distributions of geographic components, and to estimate the ranges of climatic characteristics within which these components' balance can be realized. Among the tundra geosystems, the most cold-resistant are the cobresia and dryad ones; herbal species are the least dependent on air temperature, while dwarf shrub tundra tends towards warmer environment. However, grass-sedge and dwarf shrub tundra spreading evidences the critical contribution of geological and geomorphological impacts. Ecological and climatic niches of the normal geosystem functioning can be used in forecasting the landscape transformation under climate changes. The application of statistical methods allows evaluating the contribution of landscape genesis' various factors to the formation of the high-altitude territory structure.

Keywords Landscape dynamics · Anthropogenic load · Climatic changes

K. V. Chistyakov · S. A. Gavrilkina (✉) · G. N. Shastina · M. I. Amosov
St. Petersburg State University, St. Petersburg, Russia
e-mail: s.gavrilkina@spbu.ru

E. S. Zelepukina
St. Petersburg State University, St. Petersburg, Russia

Federal State Budget-Financed Educational Institution of Higher Education, The Bonch-Bruevich Saint-Petersburg State University of Telecommunications, St. Petersburg, Russia

© Springer Nature Switzerland AG 2020
A. V. Khoroshev, K. N. Dyakonov (eds.), *Landscape Patterns in a Range of Spatio-Temporal Scales*, Landscape Series 26,
https://doi.org/10.1007/978-3-030-31185-8_21

347

21.1   Introduction

One of the tasks of landscape research at the current stage is to study the dynamics and evolution of ecosystems under a changing climate and growing man-induced impact. The relevance of spatiotemporal landscape dynamics research under the influence of natural and anthropogenic factors is caused by the increased concern of mankind with the intensification in the rates of global changes, thus entailing a noticeable accumulation of dangerous natural disasters provoking serious damage to the population and the economy.

Mountainous landscapes are characterized by high sensitivity to environmental impacts. Responses of these landscapes to the effect of spontaneous and anthropogenic factors do not reflect just random variations, but directional changes in geographical environment, which makes it possible to justifiably develop forecasting and projecting sustainable development of vast areas, such as Inner Asia. By Inner Asia we consider the landscapes of the Altai-Sayan mountainous country (Southern, Central and Mongolian Altai, Western and Eastern Sayan, Tsagan-Shibetu, Western, Eastern and Peaked Tannu-Ola), as well as other neighboring ridges and orographic formations belonging to the drainless areas of the continent (Fig. 21.1). A characteristic feature of the regional landscape is the wide spread of intermountain basins (the Great Lakes Depression, the Valley of Lakes, etc.) and relatively low isolated mountain massifs.

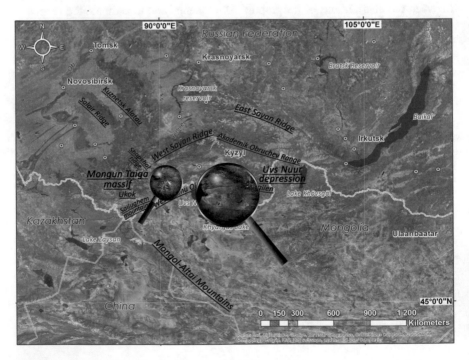

Fig. 21.1 Location of key sites in the research region

Being sparsely populated and hard-to-reach, the territory of Inner Asia still remains underexplored as far as its physicogeographical, paleogeographic, glaciological, and permafrost features are concerned.

The ground-based observation data in Inner Asia are lacking due to the sparsity of hydrological and weather stations and the insufficient duration of observation series. The use of contemporary equipment (automatic hydrometeostations and loggers connected to sensors, etc.) makes it possible to shift from unsystematic records of landscape states to continuous series of universal characteristics reflecting the functioning modes of geosystems. However, no remote studies can solve all the problems of regional ecogeographical forecasting. These circumstances increase the importance of expeditionary research; otherwise the interpretation of rapidly increasing remote survey data remains problematic.

21.2 Study Area

Geographers of St. Petersburg State University conduct research in various areas of Inner Asia following the best expeditionary traditions of the Russian Geographical Society dating back to the middle nineteenth century. When choosing research methods, semi-stationary observations of monitoring type are preferred, at which testing grounds are denoted with benchmarks. Repeated measurements are performed at model areas according to a standard procedure. The most detailed works were carried out on the massifs of Mongun-Taiga (Republic of Tuva, Russia), Tabyn-Bogdo-Ola (Republic of Altai, Russia; Mongolia), and Turgani-Nuru, Kharhira-Nuru, Tsambagarav (Mongolia). Climate, glacier balance, runoff, and other attributes of high and mid-mountain landscape component were obtained in research expeditions during the last four decades. They provide opportunity for specifying the values of regional gradients to identify various aspects of geosystem activity and, therefore, contribute to a relevant forecast of space time landscape changes.

The formation of the climate in a region located in the continental sector of the mid-latitude zone is greatly influenced by a number of factors as follows: the western air mass transport prevailing throughout the year at altitudes of 1000–2000 m; the circulation processes developing over West Siberia; remoteness from the oceans, and the proximity of the territory to arid spaces of Central Asia. The peak annual values of the radiation balance were recorded in intermountain basins. The total input of radiation on mountain slopes is determined by their aspect.

For most of the year, the region is dominated by the continental air, characterized by low temperatures, dryness, and a stable meteorological regime in winter. The vertical temperature gradient in January does not exceed -0.3 °C/100 m, but due to inversions in the depressions, the temperature drops below -50 °C. In spring, foehns in the mountains and sandstorms in the depressions are common.

Rainfall distribution in the region is in extreme contrast: in the western Altai, annual precipitation reaches 2000 mm and in the Tuva and Ubsu-Nur depressions it is less than 200 mm. In the annual course, precipitation peak is observed commonly in July; however, in the mountains, much of the winter precipitation is caused by the intervention of cyclones of the Arctic front, while the summer rainfall is mostly influenced by the cyclones of the polar front. The snow accumulation decreases southeastward. For instance, in the southeastern Altai, locally it hardly ever exceeds 5 cm. The growth of the snow accumulation with the altitude is observed only up to the altitude of 2000 m.

The considerable longitude-latitude outstretch of the region results in a wide variety of local climates, which, in its turn, leads to the formation of specific landscape structure due to the inherent diversity of edaphic mountainous conditions, such as slope gradients and Sun exposure, tectonic structures, geologic substrate, matter migration modes, etc.

Geosystems of Inner Asia are represented by boreal semi-humid, semi-arid, and arid landscapes, which differ dramatically from spectra of high-altitude belts (Ogureeva 1980; Kuminova 1960; Sobolevskaya 1950). Glacial-nival and mountain-tundra landscapes prevail in highlands; mid- and low mountains are covered by forests, mainly by taiga, as well as by various steppes; intermountain depressions and hollows are dominated by steppes and semi-deserts. The boundaries between altitudinal belts are frequently indistinct; numerous ecotone formations appear, such as forest-steppe, tundra-steppe, forest-tundra. Southward and eastward gradients in heat and water availability affect the snowline position, the upper and lower boundaries of the forest belt, and also the altitude belt range related to the increasing climate continentality.

Modifications in mountain landscapes, determined by the joint action of natural and anthropogenic factors, manifest themselves in different ways, depending on the sustainability of geosystems and on the type, intensity, and duration of the environment impact. Climate change is one of the most important natural factors of landscape dynamics.

Some global climate scenarios, based on estimations of anthropogenic greenhouse gases emissions into the atmosphere, suggest a significant increase in the global temperature of the Earth's surface. The EGISS model predicts a rise of 1–1.5 °C in the average July temperature in the region by mid-twenty-first century (Schmidt et al. 2006). This corresponds to the computation data of the RCP 8.5 scenario, widely applied for prediction calculations in numerous glaciological, meteorological, and other studies (Riahi et al. 2011). It should be noted that the dissected mountain relief always gives a much more diverse picture of potential climate change. The results of long-term weather data analysis in the study area show that, since the mid-1980s, there has been a significant increase in the average summer air temperature. Precipitation trends in the region have also been revealed. In low-mountain areas (below 900 m a.s.l.), the average annual and average summer precipitation remains stable or changes insignificantly over half a century (Kyzyl, Toora-Khem, Sosnovka weather stations), while in the mid-mountains, the precipitation is gradually decreasing (Erzin, Mugur-Aksy weather stations). However, it would be incorrect to infer decrease in moisture from this data, since observations in adjacent territories often show different trends. Thus, if the climate change remains the same in future, the most probable scenario is an increase in the average summer temperature against a slight decrease in precipitation.

The cumulative data of integrated study describing the functioning mechanisms of mountain geosystems make it possible to assess the landscape transformation trend and amplitude due to the environmental impact, first and foremost, due to the observable climatic changes. Contemporary transformations of intact arid high-mountainous landscapes are considered through an example of the Mongun-Taiga massif, the Ubsunur Biosphere Reserve cluster.

The Mongun-Taiga massif (3970.5 m a.s.l.) is located to the southeast of the linkage between the ridges of the Altai and Tannu-Ola system. The massif is surrounded by multilevel intermontane troughs. The stepped relief results from the wide occurrence of areas with gentle (2–3 °) slopes – remnants of planation surfaces lifted to different heights due to a considerable amplitude of vertical tectonic movements. Traces of glacial activity are wide-spread here: kars, cirques, moraine-pondage lakes, trough river valleys, etc. A unique feature of the massif glaciation is a diversity of glaciers morphology: dome-shaped peak complexes, sloping, hanging, cirque, and valley glaciers. The climatic snow line corresponds approximately to the level of 3600 m a.s.l.

Based on long-term expedition studies, thematic maps, and space images, we composed a unique landscape map (1: 100,000) covering an area of more than 1500 km².

The total area of the massif glaciation amounts to about 20 km². Above 3200 m plateau geosystems with a scarce soil-vegetation cover (*Waldheimia tridactylites, Saxifraga oppositifolia, S. melaleuca, Crepis nana, C. chrysanta*, etc.) dominate. The soil cover consists of nutrient-poor Cryosols (cryozems) and Leptosols (lithozems), where the average depth of permafrost is 0.3–0.4 m.[1]

Mountain tundra, occupying almost a third of the massif area, is represented by several types: kobresia (*Kobresia myosuroides* prevailing); dryad (*Dryas oxyodonta* with cryophyte herb species of *Bistorta viviparum, Papaver nudicaule, Pedicularis uliginosa*, etc.); moss-dwarf shrub (*Betula rotundifolia*, green and *Polytrichum* mosses); grass-sedge tundra (*Carex melanantha, C. orbicularis, C. sempervirens, C. stenocarpa, Eriophorum polystahion*, etc.). The tundra soil cover consists of Cryosols, Leptosols, and (in hollows) taiga Gleysols.

Alpine meadows (*Saxifraga sibirica, Draba sibirica, Aster alpinus, Trollius asiaticus, Dracocephalum grandiflorum, Gentiana grandiflora*, etc.) are distributed sporadically on Mollic and Folic Leptosols. Meadows with dominance of *Kobresia myosuroides* are combined with cryophyte mixed herb sedge meadows in moister habitats (with the predominance of *Carex melanantha*). Mixed grass meadows on Umbrisols are characterized by a wide variety of species (Festu*ca ovina, F. kryloviana, Poa altaica, Galium verum, Bistorta viviparum, Dracocephalum grandiflorum, Sajanella monatrosa, Tephroseria praticola, Campanula rotundifolia*, etc.),

---

[1]The Latin names of plants are summarized in S.K. Cherepanov (1995) classification of soils according to Shishov (2004).

and significant presence of shrubs and small shrubs (*Betula rotundifolia, Juniperus pseudosabina, Spiraea alpina, S. media, Salix reticulata, S.glauca*).

Mixed herbs larch forests are characterized by high stand density of *Larix sibirica* and diversity in the grass-shrub layer, which consists not only of the boreal herbs but of tundra species as well. *Laricetun nano-betulosum* forests (*Betula rotundifolia, Lonicera altaica, Juniperus sabina, Dryas oxyodonta, Salix sp.*), are characterized by a sparser stand (the projective cover does not exceed 30%). At altitudes below 2000 m, the presence of steppe species among boreal herbs in the soil cover is noticeable (*Artemisia tanacetifolia, Eritrichium pectinatum, Pentaphylloides fruticosa, Potentilla pensylvanica*). A few species of *Salix* brushwood with hygrophytic herbs are distributed in the river floodplains.

Mountain steppes, occupying about a third of the massif area, are represented by several types as follows: forbs steppes (*Festuca valesiaca, F. kryloviana, Setaria viridis, Poa attenuate, Erigeron krylovii, Potentilla astragalifolia* etc.), wormwood steppes (*Artemisia depauperata, Festuca lenensis, Ephedra fedtschenkoae,* etc.), and shrubby steppes (genera *Caragana, Berberis, Dasiphora fruticosa, Artemisia* types).

21.3 Methods

High-resolution (30 m) digital terrain model enabled us to use a novel approach for analyzing the present-day landscape structure. We established the probability correlation between the occurrence of geographic components and estimated the ranges of climatic attributes which allow the existence of these components and their combinations.

The vegetation cover patchiness assessment was carried out based on the concept of *landscape site*[2] determined by mesorelief homogeneity, surface deposits, matter, and moisture migration modes (Isachenko and Reznikov 1996). The accuracy of classifying into landscape sites at the selected scale is confirmed by good correspondence of the GIS thematic layer with the digital slope model. This is important when analyzing the conditions for geosystem formation and for modeling their potential transformations.

The comparison of binomial distributions for the phytocoenosis groups using Wald's method (Cox and Snell 1981) revealed the dependence of their formation on altitude and soil factors. To estimate the determination rate, the criterion χ^2 (Volkenshtein 2006) was applied. Ranking particular factors, which drive the development of phytocoenosis groups, was carried out by comparing the calculated deviations of the dependence measures.

---

[2] See Chap. 17, this volume, for details.

21.4 Results

Our estimates showed that distribution of dryad tundras and forb-bunchgrass steppes is determined mainly by the climatic factor, with altitude being an integrated climatic driver, while the distribution of grass meadows, floodplain forests, and shrub steppes is influenced by geologic and geomorphologic factors. In general, it was shown that the lower is the sensitivity of phytocoenosis groups to one of the considered factors, the higher is its sensitivity to another one.

The combined effect of various landscape genesis factors determines the formation of an ecotone in the altitude range of 2200–2800 m with approximately equal proportions of mountain tundra, mountain meadow, and mountain steppe geosystems. Similar vegetation cover of neighboring altitudinal steps, confirmed by close values of the integrated occurrence index, illustrates the mutual adaptation and relative equilibrium between vegetation communities. In case of a high patchiness of vegetation cover due to contrast topography in the ecotone, a diffuse type of phytocoenosis contact prevails. This type is characterized by blurring boundaries of the territorial units, gradual transitions between geosystems in space and by the increase in phytocoenosis species diversity (Zelepukina et al. 2018). A similar phenomenon has been observed in the adjacent arid regions of the southeastern Altai and southwestern Tuva: in the altitude range from 2100 to 2400 m meadow and gramineous steppes are enriched with cryophilic species (Makunina 2014).

To detail and analyze the conditions for the geosystems formation, we performed simulation of fields of spatial patterns for the main meteorological indices. The simulation was based not only on time series obtained from the weather station Mugur–Aksy (1830 m), which is the nearest to the massif under study, in the years of 1963–2012 (FGBU VNIIGMI-WDC), but also on data collected from seasonal field weather observations at altitudes of 2260, 2620, 3140 m. At each pixel of DEM, we calculated the average summer air temperature and summer precipitation with due regard to vertical gradient values: −0.69 °C/100 m, as well as 7 mm/100 m and 12 mm/100 m for altitudes above and below 2200 m, respectively (Chistyakov and Kaledin 2010).

The imposition of spacing areas with calculated meteorological data values on the landscape map made it possible to identify the ranges of climatic features with the highest and the lowest geosystem occurrence rate. Distribution analysis of different geosystem types depending on their heat and moisture availability regime revealed that several plant communities can exist under similar conditions.

The climatic niches (ranges of climatic feature values with occurrence of affine geosystems) characterize optimal conditions for the functioning of geosystems. Dense intersection of geosystem climatic niches in the altitude interval of 2200–2800 m, indicating equiprobable development and coexistence of communities belonging to different ecological groups under similar climatic conditions, illustrates not only the absence of prominent dominating landscapes but also the priority of geological and geomorphological factor in vegetation dispersal. This is most clearly manifested through the availability in the altitude zonation structure of arid highlands with larch forests and swamped sedge tundras at extremely low precipitation totals (about 110–150 mm per summer) in overmoistened locations due to surface runoff and seasonal thawing of frost soil.

The study of factors affecting the dynamics of environmental conditions, and trigger mechanisms of these factors, is based on the concept of long-lasting landscape states, i.e. such properties "of its structure that are being preserved for more or less prolonged period of time" (Mamay 1992). The reasons for changing the states can be both, factors external to the system and the circumstances of its self-development and self-organization. The life span of geographic complexes is comparable with the lifetime of the core system elements, first of all, of plant communities. Within the homogeneous geomorphological contours, significant transformations of the dominant plant communities of mountain steppe and alpine landscapes occur within the period of 40–50 years, and in mountain taiga – for about 100–150 years (Puzachenko and Skulkin 1981).

The similarity of the climatic niches for different geosystems implies not only the possibility of transmutation, but also indicates the transformation trend under a definite change in climatic conditions. For instance, with rise in temperature availability, the processes of increasing in the proportion of mesophytic species in the structure of tundra seem to be the most expected, resulting in transformation of cryophylic cushion plants into graminaceous or *Dryas* and *Cobresia* associations, and also a growth in the proportion of *Festuca*-forbs species groupings in *Cobresia* tundras, as well as prairiefication in the structure of high mountain meadows (Dirksen and Smirnova 1997).

However, it should be remembered that the formation of contemporary combinations of meso-, hygro-, xero-, and cryophyte plant communities took a long period of time due to the evolution of both environmental conditions and plant communities with their various adaptive capacity. Therefore, modifications in the ratio of the heat-moisture supply parameters expected by the middle of the twenty-first century seem to be not sufficient to change the landscape invariant. High sensitivity of plant communities to climate variations (dryad tundras and forb-graminaceous steppes especially) is somewhat leveled out by the influence of orographic and geomorphological factors, thus manifesting in a certain geosystem inertia.

Nowadays we face a reduction in the areas of snow and ice formations and a rise of their lower boundary. The glaciation degradation process is taking place in an irregular manner: on the one hand, there are significant differences in the glacier retreat rates, depending on their size, morphological type, exposure timing, etc.; however, since 2009, there has been some slowdown in the retreat of glaciers. During the last 40 years most of the valley glaciers retreat with an average speed of 7–11 m/year (Ganyushkin et al. 2017).

To estimate the scope of potential changes in the massif landscapes, a scenario-based space modeling of meteorological data values was carried out, with an expected increase by the year of 2050 in the mean summer air temperature by 2 °C versus a decrease in precipitation by about 15%. The consequences of such changes are thought to be most noticeable at the foot of the massif and in its highest parts, where glaciation at present is formed. Calculations predict a gradual increase in temperature resulting in a two-fold snow and ice formation area reduction (Ganyushkin et al. 2015).

Based on the concept of spatiotemporal analysis and synthesis of geographical landscapes (Beruchashvili 1986), in a general way and with regard to transition periods, conceivable trends in massif landscape transformations can be viewed as follows:

- Occupation of new locations (deglaciated rockbeds, rockslides, and moraine) by discrete petrophyte clusters
- An increase in vegetation cover shading of cobresia tundra on flat parts of the highlands, with their partial replacement by ground-shrub communities
- A gradual replacement of dryad tundras on mid-steep slopes by some kinds of mountain steppes or Cobresia associations and on undulating moraine surfaces by grass-mixed meadows
- A small reduction of moss-dwarf shrub tundras on morainic deposits cemented by permafrost, and a more noticeable one on slopes
- Increase in the proportion of xerophytic sod grasses in meadows on river valley terraces and drain slopes at the bottom stages of the massif; however, in some hollows, mixed-grass communities are expected not only to maintain their availability, but to raise their productivity as well
- Embedment and spread of larch new growth at the top of the forest boundary; strengthening of the xerophytic species in the undergrowth at the bottom forest boundary, followed by the separation of larch forests into areas or groups with patches of steppe meadows in-between
- Extended spaced grass and sagebrush steppes dispersal at the foot of the massif, followed by semi-desertic steppe replacement.

Thus, the main trends in the dynamics of high-altitude landscapes in the region under consideration at the present stage can be reduced to a decrease in the total area of glaciers, the disappearance of small glaciation forms, the release of rocky areas and loose sediments from ice cover, activation of thermokarst processes and permafrost degradation. In the long term, despite the low diversity of physicogeographical environments at the place of contact with the glacial-nival belt, one can assume some amplification of the highland belt structure resulting from glacier retreat and soil-vegetation cover formation on graded surfaces.

In the altitude interval 2400–2800 m, where the greatest values of the vegetation cover entropy are observed, the overall stability of the whole complex of various elements (emergence manifestation) is significantly increased; therefore, the consequences of climatic changes are expected to result in a certain structure transformation, for instance, an insignificant change in the reproportioning of the main geosystem areas forming the ecotone, as well as the diffusion of closely located systems and the acquisition of features specific for other ecological group communities. Thus, it was noted that cryozems in tundra areas with pronounced steppe features are characterized by a decrease in the humidity of the cryoturbated horizon and by a greater development of the humus one (Lesovaya and Goryachkin 2007). In the landscape structure of the lower parts of the massif, aridization occurring in the continental part simultaneously with a rise in temperature and a drop in the amount of precipitation is not likely to influence the structure complexity. However, the latter can acquire a more deserted view similar to the southern foothills of the Tannu-Ola ranges with nanophyton communities (Volkova 1994).

A closer look at the changes in low- and mid-mountain landscapes of Inner Asia is represented via data obtained in the Ubsu-Nur depression A bitter-saline drainless Ubsu-Nur (Ubsa) lake is located in the southwestern part of the basin at an altitude of 759 m. The height of the basin mounting reaches 4000 m and more: the depression is surrounded by sub-latitudinal ridges with small glaciers and snowfields coupled with stony placers and rock outcrops at the height of more than 3000 m. The landscape map on a scale of (1: 500,000), compiled from the fieldwork data obtained during the geographic expeditions organized by the St. Petersburg state University, covers an area of about 70,000 sq. km. Geosystem distributive characteristics depending on hypsometric, geological-geomorphological, expositional, and climatic differences formed the basis for identifying the leading factors of geosystem spatial differentiation.

The determining factor in the formation of psammophytic, petrophytic, halophyte, and semi-desert geosystems (*Cleistogenes, Stipa, Ptilagrostis, Artemisia frigida, Kochia*, etc.) is the geological and geomorphological conditions of the basin bottom, i.e. a wide distribution of soils with light mechanical composition and sandy massifs, saline soils, etc. Thus, solonchak geosystems (sedge and grass-sedge meadows) are distributed on saline substrates with close-up groundwater on rivers and lake terraces, lake-river and continental deltas. The distribution of petrophyte and psammophyte (a combination of loose sands and sparse grass and shrub) geosystems is mainly associated with a crystalline basement daylight surface emergence, a large amount of detritus in river deltas, alluvial cones, etc., as well as the spread of eolian relief forms.

The landscape shape of the low-mountain part of the basin (up to 1000 m) is characterized by the predominance of feather grass steppes, solonchaks (mainly on the lakeside hollows), and semi-deserts with insignificant participation of small-leaved valley forests. Steppe geosystems are widely spread throughout the altitudinal basin mounting profile. In the middle altitude the taiga belt with its steppified larch forests, *Pineto(sibirica)-Laricetum herbosum* forests, *Pineto(sibirica)-Laricetum* and *Piceeto-Laricetum herboso-hylocomiosum* forests is quite remarkably distinct. The upper forest boundary (about 2400 m) is represented by sparse larch woodlands coupled with moss-lichen dwarf shrubs. However, above 2200 m, the tundra with lichen and dwarf shrubs forests predominates.

The central and eastern parts of the basin are rich in barchan, hilly, shallow-cellular, and coarse-grained sands, covered with sparse feather grass (*Stipa*), wheat grass (*Agropyron*), oxytropes (*Oxytropis*) and pea shrub (*Caragana*) communities. The traditional model of nature management (grazing) alongside with the increase over the last decades in the air temperature and moisture lack contributed to the drying out of the mantled layers with their feeble cementation, as well as partial or complete root system separation, which resulted in a reinforced blowing, denudation, and desertification. However, to date, we are facing a pronounced tendency to

Fig. 21.2 A case of sand fixation by turf forming grasses and caragana in the vicinity of the Tere-Khol lake (at the top – NASA Blue Marble world imagery, 2016, below – Gavrilkina S. photo, 2013)

consolidate sandy ridges and dunes with expanding psammophyte groups (Fig. 21.2). In our opinion, the development of psammophyte communities under conditions of growing aridization is likely to be the consequence of a significant decrease in the unfavorable overgrazing impact owing to the fact that the territory acquired the SPNT (Special Protection National Territory: natural biosphere reserve) status which brought serious changes in the mode of anthropogenic activities.

To evaluate the importance of the climatic factor for the depression landscape differentiation, the ecological and climatic geosystem niches were identified with regard to plant community occurrence frequency within certain limits of estimates for the sums of active temperatures (above 10 °C) and summer precipitation at different types of locations. Simulation of the spatial distribution fields for the main climatic characteristics was carried out on evidence from 15 weather stations through a digital relief model. Due to the sizeable depression, the height difference, as well as the temperature inversions and barrier effects, both vertical and longitude gradients were used in the calculations (Chistyakov et al. 2009).

The patchiness of landscape structure observed in the middle altitude depression, determines numerous overlapping, intersecting, and sometimes, localization of ecological and climatic niches of various geosystems, is above all caused by orographic, geological, and geomorphological factors.

According to the results of expectable landscape transformation modeling under a given climatic scenario (warming versus atmospheric precipitation reduction), a significant increase in semi-desert, halophytic, and psammophyte geosystem areas in the low-mountain belt seems to be quite possible. In the middle altitude, a considerable expansion of the forbs-gramineous steppes seems possible to occur, resulting in a sharp decrease in the areas of cedar-larch and spruce-larch forests. Despite the increasing air temperature at the upper forest boundary (about 2400 m), present-day existing climatic conditions are quite favorable for the reinforcement and spread of the undergrowth, still we can hardly expect the upslope forest movement in at least one stand generation change. Moreover, the potential for tree upgrowth is limited by highland specific features, such as steep slopes, explicit cryogenic processes, strong wind, etc.

For the territory of Inner Asia, nature-use modes and landscapes interaction are of paramount importance. This is due to the extreme living conditions of ethnic groups, their struggle for survival. Generally speaking, landscapes of the region went through weak modifications, which is typical for areas with indigenous nomadic culture. It should be recognized that anthropogenic impact is the main factor limiting the spread of forest geosystems in the areas with a continental climate. Thus, in the lowlands the acute problem of wood shortage for household needs leads to the deforestation of valley forests (*Populus, Salix, Betula*), which are sensitive to external impacts and exist at a very poor atmospheric humifying (below 150 mm of precipitation over summer) only due to auxiliary groundwater feeding. Moreover, the natural renewal of small-leaved forests is hindered by the constant intensive undergrowth trampling and consumption by livestock. In the mid-altitude taiga belt of the continental sector, due to changes in meteorological conditions, a considerable reduction in mesophytic species in the ground cover (prairiefication) is observed leading to a decrease in the habitat forming capacity and, consequently, to forest sustainability, thus, changing their fire resistance and resulting in a more frequent occurrence of both anthropogenic and natural large-scale fires spreading over tens of kilometers and causing significant damage (Fig. 21.3). As for industrial logging, it is more relevant for the semi-humid areas of the Altai and Western Sayan. In this case, the renewal of dark coniferous forests occurs through a series of recovering successions with small-leaved stages (high-grass, small-leaved, small-leaved-coniferous), as well as with the change of the main forest-forming species: in the overwhelming majority of cases *Pinus sibirica* is displaced by *Abies sibirica* (Gavrilkina and Zelepukina 2017). In semi-arid climate, where the light-coniferous taiga predominates, the *Larix* regeneration usually takes place avoiding a small-leaved stage and forest-forming species change. Thus, the taiga belt of the region has been subject to fundamental complication of its structure and growth of its patchiness under the anthropogenic activity impact during the last 50 years. Strongly disturbed landscapes in the region occupy rela-

Fig. 21.3 Larch forests after fire disturbance in the Naryn River valley, Erzinsky district, Republic of Tuva. At the top – Landsat 8 imagery, 2017, below – photo Amburtseva N., 2011)

tively small areas and are related to water management and mining activities, occasionally to agricultural industry.

21.5   Conclusion

The research data obtained from long-term field studies of high-mountain landscapes of Inner Asia do not give enough ground for being convinced in the inevitability of global warming and regional landscape transformations resulting from it. All the changes in Inner Asia are of implicit nature: the winter season warming is observed in reality; but the warm half-year does not demonstrate the apparent trends in both: temperature and precipitation changes. Moreover, the multidirectional nature of precipitation fluctuations in the region does not allow one to unambiguously associate the processes of landscape dynamics with changes in external factors. On the one hand, in the highlands glacier and permafrost degradation processes, thawing of ice lenses and veins occur, consequently causing a level decline by 1–1.5 m in most mid-mountain lakes of moraine genesis; and, on the other hand, many large lakes located in the Great Lakes Depression, at the turn of the century, exhibited a rising trend in their level. To sum it up, one can affirm that the dynamics of slightly disturbed high altitude landscapes to a greater degree depends on environmental factors, while the dynamics of low- and mid- altitude landscapes results from anthropogenic impact of various intensity.

The continuation of the investigation depends on the interregional comparison and correlation of the research results, carried out in other mountain regions, as well as using modern remote methods (satellite sounding, electronic tacheometry, aerial survey, and pilotless vehicles etc.) followed by the obligatory verification during field observations.

Acknowledgements The work of the geographers of St. Petersburg State University in Inner Asia is supported by RFBR grant 18-05-00860 and Russian Geographical Society (18.63.99.2017).

References

All-Russian Scientific Research Institute of Hydrometeorological Information – World Data Center (FGBU VNIIGMI-WDC). http://meteo.ru/. Accessed 04 Aug 2018.

Beruchashvili, N. L. (1986). *Four dimensions of a landscape*. Moscow: Mysl'. (in Russian).

Cherepanov, S. K. (1995). *Vascular plants of Russia and neighboring countries (within the former USSR)*. St. Petersburg: Mir i sem'ya. (in Russian).

Chistyakov, K. V., & Kaledin, N. V. (Eds.). (2010). *Mountains and people: Changes in landscapes and ethnoses of the inland mountains of Russia*. St. Petersburg: VVM. (in Russian).

Chistyakov, K. V., Moskalenko, I. G., & Zelepukina, E. S. (2009). Climate of Ubsu-Nur depression: Spatial model. *Proceedings of Russian Geographical Society, 141*(1), 44–61. (in Russian).

Cox, D., & Snell, E. (1981). *Applied statistics. Principles and examples*. London: Chapman & Hall.

Dirksen, V. G., & Smirnova, M. A. (1997). Characteristics of the vegetation of the northern macroslope of the high-mountain massif Mongun-Taiga (Southwestern Tuva). *Botanical Journal, 82*(10), 120–131. (in Russian).

Ganyushkin, D. A., Zelepukina, E. S., & Gavrilkina, S. A. (2015). Forecasts of landscape dynamics of the Mongun-Taiga massif (southwestern Tyva) under given scenarios of climate change. *In the World of Scientific Discoveries, 4*(64), 273–307. (in Russian).

Ganyushkin, D. A., Chistyakov, K. V., Volkov, I. V., Bantcev, D. V., Terekhov, A. V., & Kunaeva, E. P. (2017). Present glaciers and their dynamics in the arid parts of the Altai mountains. *Geosciences, 7*(4), 117. https://doi.org/10.3390/geosciences7040117.

Gavrilkina, S., & Zelepukina, E. (2017). Dynamics of mountain forest ecosystems in the continental sector of Siberia: Patterns and reasons. *Proceedings of 17th International Multidisciplinary Scientific Geo Conference SGEM, 17*(32), 797–804.

Isachenko, G. A., & Reznikov, A. I. (1996). *Dynamics of landscapes of the taiga of the North-West of European Russia*. St. Petersburg: Russian Geographical Society Publishing. (in Russian).

Kuminova, A. V. (1960). *Vegetation cover of Altai*. Novosibirsk: SB AS USSR. (in Russian).

Lesovaya, S. N., & Goryachkin, S.V. (2007). Cryogenic soils of Altai highlands: Morphology, mineralogy, genesis, classification problems and connection with soils of polar regions. In *Proceedings of the international conference "cryogenic resources of the polar regions"* (Vol. 2, pp. 96–99). Salekhard. (in Russian).

Makunina, N. I. (2014). Mountainous forest-steppe of the South-Eastern Altai and South-Western Tuva. *Vegetation of Russia, 24*, 86–100. (in Russian).

Mamay, I. I. (1992). *Landscape dynamics. Research methods*. Moscow: MSU Publishing House. (in Russian).

Ogureeva, G. N. (1980). *Botanical geography of Altai*. Moscow: Nauka. (in Russian).

Puzachenko, Y. G., & Skulkin, B. C. (1981). *Structure of vegetation in the forest zone of the USSR*. Moscow: Nauka. (in Russian).

Riahi, K., Rao, S., et al. (2011). RCP 8.5 – A scenario of comparatively high greenhouse gas emissions. *Climatic Change, 109*(1), 33–57.

Schmidt, G. A., et al. (2006). Present-day atmospheric simulations using GISS Model E: Comparison to in situ, satellite, and reanalysis data. *Journal of Climate, 19*, 153–192.

Shishov, L. L., et al. (2004). *Classification and diagnostics of soils in Russia*. Smolensk: Oykumena. (in Russian).

Sobolevskaya, K.A. (1950). *Vegetation of Tuva*. Novosibirsk. (in Russian).

Volkenshtein, M. V. (2006). *Entropy and information*. Moscow: Nauka. (in Russian).

Volkova, E. A. (1994). *Botanical geography of the Mongolian and Gobi Altai*. St. Petersburg: BIN. (in Russian).

Zelepukina, E. S., Gavrilkina, S. A., Lesovaya, S. N., & Galanina, O. V. (2018). Landscape structure of the high-altitude ecotone band of the high-mountain massif Mongun-Taiga. *Proceedings of Russian Geographical Society, 150*(2), 33–47. (in Russian).

Part VII
How Patterns Affected Land Use in the Past

Chapter 22
Initial Stages of Anthropogenic Evolution of Landscapes in Russia

Viacheslav A. Nizovtsev and Natalia M. Erman

Abstract Our research is aimed at establishing the patterns of formation and evolution of landscapes at the initial stages of anthropogenically-induced development, when human activity becomes an important landscape-forming and landscape-transforming factor. Of particular importance in solving this problem is identifying the role of the anthropogenic component in the formation of structure, dynamics, and evolution of landscapes at the regional and local levels. Despite extensive historiography concerning contribution of the anthropogenic factor to the formation of the structure, functioning, and dynamics of the landscape, the initial stages of the man-induced evolution is poorly studied. This work presents the results of long-term research, focusing on the history of the relationship between society and nature, man and landscape. The research allowed us to establish some patterns of anthropogenic landscape genesis in Russia. To determine anthropogenic changes in landscapes in the past and emerging environmental problems, the landscape-historical-archeological approach is believed to be the most effective. It is based on a set of interrelated methods (landscape-archeological, landscape-paleo-pedological, landscape-edaphic, etc.).

Keywords Landscapes · Anthropogenic Landscape genesis · Landscape-economic systems · Producing economy

V. A. Nizovtsev
Lomonosov Moscow State University, Moscow, Russia

N. M. Erman (✉)
Vavilov Institute for the History of Science and Technology, e Russian Academy of Sciences, Moscow, Russia

© Springer Nature Switzerland AG 2020
A. V. Khoroshev, K. N. Dyakonov (eds.), *Landscape Patterns in a Range of Spatio-Temporal Scales*, Landscape Series 26,
https://doi.org/10.1007/978-3-030-31185-8_22

365

22.1 Introduction

Studies of the formation of anthropogenic landscape genesis (anthropogenic evolution of landscapes) in Russia were carried out at two hierarchical levels: (1) analysis of the data sources for identifying zonal patterns of the structure and dynamics of anthropogenically changed landscapes of forest, forest-steppe, and steppe zones of Russia at the regional level; and (2) detailed large-scale studies on key sites. The analysis of the extensive source base, including complex landscape, paleo-geographic and archaeological, historical and geographical published and archive cartographic and text materials made it possible to compile maps showing distribution of the main material cultures at the initial stages of the producing economy establishment in the forest and southern non-forest areas within the framework of the landscape zonality concept.

22.2 Methods

Original maps made it possible to establish landscape-zonal features of the periodization of the process of initial economic development, nature use, and anthropogenic landscape genesis for the entire territory of Russia on the basis of a comparative geographical method (Nizovtsev et al. 2014). Large-scale field research was performed at key sites in the Smolensk, Tver, Moscow, Chelyabinsk, Ryazan, and Krasnodar regions. In key areas, landscape mapping and transect studies were carried out with focus on present-day landscape structure, anthropogenic loads, and the degree of man-induced diversity of landscape units (Graves et al. 2009).

At the next stage for key areas, we performed paleo-reconstruction of landscape features affecting the living conditions of settlers in different material cultures and retro- reconstruction of landscape-economic systems (LES) for the initial stages of anthropogenic landscape genesis. This allowed us to reveal differences in intra-landscape structure for the main zonal types of forest, forest steppe, and steppe landscapes, as well as the associated features of the settlement structure development and the formation of the producing economy. A series of large-scale landscape, historical, and archaeological maps and profiles were compiled, which became the basis for the landscape and historical geographic information systems. In addition, basic reconstruction models were developed, which reflected the spatial and time structure of formation of the producing economy (maps and schemes of the structure for the most typical cultural and historical landscapes) for the Eneolithic-early Bronze Age and early and late Iron Age for the territory of Russia (Nizovtsev et al. 2014).

The basic schemes of the structure of landscape-economic systems (LES) for key areas have been drawn up. We revealed the dependence of economic development of steppe and forest landscape complexes on their spontaneous development during the formation of the producing economy.

An analysis of paleogeographic materials, together with research on key areas, makes it possible to establish differences in the spontaneous development of landscapes in the Holocene in different natural zones, different glaciation regions, and extra-glacial regions.

22.3 Results

Features of Natural Development of Landscapes Rhythmic variations in climate had crucial importance in the evolution of the landscapes, leading to repeated shifts of their types (zones). The specific pattern of landscapes occurrence, both at regional and local levels, is determined by structural and geomorphological factors. This is especially evident in glaciation-affected regions. The areal of Moscow (analogous to Riss) glaciation has limited diversity of the morphological landscape structures as compared to the area of the Valdai (Wurm) glaciation with characteristic smaller landscape complexes at the local level. In Holocene, the landscape types in the mixed forests zone changed four times (taiga – mixed forests – deciduous forests – mixed forests), while landscape sub-types changed six times. The southern taiga landscapes and northern taiga landscapes experienced four changes of types and from 5 to 11 shifts of zone subtypes. Landscapes of the steppe zone are more conservative in terms of evolution since they depend on the provincial conditions. They experienced from two to four changes of zonal affiliation (3–5 subtypes were replaced). The duration of existence of each landscape type averaged about 3000 years for forest landscapes and 3000–4000 years for steppe landscapes (Nizovtsev 2014).

Forms of Anthropogenic Transformation of Landscapes The social and natural history of contemporary landscapes of Russia was progressive, and sometimes, was cyclical in nature. At almost any historical period in similar landscape conditions, the settlers were involved in similar types of economy, under which analogous types of natural and economic systems were formed. With the development of production and production relations, gradual complication of the emerging natural-economic systems took place. During the period of the appropriating type of economy, there was a balanced man-nature equilibrium system (Nizovtsev 1999). Since the first settlers practiced the appropriating type of economy, engaging in hunting, fishing and gathering, the human impact on landscapes was minimal and limited to the biota. Human settlements were clearly confined to the resource base of certain types of *natural-territorial complexes* (NTC). A man could not "leave" their frontiers. Therefore, during the meso-neolithic stage, only anthropogenic modifications of the NTC were formed (Nizovtsev 2005). The formation of the first anthropogenic (ALC) and cultural landscape complexes (CLC) was caused by the development of types of farming and pasture economy, with the formation of permanent settlements. In the Bronze Age, with the transition from the appropriating type of economy to the producing economy, anthropogenically-transformed landscape complexes and even anthropogenic landscape complexes (ALC) were formed. Actually, ALC and cultural landscape complexes (CLC) emerged as late as in the Iron Age with the advent of developed agriculture and a permanent, long-existing settlement structure.

The Initial Stages of the Socio-natural History of Landscapes in Russia The economic human activity as a factor of differentiation and development of landscapes manifested itself since the Eneolithic (the Atlantic period) in the southern steppe regions (on the northern periphery of the main centers of the producing economy), and since the Bronze Age, in the forest areas. At this time, in IV (Eneolithic) – III (early Bronze Age) millennium BC, there was a change from the appropriating type of economy to the producing economy (Fig. 22.1).

This was the period when the first center of the producing economy in Russia, connected with the tribes of the Maykop culture, was developing. They formed a combination of hoe farming and cattle-breeding, developed in the valley and near-valley plains and foothill steppe landscapes of the North Caucasus in the areas of settlements (Nizovtsev et al. 2015). Also, in Siberia in the Eneolite, it was in the riverside stands, where the emergence of the producing economy was detected (Nikolaev 1999). Long-existing settlement and agricultural structure resulted in development of one of the first landscape complexes that experienced anthropogenic transformation, i.e. natural-anthropogenic landscape complexes that included from 4 to 6–7 landscape-economic systems (LES) (Fig. 22.2a): residential and agricultural (terraces and foothills of interfluves), pasture-meadow (floodplains and terraces), and pasture-steppe (inland interfluves). At the end of the Eneolithic, the resource base of the surrounding landscapes was severely exhausted, and local tribes had to bring new land into cultivation, namely, both river valleys and flat areas of interfluve steppe landscapes. The rudiments of the producing economy were introduced into the steppe and forest-steppe landscapes in the vast areas of the European part of Russia and Siberia.

In the steppes of Eurasia, there were two modes of the producing economy with the predominance of cattle breeding: mobile nomadic cattle breeding and homestead cattle breeding, with significant role of farming in the economy and long-term settlements (Buganov 1987). The formation of the two economic-cultural types and their further distribution was completely controlled by local landscape features. Migrating cattle breeders preferred steppe landscapes, however, given scarce population and a relatively limited number of cattle, their movements were relatively limited to the valleys of large rivers and their tributaries. Settled herders and farmers preferred the valley forest-steppe landscapes. It should be noted that in both cases, hunting, fishing, and gathering were still an essential part of the economy. The fact that the producing economy in the Eurasian steppe and forest-steppe landscapes was oriented mainly on cattle breeding can be explained by the natural properties of steppe landscapes with their fodder resources almost inexhaustible for that time. In the conditions of feather-grass and dry-steppe wormwood landscapes, the possibilities of their agricultural reclamation without developed arable tools with iron tips

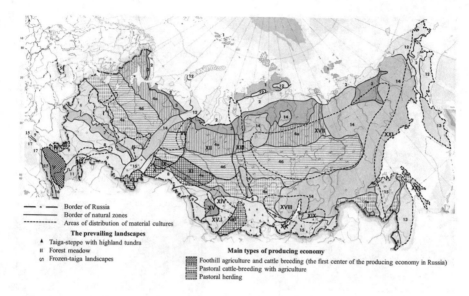

— · — Border of Russia
———— Border of natural zones
- - - - - Areas of distribution of material cultures
The prevailing landscapes
▲ Taiga-steppe with highland tundra
II Forest meadow
ᴄɴ Frozen-taiga landscapes

Main types of producing economy
▨ Foothill agriculture and cattle breeding (the first center of the producing economy in Russia)
▤ Pastoral cattle-breeding with agriculture
▦ Pastoral herding

Fig. 22.1 The main types of productive economy. Eneolithic, Early Bronze Age. Scale 1:30,000,000. On the map, the Arabic numerals indicate the prevailing landscapes and Roman numerals indicate the areas of distribution of material cultures

The prevailing landscapes

Latitudinal-zonal plain areas:

1. Arctic deserts; 2. Tundra; 3. Forest-tundra woodlands; 4. Taiga forest (including permafrost-taiga); 4a. North Taiga; 4б. Middle Taiga; 4в. South Taiga; 5. Mixed forest, broadleaf-coniferous; 6. Small-leaved forest; 7. Broad-leaved forest; 8. Forest steppe; 9. Steppe; 10. Semi-desert; 11. Deserted

Mountain areas (types of altitudinal zoning):

12. Arctic Tundra-desert-arctic; 13. Near-oceanic tundra forest; 14. Taiga and tundra-taiga; 15. Taiga forest-meadow, taiga-steppe; 16. Forest meadow-steppe; 17. Subtropical and wet forest-meadow

Material culture:

I. Volosovskaya; II. Garinskaya; III, III.I. Khvalynsko-Srednestogovskaya community; IV. Novotitorovskaya; V. Maikopskaya; VI. Sintashtinskaya; VII. Sartyninskaya; VIII. Polymyatskaya; IX. Tashkovskaya; X. Petrovskaya; XI.Odintsovskaya and Alexandrovskaya; XII. The area of interaction of otstypaushche-nakolchatoj, grebenchatoj, and grebenchato-yamochnoj traditions; XIII. Contact area of East Siberian and West Siberian cultures; XIV. Krokhalevskaya; XV, XV.I. Eluninskaya; XVI. Afanasevsko-Okunevskiy area; XVII. Ymyakhtakhskaya; XVIII. Glazkovskaya; XIX. Omolonskaya; XX. Fofanovskaya; XXI. Culture of the Perezhitochnogo Neolithic

were extremely limited. The most suitable sites for farming with primitive hand-cultivating tools were located in valley (floodplain) landscape complexes and wood-side of forest-steppe landscapes, riverine landscapes of the southern forest zone and forest-steppe foothill landscapes, and intramontane depressions (Nizovtsev et al. 2015).

To the north of the steppe zone, the belt from the Volga region to the Seversky Donets was inhabited by the tribes of the Eneolithic cultures belonging to the

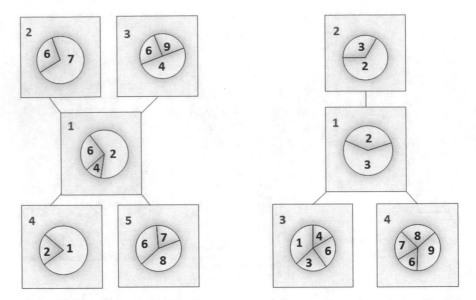

Fig. 22.2 Basic diagrams of the structure of natural-anthropogenic and anthropogenic landscapes (landscape-economic systems) (Aeneolithic, Early Bronze Age)
(**a**) – Aeneolithic, Early Bronze Age (steppe zone). Agricultural and cattle-breeding tribes
1. Residential (permanent settlements of pastoralists and agriculturists); 2. Agricultural (hoe agricultural); 3. Pasture and agricultural; 4. Pasture and meadow; 5. Pasture-steppe
(**b**) – Aeneolithic, Early Bronze Age (steppe zone). Cattle-breeding tribes
1. Residential (settlements of steppe nomads); 2. homestead economy; 3. Pasture and meadow; 4. Pasture-steppe
In the squares, there are numbers of landscape-economic systems. In the circles, there are numbers of landscape complexes that make up these natural-economic systems in the key areas (see captions to Fig. 22.1)

Khvalyn-Stogov community. In Siberia, the Eneolithic cultures were localized in the areas adjacent to the mining and metallurgical regions. Later, representatives of the *ancient yam* culture conquered the areas from the Prut to the Trans-Urals (Borodko and Vedenin 2008), where their main occupation was cattle breeding and hunting; however, they were already familiar with primitive agriculture and metallurgy (Merpert 1974). Representatives of this culture inhabited not only extensive forest-steppe and steppe landscapes, but also the foothills of the Altai, and later penetrated into the Minusinsk intramontane depression and further to Tuva and Mongolia (Shnirelman 2012). At the same time, on the Yenisei in the Krasnoyarsk region, in the Baikal area and in some areas of Trans-Baikalia, the *Glazkovsk* culture of hunters and fishermen was widespread. Apparently, the landscapes of these areas were less favorable for cattle breeding and agriculture of that time (Rumin 1995).

Natural-anthropogenic landscapes, commonly, have an extremely simple structure based on the following LESs (Fig. 22.2b): residential (temporary sites in the near-valley and valley complexes), pasture-steppe (valley steppe tracts of watershed spaces), and pasture meadow (in river valleys) (Nizovtsev et al. 2015). Despite the fact that on the vast interfluvial steppe areas cattle breeding became the most wide-

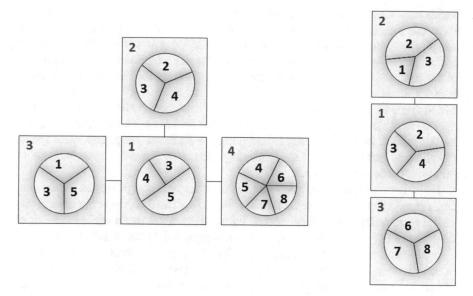

Fig. 22.3 Basic diagrams of the structure of natural-anthropogenic and anthropogenic landscapes (landscape-economic systems) (Bronze Age, steppe zone)
In the squares, there are numbers of landscape-economic systems. In the circles, there are numbers of landscape complexes that make up these natural-economic systems in the key areas (see captions to Fig. 22.1)
(**a**) – Agricultural and cattle-breeding tribes.
1. Residential (settlements of .steppe nomads); 2. Arable; 3. Pasture and meadow; 4. Pasture-steppe
(**b**) – Cattle-breeding tribes
1. Residential (settlements of steppe nomads); 2. Pasture and meadow; 3. Pasture-steppe

spread form, traces of meat and dairy farming with a predominance of cattle and pigs are clearly traced along river valleys around permanent settlements.

The forest zone was inhabited by the tribes of hunter-fishermen that had not yet overcome the stage of conducting an appropriating economy. In the river valleys, hunter-fishermen penetrated far into the taiga zone and began to move to a settled way of life with the spread of the fishing economy (Borodko and Vedenin 2008). In the forest-steppe zone, slash-and-fire and land-based agriculture were developing initially on the European territory of Russia and later in Siberia,

In the Bronze Age (II-I millennium BC), the producing economy spread over the new territories. In the steppe zone, there was a rapid and ubiquitous spread of mobile nomadic or semi-nomadic cattle breeding with complete unification of the economy on vast areas. The settled farming with cattle-breeding was preserved only in the forest-steppe zone (Nizovtsev et al. 2015) (Fig. 22.3a). In the steppe zone of the European part of Russia, a vast set of catacomb cultural and historical community was formed (the *Donets, Azov, Manych, Volga* and other cultures), and on its northern border entered the forest-steppe. In Siberia, first, the *Afanasiev* culture in the south was replaced by the *Okunev* culture with developed metallurgy, and close to it the *Serov* culture in the upper part of the Angara and Lena basins (Borodko and

Vedenin 2008). At this time, climate entered the xerothermic phase of the subboreal Holocene period. A sharp aridization of climate resulted in drying up of steppe lakes and disappearance of small rivers, especially in the Asian part of Russia. Throughout almost all the territory of the steppes a cattle-breeding economic and cultural type of the producing economy was developing with a simplified structure of natural-anthropogenic and anthropogenic landscapes, including 3 or a maximum of 4–5, LESs) (Fig. 22.3b). It should be noted that this structure existed in a number of regions until the late Middle Ages. Rare settled farming with cattle-breeding was limited to river valleys. They started migrating to the forest-steppe and forest zones. However, as early as at this time, sedentary tribes experienced a constant press of predatory cattle breeding tribes.

In the Bronze Age in the forest zone of the European part of Russia (in the sub-areas of mixed and broad-leaved forests), the *Fatyanovo-Balanovo* cultural and historical community was widely spread. With the extensive forest cattle-breeding of the tribes of Fatyanovo culture, one of the first significant human-nature conflicts in Central Russia is associated. They practiced forest cattle-breeding with grazing pigs first, and after that cows, sheep, and goats; hunting and fishing was a second-important occupation. Pastures for pigs and cattle were localized mainly at flood-plains and lacustrine lowlands, which had more open areas, such as meadow clearings and glades. Long-term grazing at the same site caused the complete destruction of vegetation and forced migration to the other sites looking for pastures and involving new sites into the economy. The beginning of deforestation at the floodplains can be linked to the extensive nature use of the *Fatyanovo* tribes. In terms of the morphological structure of landscapes, in floodplain tracts and localities the first stable anthropogenic elements appeared, namely, floodplain meadows. After the extinction of the Fatyanovo tribes from the region, the established natural and economic system with floodplain cattle-breeding (agro-geo-systems of pasture type with floodplain meadows and light forests) existed for a long time (Fig. 22.4),

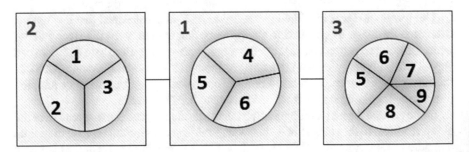

Fig. 22.4 Basic diagrams of the structure of natural-anthropogenic and anthropogenic landscapes (landscape-economic systems) (Bronze Age, forest zone)
In the squares, there are numbers of landscape-economic systems. In the circles, there are numbers of landscape complexes that make up these natural-economic systems in the key areas (see captions to Fig. 22.1). Cattle-breeding tribes with rudimentary farming
1. Residential (permanent settlements); 2. Pasture-floodplain; 3. Rudimentary slash-and-burn hoe farming

as local tribes adopted these efficient forms of farming from the *Fatyanovo* tribes (Nizovtsev 1997). In a number of regions, apparently, the elements of slash-and-burn farming appeared.

In Siberia, as a result of climate aridization, forest-steppe landscapes were the most widely-spread and provided the opportunity for nonirrigation farming. The main migration flows passed through this area. The main settlements emerged at the first terraces of large rivers that preserved runoff. Their inhabitants were engaged in pastoral cattle-breeding and hoe farming. Thus, cattle-breeding was of a near-valley (sedentary) character, and agriculture was possible only in floodplain landscapes (Nizovtsev and Erman 2014). Forced reduction of agricultural land led to a significant environmental crisis. A sharp increase in pasture loads in riverine landscape complexes led to the strongest pasture digression of psammophytic steppes in floodplain-terraced complexes, accompanied by deflation of soils in several regions (Nikolaev 1999). As a result, tribes of the *Andronovo* culture were forced to migrate to new lands northward within the West Siberian plain (Kosarev 1984).

The formation of the producing type of economy, with the formation of basic landscape-economic systems, depending on the zonal conditions, is presented in the tables (Tables 22.1 and 22.2).

Table 22.1 Initial stages of socio-natural evolution of landscapes (forest zone)

Types of anthropogenically altered landscapes	Types of landscape-economic systems	
	Bronze Age	Eneolithic-Bronze Age
Natural-anthropogenic modifications of landscapes	Pasture and forest	Pasture and forest
Nature-anthropogenic landscape	Pasture and meadow	Pasture and meadow
	Agricultural and forest	
Anthropogenic landscape	Residential (permanent settlements)	Residential (permanent settlements)

Table 22.2 Initial stages of socio-natural evolution of landscapes (steppe zone)

Types of anthropogenically altered landscapes	Types of landscape-economic systems	
	Bronze Age	Eneolithic-Bronze Age
Natural-anthropogenic modifications of landscapes	Pasture-steppe	Pasture-steppe
		Pasture and hay
Nature-anthropogenic landscape	Pasture and meadow	Pasture and meadow
	Residential (settlements of steppe nomads)	Pasture and agricultural agricultural (hoe agricultural)
		Residential (settlements of steppe nomads)
Anthropogenic landscape	Residential (permanent settlements of pastoralists and agriculturists)	Residential (permanent settlements of pastoralists and agriculturists)
	Arable (among pastoralists and agriculturists)	

The transition from the Bronze Age to the Iron Age falls on the boundary of the II-I millennium B.C. Climatic conditions became continental with significant drying up of the steppes. In the steppe landscapes, mobile nomadic or semi-nomadic cattle-breeding with a complete unification of the economy on vast areas was spreading rapidly and everywhere. The sedentary stock-farming economy was preserved only in the forest-steppe (Buganov 1987). Climate change was accompanied by an ecological crisis that led to a "great migration of peoples" (Gumilev 1987). Huge masses of nomads joined the movement in the opposite direction from East to West. The nature of the relationship and coexistence of the nomadic and agricultural population was becoming completely hostile. At the end of the Bronze Age – the beginning of the Iron Age, there was a separation of taiga fishermen and hunters from cattlemen of the Siberian forest-steppe and steppe (Borodko and Vedenin 2008).

22.4   Conclusion

Thus, the emergence of producing economy on the territory of Russia dates back to the IV (Eneolithic) – III (early Bronze Age) millennium B.C. In the steppe, forest-steppe, and forest zones, the shepherd-agricultural economy developed with a pronounced trend to a complex diversified economy that organically combined producing industries (pastoralism and agriculture) and appropriating industries (hunting, fishing, and gathering). In the forest zone, mainly in the landscapes of broad-leaved, mixed, and southern taiga forests, specific forest cattle-breeding was developed along the river and lake floodplains. In all landscape zones at the early stages of the socio-natural history, formation of land use systems and, accordingly, settlement structures was strongly related to the local landscape structure. In similar landscape (zonal) conditions, the settlers had the same type of economy, which determined the formation of the same type of natural economic systems and the same type of anthropogenically transformed landscapes.

Acknowledgments The research was financially supported by Russian Foundation for Basic Research (project No. 19-05-00233).

References

Borodko, A. V., & Vedenin, Y. A. (Eds.). (2008). *The National Atlas of Russia. History and culture, Vol. 4.* Moscow: PKO "Kartographiya". (in Russian).

Buganov, V. I. (Ed.). (1987). *The history of the peasantry in the USSR from ancient times to the Great October Socialist Revolution, Vol. 1.* Moscow: Nauka. (in Russian).

Graves, I. V., Galkin, Y. S., & Nizovtsev, V. A. (2009). Landscape analysis of formation of settlement structure of the Moscow region. In *Archaeology of Moscow Region: Proceedings of Scientific Seminar* (Vol. 5, pp. 43–55). Moscow: Institute of Archaeology RAS. (in Russian).

Gumilev, L. N. (1987). People and nature of the Great Steppe: Experience of explanation of some details of nomad history. *Questions of History, 11*, 64–77. (in Russian).

Kosarev, M. F. (1984). *Western Siberia in ancient times*. Moscow: Nauka. (in Russian).

Merpert, K. Y. (1974). *The oldest cattle breeders of the Volga-Ural interfluve*. Moscow: Nauka. (in Russian).

Nikolaev, V. A. (1999). *Landscapes of Asian steppes*. Moscow: University Press. (in Russian).

Nizovtsev, V. A. (1997). The history of the formation of the first natural-economic systems of the Moscow region. In *The history of studying, using and protecting the natural resources of Moscow and the Moscow region* (pp. 72–81). Moscow: Yanus-K. (in Russian).

Nizovtsev, V. A. (1999). Anthropogenic landscape genesis: The subject and research problems. *Proceedings of Russian Geographical Society, 1*, 26–30. (in Russian).

Nizovtsev, V. A. (2005). The history of the formation of anthropogenic and cultural landscapes of Central Russia. In *Geospatial systems: Structure, dynamics, interrelations. Proceedings of the XII Congress of the Russian Geographical Society, Vol. 2* (pp. 54–59). St. Petersburg: Russian Geographical Society. (in Russian).

Nizovtsev, V. A. (2014). Zonal features of anthropogenic landscapes structure and dynamics at European Russia area in historical aspect. In *Geology, geo-ecology, evolutionary geography, Vol. XII* (pp. 52–58). St. Petersburg: Herzen RGPU. (in Russian).

Nizovtsev, V. A., & Erman, N. M. (2014). Zonal and landscape features of development of manufacturing economy in Siberia. In *Geology, geo-ecology, evolutionary geography, Vol. XIII* (pp. 168–171). St. Petersburg: Herzen RGPU. (in Russian).

Nizovtsev, V. A., Snytko, V. A., & Erman, N. M. (2014). Methodological aspects of research of the formation of anthropogenic landscape genesis. In *Geology, geo-ecology, evolutionary geography, Vol. XIII* (pp. 161–164). St. Petersburg: Herzen RGPU. (in Russian).

Nizovtsev, V. A., Snytko, V. A., Erman, N. M., & Graves, I. V. (2015). Features of the producing economy in the forest, forest-steppe and steppe landscapes of Russia. In *Landscape and ecological state of Russian regions* (pp. 145–153). Voronezh: Istoki. (in Russian).

Rumin, V. V. (1995). The main stages of the development of nature in Southern Siberia. In *Historical geography: Trends and prospects* (pp. 14–25). St. Petersburg: Russian Geographical Society. (in Russian).

Shnirelman, V. A. (2012). *The emergence of the productive economy*. Moscow: Librokom. (in Russian).

Chapter 23
How Natural and Positional Factors Influenced Land-Use Change During the Last 250 Years in Temperate Russia

Victor M. Matasov

Abstract The analysis of spatial and temporal dynamics of the land-use structure in the Meschera Lowland since the late eighteenth century has been realized considering the contributions of natural and positional factors. The "natural" factors refer to the land's agricultural suitability characterized by drainage, soil fertility, erosion, etc. The "positional" ones mean distance to rivers, villages, roads. Three study sites, located in the Ryazan region (central European Russia) in differing natural conditions (same climate, but different sediments and waterlogging) were selected for study. The land-use structure was analyzed using the maps created from General Land Survey maps (the eighteenth century), Atlas Mende maps (the nineteenth century), satellite imagery Corona (the twentieth century) and modern satellite images (the twenty-first century). These were matched up in GIS with landscape maps bound to georeferenced topographic maps considering features of the landscape. Quantitative analysis of land-use distribution across natural landscape patterns and its remoteness from social infrastructure showed that in the regions where economic activity is strongly limited by natural conditions, the structure of land use remains practically unchanged for a long time.

Keywords Long-term land-use change · Historical maps · Landscape structure · Natural and positional factors

V. M. Matasov (✉)
Lomonosov Moscow State University, Moscow, Russia
e-mail: victor.matasov@geogr.msu.ru

© Springer Nature Switzerland AG 2020
A. V. Khoroshev, K. N. Dyakonov (eds.), *Landscape Patterns in a Range of Spatio-Temporal Scales*, Landscape Series 26,
https://doi.org/10.1007/978-3-030-31185-8_23

23.1 Introduction

One of the most criticized questions in geography is the question of the determination of some phenomena by others (Sluyter 2003; Engerman and Sokoloff 2011). Most often, scholars talk about the influence of environmental factors on the development of society, economy, and culture. But this issue is somewhat wider. Rather, it is worth talking about how much mobile, rapidly changing components (with a small characteristic time scale) are determined by stable invariant properties of the environment (with a large characteristic time scale). This may be the dependence of vegetation on relief or it can also be found in how the different sediments in a similar climate determine the structure of the land use of only one society. On the other hand, this geographical determinism can be manifested not only in different characteristic time scales of phenomena but also by their impact intensity.

Many modern studies show that the environmental impact of human activities is now rapidly increasing due to population growth. Already more than half of the land area has been transformed by human activity (Ellis et al. 2010; Ellis 2011). These changes lead to a decrease in biodiversity, a violation of global geochemical cycles (Foley et al. 2005), and are considered to be one of the main challenges for the future stable humanity development (Steffen 2005). The degree of human impact on nature grows owing to the technical progress, and many scientists believed that the Industrial Revolution marked the beginning of a period called Anthropocene (Ellis et al. 2013). Before the advent of widespread industrial production, the agricultural development heavily depended on natural factors. Exploring the contribution of various natural and socioeconomic factors to the dynamics of land transformation is an area of investigation for many researchers (Verburg et al. 2015; Plieninger et al. 2016). With regard to spatial data, natural factors usually include climate, soil maps, relief models, etc. Studies for agricultural development usually focus on land suitability evaluation, in which these factors are examined (Kiryushin and Ivanov 2005). Such socioeconomic factors as accessibility to markets, production centers, trade routes are explored based on Ricardo's and von Thünen's land-use theories. Exploring land use, it is also important to consider institutional changes and major political shifts, technological innovations, and teleconnections that affect economic structure, therefore change the land use (Prishchepov et al. 2012; Jepsen et al. 2015; Schröter et al. 2018).

We also need to learn from the past in order to gain a better understanding of influence of different socioeconomic legacies on land transformation (Rhemtulla et al. 2009; Price et al. 2017). The Landsat program has been in operation for the last 40 years – the time of great acceleration of the post-war period, when human-nature interaction quickly became more complicated. At the same time, we have a large number of historical materials (maps, descriptions) that allow us to reconstruct the structure of the land use during the last several centuries (Petit and Lambin 2002; Feurdean et al. 2017). An intermixing of history and geography is continuing – more and more historians turn to the study of the environment, and an increasing number of geographers use historical materials in their studies (Josephson et al.

2013; Moon 2013). Historical data are used to test global land-use patterns, and often form the basis for these models (Hurtt et al. 2011). The use of such data, on the one hand, is problematic because of a large number of assumptions and errors, and, on the other hand, opens a lot of opportunities in today's spatial planning for sustainable development. Many studies successfully combine historical maps, paleoecological data, modern satellite images in order to reconstruct the land-use dynamics (Veski et al. 2005; Poska et al. 2014).

Russia, as a country with a huge territory, and, therefore, with a wide range of natural conditions, for a long time remained agrarian. Underlying drivers of land-use change were specific due to political and economic development in the twentieth century and has been considered in many studies focused on forest dynamics (Tsvetkov 1957; Kuemmerle et al. 2015; Alix-Garcia et al. 2016), agriculture (Lyuri et al. 2010; Prishchepov et al. 2012, 2013), industrial development (Kalimullin 2006), etc. But there are few investigations related to the study of agricultural development in the eighteenth–nineteenth centuries, in spite of the fact that there is a magnificent cartographic material, unique in detail and territorial coverage, namely, General Land Survey (Kusov 1993; Golubinsky et al. 2013). Multiple historical studies have been devoted to exploring these materials (Gedymin 1960; Milov 1965; Goldenberg and Postnikov 1985), while the question of how the Russian peasant adapted to such various natural conditions of the vast Russian Empire with a fairly low level of technological capability is still an issue of concern for many researchers (Milov 2006; Alyabina et al. 2015; Matasov 2016). How strongly natural factors limited the agricultural development at that time? And how local natural and positional patterns redistributed land use associated with regional or global socioeconomic changes?

The purpose of our work was to find out the dependence of agricultural development on the landscape structure in different natural conditions. To this end, the tasks were as follows:

– Reconstruction of land use based on historical maps and modern satellite images
– Analysis of the land use and settlements distribution across different landscapes
– Assessing the contribution of natural and positional factors to land-use spatial distribution

23.2   Materials and Methods

23.2.1   Study Area and Period

To reconstruct long-term land-use change, we selected three study sites in the northern part of Ryazan province (oblast, administrative level-2 unit in Russia) in European Russia: site 1 "Prudki"; site 2 "Lesunovo"; site 3 "Shulgino". These study sites cover much of the local landscape variation in terms of soil wetness and edaphic conditions. They represent typical composition of the landscapes in the northern edge of temperate Russia and are similar to many common landscapes in Belarus, northern Ukraine, and eastern Poland. Our selection of study sites represents variations in the proportion of cultivated croplands from large clusters of fields on loess-like sediments that had been in use for a long time (*Opolje*) to scattered agricultural fields in the marshy woodlands on glaciofluvial sands (*Polessje*) (Gvozdetskii 1968; Krivtsov et al. 2011).

This study covers the period from 1770 to 2010. During this period, three main political and economic regimes replaced each other: (1) the traditional feudal economy that existed from 1721 until the revolution in 1917; (2) the planned economy, which lasted from 1917 until 1991; (3) a market-based economy since 1991. The population of study area first increased from approximately 65,000 people in 1770 to 190,000 in 1900. Then, coinciding with perturbations caused by World War I, the revolution in 1917, and World War II, the population declined to approximately 50,000. Population slightly stabilized only in the middle of the twentieth century and reached 71,000 by 2010. The urbanization has become one of the primary causes of rural population decline. The percentage of the rural population dropped from 95% in the late eighteenth century to approximately 35–40% in 2010 (Selivanov 1890; Rosstat 2010).

23.2.2   Landscape Mapping

We got landscape maps of study area at the university archives. These maps were made in the 1960s at a scale of 1:50,000, with a second hierarchical level geosystems (*urochische*[1]) as the main mapping unit following the traditions of the Russian school of landscape research and landscape ecology (Bastian et al. 2015). These maps were scanned, georeferenced to modern topographic maps and satellite images using QGIS software. Then we digitized the units, each of which being associated with the attribute table derived from the map legend. To renew these maps, we conducted fieldwork with observations at the key points within boundaries of the dominant geosystem level-1 (*facies*[2]) with description of relief, the composition of the vegetation and soil types. In addition, all field observations, archival maps, and modern satellite images were combined into a single GIS, and final landscape maps were compiled. The following characteristics were assigned for each contour of geosystem level-2 in the attributive table: relief, sediments, soil type, vegetation composition, nutrient supply level ("*trophotope*"), and moisture content level ("*hygrotop*"). We assumed that the relief and sediments structure remained unchanged for the last 250 years as the geotope framework of the territory, which determines the development of soils and vegetation (Isachenko and Reznikov 1996). Since the study area is mostly flat, erosion activity here is almost insignificant.

---

[1] See Chap. 1 and Glossary for definition.
[2] See Chap. 1 and Glossary for definition.

23.2.3 Reconstruction of Land-Use Change

To reconstruct land use for our three study sites, we used historical land use and topographic maps as well as satellite images. The earliest available cartographic data were General Land Surveying (GLS) results from the 1770s. We obtained the GLS data from the Russian State Archive of Ancient Documents (RSAAD, www.rgada.info). We made photos of the 1:8400 maps, including individual land tenures (*dacha*). Each map contained boundaries, angles, and distances recorded in the field, and was supplied with additional descriptive information about the land ownership and its common land-use types (Milov 1965; Kusov 1993). We complemented GLS maps with the 1:42,000 Mende Atlas maps dated 1850–1860 (www.etomesto.ru). For assessment of more recent land cover and land-use change, we relied on products derived from satellite imagery. We used declassified 3-m Corona imagery for 1970 (glovis.usgs.gov) and 10-m Spot-5 HRG for 2011 (search.kosmosnimki.ru).

We reconstructed land use from GLS maps via georeferenced land tenure plans (Matasov 2016). Maps from Atlas Mende and SPOT-5 satellite images were already georeferenced and we just co-registered the Corona satellite images to these data. Land-use types were digitized from satellite imagery by visual interpretation. Atlas Mende maps from 1860 matched the GLS maps very well because both maps used the same legend. To identify land-use change, we decided to generalize the legend to the following classes: forest (including peatland), arable land, grassland (including pasture, hay, and fallow land), settlement, industry (quarries and gas storage constructions), and water.

23.2.4 Analysis of Land-Use Distribution Factors

We divided explanatory factors into two main groups – natural and positional ones. With an assumption based on the von Thünen's model of land-use distribution around the settlements, distances to the nearest villages were calculated in SAGA-GIS™ (Conrad et al. 2015). Maps of distances to the main possible trade routes such as nearest navigable river, provincial tracts (up to the twentieth century), or paved roads (since the twentieth century) were also obtained. We also assumed that grasslands would cluster in proximity to rivers and streams. For example, the floodplain of Oka River was famous for its high-quality hayfields utilized during the Soviet time for ь and livestock grazing. Therefore, the distance to rivers and streams was one of the important parameters we calculated.

We assumed that geotope is the most important control over agricultural development due to the sediments quality type and the degree of moistening, which are the main limiting factors in this region (Vidina 1962). Therefore, maps of hygrotopes (with values from 1 to 5, that is, from dry to wet habitats) and trophotopes (from A to D – i.e. from nutrient-poor sandy to nutrient-rich loamy soils) were subsequently constructed. Geotopes were established as combinations of hygrotopes and trophotopes (A1...D5). All landscape maps were transformed into a raster format with 30 m resolution.

Thus, for each grid cell, we obtained information about land-use type for each time-step and variables characterizing positional and natural factors. Then probability density plots were made for the three types of land use (forest, arable lands, and grassland), depending on the positional factors in R software (R Core Team 2013). Also, the proportions of the land-use types within geotope types were calculated.

It is well known that for the nonchernozem zone, since the seventeenth until the nineteenth centuries, when a modern system of settlement developed, organic fertilization of soils was very important for agriculture, and that duration of manure input correlated positively with the distance to a village (Ioffe et al. 2004; Trapeznikova 2014). Thus, we assumed that the choice of the village location was also not accidental, but depended on natural and positional features. Therefore, we built maps with 1-, 2-, and 3-km-wide buffer zones around the villages in order to check in which geotope villages and their buffer zones were located in 1860.

23.3 Results

23.3.1 Landscape Structure

Prudki (site 1) is a typical *Polessje* landscape with sandy soils of low fertility (Rustic Podzols) and small fields in a matrix of wooded wetlands. Prudki is located on the border of the fluvial (with numerous lakes) and outwash plain landscapes. Thirty-five percent of the total site area is dominated by geotope D5 and 22% by geotope A3. Other three subdominant geotopes (A4, B3, B4) occupied about 25% of the territory in equal share. These geotopes (together with A3) are the most suitable for agriculture due to normal or relatively redundant soil wetness. A5, C5, D5 geotopes correspond to various types of peatlands.

Lesunovo (site 2) is located in the transition zone between moraine and outwash plain landscapes with a large field cluster in a forest matrix. Lesunovo is characterized by the greatest diversity of geotopes, but with a predominance of nutrient-poor and dry sandy terraces and outwash plains with eolian relief (A1) with significant patches of floodplain fens in river valley (D5), interfluvial morainic plain (C3), and outwash plains (B2 and B3) with suitable-for-agriculture soil moisture.

Shulgino (site 3) is located on the border of the Oka river valley and high moraine karst plateau that is characterized by fertile grey forest soils (Greyic Phaeozems) that developed in carbonate loess and loam sediments. This karst plateau consists from geotopes B3 (36%) and C3 (30%) with normal soil wetness and suitable for agriculture soil nutrients supply. Nutrient-poor and dry (A1), the Oka terraces, as well as nutrient-rich and humid (D4) floodplains occupied 18% and 10% of total area. Shulgino looks like a huge field matrix with several forest patches.

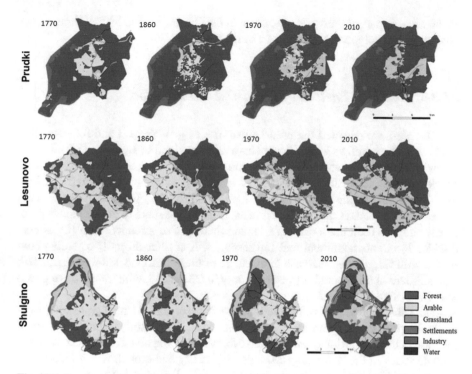

Fig. 23.1 Land-use reconstructions

23.3.2 Reconstruction of Land-Use Change

Our reconstructed trajectories of LULCC showed (Fig. 23.1) that the maximum cultivated area dates back to 1770, comprising 12% of total arable land (130 km²) for *Prudki*, 29% for *Lesunovo* (320 km²), and 51% for *Shulgino* (200 km²). In the nineteenth century, the arable land increased slightly only in Lesunovo (by 3%), while it declined in *Shulgino* (from 51% to 48%) and *Prudki* (from 12% to 8%), accompanied by an increase in forest and grassland areas.

During the second half of the twentieth century, the share of forest grew everywhere, but the share of grassland also increased. In the early twenty-first century, there was a sharp reduction in arable land, accompanied by an increase in grassland. Arable land in *Lesunovo* and *Prudki* vanished completely by 2010. In *Shulgino*, the extent of arable land decreased to 9% by 2010.

The most noticeable growth of arable land took place in the eastern part of *Prudki* in the twentieth century, mainly along the river and on formerly forested land. Almost all the territories around the villages were converted from arable land to grassland by 1970. *Lesunovo* was characterized by an almost unchanged structure of agricultural land, allocated mostly between the main road and the navigable river. In *Shulgino*, abandonment of arable land occurred already in the nineteenth century in the western part of the study site, close to the Oka River, and by that time, arable land remained only in the eastern part of the study site.

23.3.3 Natural Factors of Land-Use Distribution

For *Shulgino*, we detected the predominant use of geotopes C3 and A1 for arable land in 1770, as well as a significant share of arable land in B3 (Fig. 23.2). D4 are occupied mainly by grasslands in all periods. However, decline in the proportion of arable land was active already by 1860 on nutrient-poor and dry terraces (from 12% to 7%). Increase in the share of arable land in C3 (by 2%) was accompanied with decrease of grasslands, and not just forests. In 1970, the share of arable land is practically equal to 0 on the terraces (A1), and there is also a reduction in B3, almost double, due to abandonment and reforestation. A smaller drop-off of arable land occurs with C3, with an increase in the share of both grassland and forest. In 2010, a small share of arable land is preserved only in C3 and B3, with transition to grassland (first of all, fallow land).

While abandonment of arable land in *Shulgino* has started already by 1860 (especially for poor and dry terraces), in *Lesunovo*, on the contrary, there was a small increase in the proportion of arable land everywhere, even on terraces, accompanied with decrease in the share of grasslands. Especially this growth was noticeable in the most suitable for plowing geotopes, C3 and B3. By 1970, however, there was a sharp reduction in the share of arable land in all geotopes – the fastest was for the nutrient-poor dry terraces A1 and the slowest for B3 and C3 with an increase in the share of grassland, rather than forests. By 2010, we detected complete abandonment of agricultural lands and their transition to grassland (fallow land).

For *Prudki*, we observed a very small share of agricultural land in the eighteenth and nineteenth centuries. Geotopes B4 and A3 were often used here for agriculture in 1770. By 1860, there was a decrease in the share of arable land. However, by 1970, there was an increase in the share of agricultural land (especially grasslands) in geotope A5. There was an expansion (Fig. 23.1) of grasslands along a huge swampy area (drained by the Vozha River to the east of study site) due to land reclamation and peat extraction. Arable lands here remained only in geotope B4 and A3. By 2010, a complete abandonment of agricultural land was detected, similar to *Lesunovo*.

23.3.4 Positional Factors of Land-Use Distribution

All the study sites are characterized by the location of arable land near villages, this radius being within 1 km for *Prudki* and 2–2.5 km for *Lesunovo* and *Shulgino*. The meadows in *Prudki* and *Lesunovo* have a similar location, albeit more extended and with a less-pronounced peak, and *Shulgino* is characterized by the presence of

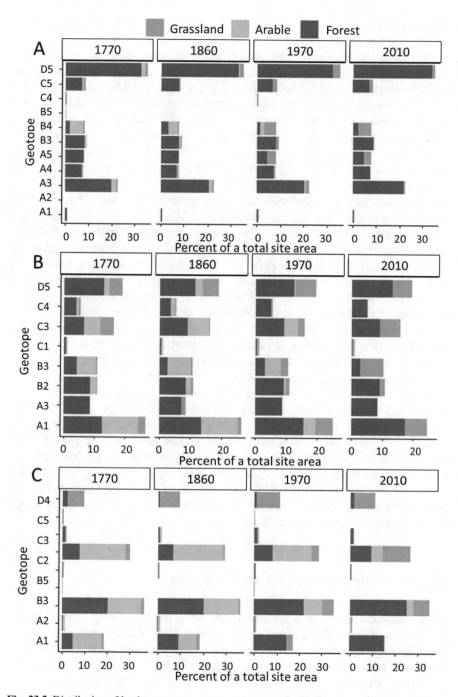

Fig. 23.2 Distribution of land-use types across geotopes. A – Prudki, B – Lesunovo, C – Shulgino

meadows as well as at maximum distance from the villages (about 4–5 km). These patterns observed in all periods.

For *Prudki*, we detected the location of arable land also close to the main road, because almost all the villages of this site were located on the main road. For grasslands, the second peak in the range of 2.5–3 km is more noticeable in different periods. We observed the placement of arable land no further than 5 km from the main road in *Lesunovo*, where only a part of the villages were located on the main road. In 1970, there was a slight increase in the density of road networks after the construction of asphalt roads, so this distance reduced to 2.5–3 km. Grasslands in the eighteenth and the nineteenth centuries were situated within 9–10 km from the road. Since the twentieth century, again, this distance reduced to 5–6 km. In *Shulgino*, in the eighteenth and the nineteenth centuries, arable lands were situated within 12–13 km from the main road, but grasslands had their maximum peak 12–13 km away from the main road. This situation changed dramatically in the twentieth century after the construction of asphalt roads, which led to the conservation of arable land in the range of 1.5–2 km from the road and within 4 km from grasslands.

The location of almost all the meadows near the large river in *Shulgino* indicates high importance of hayfields in the flood plain of the Oka River for farming. For arable land, the proximity to the river does not matter. Only by 2010, there was a situation when arable land was preserved only within the range of 1–5 km from the river. Probability density for grasslands in 2010 has two peaks. This is explained by the massive abandonment of arable land in the area of 5–6 km from the river. For *Lesunovo*, we observed the presence of arable land between the eighteenth and twentieth centuries within a wide range of distances to the river with a certain peak in the 2–3 km surroundings. Grasslands almost always have two peaks of occurrence – near the river and at some distance. *Prudki* is characterized mainly by the coincidence of the location of arable land and grasslands at a distance of 2.5–5 km from the system of navigable lakes. The probability of encountering grassland near water bodies in *Prudki* declines with an increase in distance; only for 1860, the second peak in the range of 1 km was detected. Arable land has a more noticeable peak in the area of 1 km from water bodies. For *Shulgino* and *Lesunovo*, we observed approximately the same picture.

23.3.5   Analysis of the Settlement System

Our results showed evidence that arable land is almost completely outlined by the 1 km radius of the buffer zone in *Prudki* site (Fig. 23.3). Also, almost all the settlements demonstrated preference for geotope B4. One of the villages is on the shore of the Klepiki lakes system, the other two are located far away from the main road. All the other major villages are situated on the provincial road from Ryazan to Vladimir that passes through the lowest parts of the Meshchera Lowlands, so it locally runs along the driest ridges between peatlands. Thus, here we observe a kind of "island" type of settlement, where peasants often work in handicrafts associated with forest.

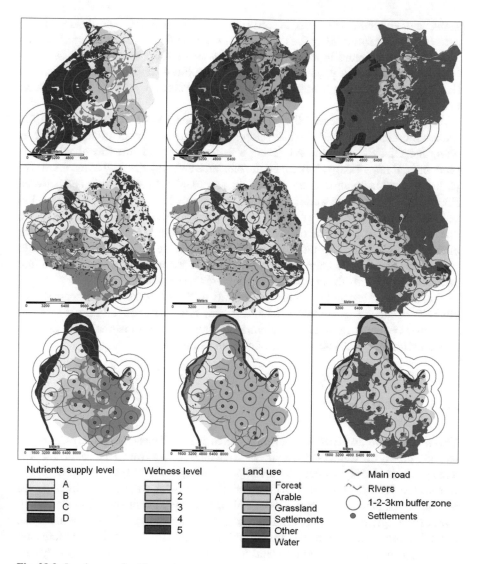

Nutrients supply level

	A
	B
	C
	D

Wetness level

	1
	2
	3
	4
	5

Land use

	Forest
	Arable
	Grassland
	Settlements
	Other
	Water

~ Main road
~ Rivers
◯ 1-2-3km buffer zone
● Settlements

Fig. 23.3 Land-use and settlements structures across geotopes

Most villages in *Lesunovo* were confined to river valleys, i.e. to areas with certain proportion of slopes and sandy sediments (often geotope A1). This, again, demonstrates that drainage here was an important, limiting factor. At the same time, all the closest territories to villages were plowed up – we see that the arable land is already delineated by only a two-kilometer radius. And within 2–3 km from villages there were different combinations of geotopes, mostly with high share of B3 and C3. The share of arable land here was growing slightly from 1770 to 1860. While population grows, this extensive development leads to transition from grasslands to arable land and then to decrease in fertilization of remote arable land. This results in a drop in yields, followed by the conversion back to arable land expansion. The type of settlement structure is rather "linear", stretched along the rivers and the main road, which defines a peculiar triangle of agricultural development between the river and the road. And all waterlogged flat moraine plains were covered with forest due to soil moisture.

In *Shulgino*, we observed the picture of "equitable distribution" type of settlement structure and agricultural development. There is a small group of villages along the Oka River within A1 geotope, but the group of interfluvial villages is predominant numerically. Moreover, the nutrients and water supply at the interfluve plain are suitable for agriculture, so the villages are fairly evenly distributed between B3 and C3 geotopes. In addition, we see that arable land is delineated by a 2 km radius, but there are a lot of forest patches close to the villages (Fig. 23.3). We assumed that the presence of the high-quality meadows on the Oka floodplain allowed peasants to keep more livestock, and therefore to fertilize more land, to choose the better lands from the surrounding. Hence, intensification of agriculture with population growth was greater as compared to *Lesunovo*.

At least the driest and nutrient-rich habitats were selected for settlements if possible. This illustrates perfectly the thesis that the two main limiting factors in Meshchera (waterlogging and poor nutrients supply of the substrate) were of decisive importance in the economic development of the territory.

23.4   Discussion

The main difference between our study sites was in landscape patterns gradient (mostly based on wetness and nutrient content in soils) from *Polessje* to *Opolje*. This gradient sharply limited the development of agriculture that led to the formation of the "center-periphery" vector, according to which the development of the territory took place (Ioffe et al. 2004, 2014). In the eighteenth century, the expansion of the agricultural transformation of the territory affected only the marginal zone (Lesunovo site), captured the central Meshchera (Prudki) in the twentieth century, and abandonment of arable land was in the reverse order in the twenty-first century.

Land-use changes across three study sites were associated with crucial sociopolitical processes and technological innovation, which via a set of positional and biophysical determinants shaped land use in the study area. The allocation of land-use types was predetermined by agro-environmental site conditions, such as soil quality and humidity and distance to rivers. The distance to the villages, as a position factor, was largely determined by the location of the village, which was also not accidental, but depended on the natural features of the territory. During the eighteenth and the nineteenth centuries, the threefold growth of the population did not cause the expansion of agricultural lands, whose development was limited to natural patterns. In the twentieth century, a sharp decline in the population, along with the development of technology and the implementation of agricultural machinery, would seem to have greatly altered the structure of the land, but the pattern has changed significantly only in *Prudki* due to drainage of marshes and extraction of peat. And only in the twenty-first century, after the transition to market agriculture, land was abandoned. This suggests that for 250 years, despite all sociopolitical changes, the spatial structure of agricultural development has not changed much, determined by natural patterns.

Acknowledgments This research was conducted according to the State target for Lomonosov Moscow State University "Structure, functioning and evolution of natural and natural-anthropogenic geosystems" (project no. AAAA-A16-116032810081-9).

References

Alix-Garcia, J., Munteanu, C., Zhao, N., et al. (2016). Drivers of forest cover change in Eastern Europe and European Russia, 1985–2012. *Land Use Policy, 59*, 284–297. https://doi.org/10.1016/j.landusepol.2016.08.014.

Alyabina, I. O., Golubinsky, A. A., Kirillova, V. A., & Khitrov, D. A. (2015). Soil resources and agriculture in the center of European Russia at the end of the 18th century. *Eurasian Soil Science, 48*, 1182–1192. https://doi.org/10.1134/S1064229315110034.

Bastian, O., Grunewald, K., & Khoroshev, A. V. (2015). The significance of geosystem and landscape concepts for the assessment of ecosystem services: Exemplified in a case study in Russia. *Landscape Ecology, 30*, 1145–1164. https://doi.org/10.1007/s10980-015-0200-x.

Conrad, O., Bechtel, B., Bock, M., et al. (2015). System for Automated Geoscientific Analyses (SAGA) v. 2.1.4. *Geoscientific Model Development, 8*, 1991–2007. https://doi.org/10.5194/gmd-8-1991-2015.

Ellis, E. C. (2011). Anthropogenic transformation of the terrestrial biosphere. *Philosophical Transactions of the Royal Society A: Mathematical, Physical, and Engineering Sciences, 369*, 1010–1035. https://doi.org/10.1098/rsta.2010.0331.

Ellis, E. C., Klein Goldewijk, K., Siebert, S., et al. (2010). Anthropogenic transformation of the biomes, 1700 to 2000. *Global Ecology and Biogeography, 19*(5), 589–606. https://doi.org/10.1111/j.1466-8238.2010.00540.x.

Ellis, E. C., Fuller, D. Q., Kaplan, J. O., & Lutters, W. G. (2013). Dating the Anthropocene: Towards an empirical global history of human transformation of the terrestrial biosphere. *Elementa: Science of the Anthropocene, 1*(18). https://doi.org/10.12952/journal.elementa.000018.

Engerman, S. L., & Sokoloff, K. (2011). *Economic development in the Americas since 1500: Endowments and institutions*. Cambridge: Cambridge University Press.

Feurdean, A., Munteanu, C., Kuemmerle, T., et al. (2017). Long-term land-cover/use change in a traditional farming landscape in Romania inferred from pollen data, historical maps and satellite images. *Regional Environmental Change, 17*(8), 2193–2207. https://doi.org/10.1007/s10113-016-1063-7.

Foley, J. A., et al. (2005). Global consequences of land use. *Science, 309*(5734), 570–574. https://doi.org/10.1126/science.1111772.

Gedymin, A. V. (1960). Experience in the use of Russian Land Survey materials in geographical research for agricultural. In *Issues in geography. Vol. 50. Historical geography* (pp. 147–171). Moscow: Geografgiz. (in Russian).

Goldenberg, L. A., & Postnikov, A. V. (1985). Development of mapping methods in Russia in the eighteenth century. *Imago Mundi, 37*(1), 63–80. https://doi.org/10.1080/03085698508592588.

Golubinsky, A. A., Alyabina, I. O., Shalashova, O. V., & Khitrov, D. A. (2013). From survey plans to land cover maps: Data generalization in the cartographic materials of the General Land Survey in Russia (1765–1800). In *Proceedings of the 26th international cartographic conference* (pp. 1–7). Dresden, Germany.

Gvozdetskii, N. A. (1968). *Physico-geographical regionalization of the USSR: The characteristics of the regional units. Moscow.* Moscow: University Press. (in Russian).

Hurtt, G. C., Chini, L. P., Frolking, S., et al. (2011). Harmonization of land-use scenarios for the period 1500–2100: 600 years of global gridded annual land-use transitions, wood harvest, and resulting secondary lands. *Climatic Change, 109*, 117–161. https://doi.org/10.1007/s10584-011-0153-2.

Ioffe, G., Nefedova, T., & Zaslavsky, I. (2004). From spatial continuity to fragmentation: The case of Russian farming. *Annals of the American Association of Geographers, 94*, 913–943.

Ioffe, G., Nefedova, T., & de Beurs, K. (2014). Agrarian transformation in the Russian breadbasket: Contemporary trends as manifest in Stavropol. *Post-Soviet Affairs, 30*, 441–463. https://doi.org/10.1080/1060586X.2013.858509.

Isachenko, G. A., & Reznikov, A. I. (1996). *Dynamics of landscapes of the taiga of the North-West of European Russia.* St. Petersburg: Russian Geographical Society Publishing. (in Russian).

Jepsen, M. R., Kuemmerle, T., Müller, D., et al. (2015). Transitions in European land-management regimes between 1800 and 2010. *Land Use Policy, 49*, 53–64. https://doi.org/10.1016/j.landusepol.2015.07.003.

Josephson, P., Dronin, N., Mnatsakanian, R., et al. (2013). *An environmental history of Russia.* Cambridge: Cambridge University Press.

Kalimullin, A. V. (2006). *Historical research of regional environmental problems.* Moscow: Prometej. (in Russian).

Kiryushin, V. I., & Ivanov, A. L. (2005). *Agroecological evaluation of lands and design of landscape-adaptive farming systems and agrotechnologies: Methodological recommendations.* Moscow: Rosinformagrotekh. (in Russian).

Krivtsov, V. A., Tobratov, S. A., Vodoresov, A. V., et al. (2011). *Natural potential of landscapes of Ryazan oblast.* Ryazan: Ryazan State University Publishing House. (in Russian).

Kuemmerle, T., Kaplan, J. O., Prishchepov, A. V., et al. (2015). Forest transitions in Eastern Europe and their effects on carbon budgets. *Global Change Biology, 21*(8), 3049–3061. https://doi.org/10.1111/gcb.12897.

Kusov, V. S. (1993). Quality surveying maps and the possibility of their use for retrospective mapping. *Proceedings of Moscow University, Series 5. Geography, 3*, 66–76. (in Russian).

Lyuri, D. I., Goryachkin, S. V., Karavaeva, N. A., et al. (2010). *Dynamics of agricultural land in Russia in the twentieth century and postagrogenic restoration of vegetation and soil.* Moscow: GEOS. (in Russian).

Matasov, V. M. (2016). Methodological aspects of the analysis of the spatial land use structure of Kasimov county in the late 18th century. *Geodesy and Cartography, 3*, 59–64. (in Russian).

Milov, L. V. (1965). *Study on "Economic Notes" to the General Land Survey.* Moscow: University Press. (in Russian).

Milov, L. V. (2006). *The Great Russian plowman and the peculiarities of the Russian historical process* (2nd ed.). Moscow: Russian Political Encyclopedia. (in Russian).

Moon, D. (2013). *The plough that broke the steppes: Agriculture and environment on Russia's grasslands, 1700–1914.* Oxford: Oxford University Press.

Petit, C. C., & Lambin, E. F. (2002). Long-term land-cover changes in the Belgian Ardennes (1775–1929): Model-based reconstruction vs historical maps. *Global Change Biology, 8*(7), 616–630. https://doi.org/10.1046/j.1365-2486.2002.00500.x.

Plieninger, T., Draux, H., Fagerholm, N., et al. (2016). The driving forces of landscape change in Europe: A systematic review of the evidence. *Land Use Policy, 57*, 204–214. https://doi.org/10.1016/j.landusepol.2016.04.040.

Poska, A., Saarse, L., Koppel, K., et al. (2014). The Verijärv area, South Estonia over the last millennium: A high resolution quantitative land-cover reconstruction based on pollen and historical data. *Review of Palaeobotany and Palynology, 207*, 5–17. https://doi.org/10.1016/j. revpalbo.2014.04.001.

Price, B., Kaim, D., Szwagrzyk, M., et al. (2017). Legacies, socio-economic and biophysical processes and drivers: The case of future forest cover expansion in the Polish Carpathians and Swiss Alps. *Regional Environmental Change, 17*(8), 2279–2291. https://doi.org/10.1007/s10113-016-1079-z.

Prishchepov, A. V., Radeloff, V. C., Baumann, M., et al. (2012). Effects of institutional changes on land use: Agricultural land abandonment during the transition from state-command to market-driven economies in post-Soviet Eastern Europe. *Environmental Research Letters, 7*(2), 024021. https://doi.org/10.1088/1748-9326/7/2/024021.

Prishchepov, A. V., Müller, D., Dubinin, M., et al. (2013). Determinants of agricultural land abandonment in post-Soviet European Russia. *Land Use Policy, 30*(1), 873–884. https://doi. org/10.1016/j.landusepol.2012.06.011.

R Core Team. (2013) *R: A language and environment for statistical computing.* http://www.R-project.org. Accessed 4 Aug 2018.

Rhemtulla, J. M., Mladenoff, D. J., & Clayton, M. K. (2009). Legacies of historical land use on regional forest composition and structure in Wisconsin, USA (mid-1800s–1930s–2000s). *Ecological Applications, 19*(4), 1061–1078. https://doi.org/10.1890/08-1453.1.

Rosstat. (2010). *Russian Federal Service of State statistics.* Regions of Russia. Socio-economic measures. http://www.gks.ru/bgd/regl/B10_14p/Main.htm. Accessed 4 Aug 2018. (in Russian).

Schröter, M., Koellner, T., Alkemade, R., et al. (2018). Interregional flows of ecosystem services: Concepts, typology and four cases. *Ecosystem Services, 31B*, 231–241. https://doi. org/10.1016/j.ecoser.2018.02.003.

Selivanov, A. (1890). *Collection of statistical data for the Ryazan province.* Ryazan: Ryazan Province Zemstvo. (in Russian).

Sluyter, A. (2003). Neo-environmental determinism, intellectual damage control, and nature/society science. *Antipode, 35*(4), 813–817. https://doi.org/10.1046/j.1467-8330.2003.00354.x.

Steffen, W. L. (Ed.). (2005). *Global change and the Earth system: A planet under pressure.* Berlin/New York: Springer.

Trapeznikova, O. N. (2014). Historical types of agrolandscapes of the forest zone of the East European Plain and the natural factors of their spatial organization. In K. N. Dyakonov, V. M. Kotlyakov, & T. I. Kharitonova (Eds.), *Issues in geography. Vol. 138. Horizons of landscape studies* (pp. 384–408). Moscow: Kodeks. (in Russian).

Tsvetkov, M. A. (1957). *The change in the forest cover of European Russia from the end of the 17th century to 1914.* Moscow: Academy of Sciences of the USSR Press. (in Russian).

Verburg, P. H., Crossman, N., Ellis, E. C., et al. (2015). Land system science and sustainable development of the earth system: A global land project perspective. *Anthropocene, 12*, 29–41. https://doi.org/10.1016/j.ancene.2015.09.004.

Veski, S., Koppel, K., & Poska, A. (2005). Integrated palaeoecological and historical data in the service of fine-resolution land use and ecological change assessment during the last 1000 years in Rõuge, southern Estonia: Land use and ecological change assessment in southern Estonia. *Journal of Biogeography, 32*(8), 1473–1488. https://doi. org/10.1111/j.1365-2699.2005.01290.x.

Vidina, A. A. (1962). *Methodical instructions for field large-scale landscape research. For the purposes of agricultural production in the forest zone of the Russian Plain.* Moscow: University Press. (in Russian).

Chapter 24
Landscape Features of the Prehistory of Moscow

Viacheslav A. Nizovtsev

Abstract The study focuses on determining the role of landscape factor in the origin and development of Moscow city. The methodological basis was a landscape-historical approach, which considers the regional and local physical and geographical differentiation of the territory with regards to both space and time. The natural properties of landscapes in many respects predetermined the development of economy and settlement networks. A diversity of landscape conditions is caused by features of abiotic template (geological substrate and topography) and variations in local climate. Moscow is located on the boundary of three physical-geographical provinces which are in correspondence with the relief of the East-European plain. Moscow, similar to the other cities of the region, initially came into existence as the center of agricultural area. The choice of a place for construction of the Kremlin was not accidental. A better place other than Borovitsky hill could hardly be found within the contemporary limits of Moscow city. This place combines the control over the transport routes with excellent defense opportunities. Kremlin is located on the cape near the junction of the rivers within the low outwash plain, which is protected by a water barrier and natural coastal slopes.

Keywords Landscape reconstruction · Moscow · Kremlin

24.1 Introduction: Physical-Geographical Features of the Historical Center of Moscow

The geographical position of Moscow is absolutely unique. The city is located at the neck of the meander near the node of the hydrographical network, which allowed settlers to move easily along numerous tributaries and manage diverse landscapes both in the vicinity of the city and in remote areas. Original landscapes on the terri-

V. A. Nizovtsev (✉)
Lomonosov Moscow State University, Moscow, Russia

© Springer Nature Switzerland AG 2020
A. V. Khoroshev, K. N. Dyakonov (eds.), *Landscape Patterns in a Range of Spatio-Temporal Scales*, Landscape Series 26,
https://doi.org/10.1007/978-3-030-31185-8_24

Fig. 24.1 Physical-geographical provinces and indigenous landscapes of Moscow (within the boundaries of the contemporary Moscow ring road) **Provinces: I – Moscovskaya, II – Moskvoretsko-Okskaya, III – Meshcherskaya**

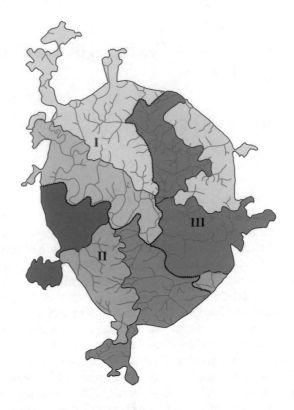

tory of Moscow were formed at the end of the late Holocene (2000–2500 years ago) during last significant climate change. The territory of Moscow in the past was characterized by a striking diversity of landscape complexes and, accordingly, soil, vegetation cover, and fauna. Despite the fact that Moscow is located in the natural zone of mixed (spruce-broad-leaved) forests, the territory was covered by not coniferous-deciduous forests only, but also by taiga spruce and pine forests, as well as by oak, lime, and elm forests. The diversity and, often, also the contrast of landscape conditions was determined by abiotic template (geological substrate and topography) and the differences in the local climate. Moscow is located at the boundary of three major physical-geographical provinces: Moscovskaya (No. I in Fig. 24.1), Moskvoretsko-Okskaya (No. II), and Meshcherskaya (No. III). They were formed within the relief mega-forms of East-European Plain: the Moskvoretsko-Okskaya erosion plain, the Klinsko-Dmitrovskaya ridge and the Meshcherskaya lowland, respectively (Fig. 24.1).

The formation of the landscape structure of the city territory was influenced by the events of the Quaternary period, namely, the glaciation and the flows of glacial waters. Moscow (Riss) glaciation left the most important legacy. Resulting different types of morainic, glaciofluvial, and lacustrine-glaciofluvial plains masked initial geological contrasts and determined landscape differentiation. The subsequent formation of the erosion network, deluvial processes, and solifluction transformed glacial relief and influenced the formation of landscapes morphological structure resulting in individual, unique features of each landscape.

We identified nine indigenous landscapes, most of which extend far beyond the territory of contemporary Moscow. A unique fact is that eight of them almost converge in the central part of the city. There is no such close neighborhood and such large quantity of diverse landscapes elsewhere in the Moscow region or in other areas of the center of the East-European Plain.

24.2 Landscape Structure of the Territory of the Modern Kremlin

The restored landscape structure of the territory of the modern Kremlin is represented by the following natural-territorial landscape complexes (NTC) (Fig. 24.2):

1. High floodplain of the Moscow river, flat and slightly undulating
2. Floodplain of the Neglinnaya river, gently inclined
3. Floodplain of the Neglinnaya river, flat with low ridges and fens
4. The first alluvial terrace of the Moscow river, imperfectly drained
5. The first alluvial terrace of the Moscow river, flat and gently inclined, well drained
6. Alluvial terrace of the Neglinnaya river, flat and slightly inclined, poorly drained
7. Slope of the alluvial terrace of the Neglinnaya river, gently inclined
8. Deluvial fan superimposed on the first terrace of the Moscow river
9. Steep south-facing slope of the Moscow river valley with complex slope profile
10. Steep north-facing slope of the Neglinnaya river valley
11. Gentle south-facing slope of the Neglinnaya river valley
12. Low valley outwash plain with gently sloping main surface
13. North-facing slope of a low valley outwash plain, gently inclined
14. Moraine and kame hills, weakly convex
15. Flat and concave hollow of drain glacial waters (local inter-basin flows)
16. *Balka*[1] bottoms, flat and concave, moist and wet
17. *Balka* slopes, gentle and steep
18. Hollows and hollow-like lowlands, concave
19. Hollow-like water collecting lowlands, concave

---

[1] *Balka* is the Russian term for flat-bottom erosional landforms with ephemeral water courses.

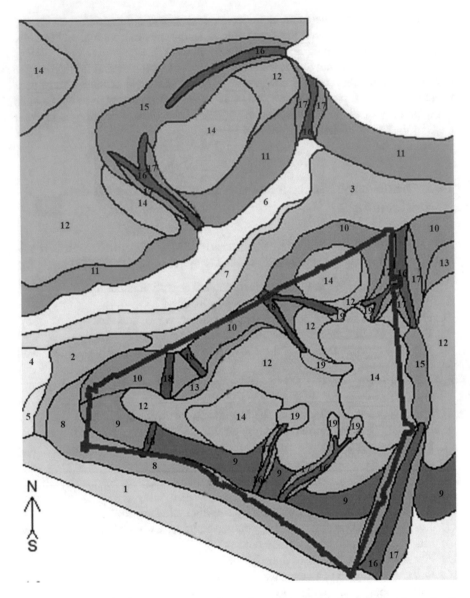

Fig. 24.2 Restored landscape structure on the territory of the modern Kremlin (numbers of units see in text)

24.3   Reconstruction of Landscape Features of Nature Use in the Kremlin Area in the Bronze Age

According to archaeological data, the first people on the territory of Moscow appeared in the IV millennium BC (Sakharov 1997). More ancient traces of human presence here have not yet been found. There are very few archaeological finds of this time discovered in the immediate vicinity of the Moscow river, in its floodplain landscapes. This indicates that the first "Muscovites" preferred water streams while choosing places for the settlements. The finds of stone-made and bone-made tools show that they were fishers-hunters related to the Lyalovskaya culture. In the Neolithic age, flourishing of the fishing-hunting-gathering economy started. Since the settlers practiced the appropriating economy being engaged in hunting, fishing and gathering, the human impact on landscapes was minimal. During the period of the appropriating type of economy, a balanced equilibrium system "man-nature" was formed (Nizovtsev 1990). While the production and the relations of production were developing, gradual complication of the emerging natural-economic systems was evident. Human settlements were clearly attached to the resource base at certain types of NTC. Humans could not ignore their boundaries. The immigration of the Fatyanovo culture representatives to the Moscow region (the Bronze Age) was the critical event which provoked a significant change in the economic development of landscapes.

The Bronze Age is associated with various substages of the subboreal period. The results of paleobotanical studies, both to the west and east of Moscow within the forest and steppe zones of the European Russia (Spiridonova 1991), allow us to reconstruct the most complex and contrasting changes in the landscapes of this period throughout the entire Holocene. A sharp climate aridization that occurred 3200–4100 years ago (the average subboreal warming phase) caused a drop in the water level in the hydrographic network and made floodplain complexes available for the economy. In the Bronze Age (second millennium BC), Fatyanovo settlers began to develop not only the valley of the Moscow river, but also its tributaries such as Kotlovka, Likhoborka, etc. In such valleys there were more grounds favorable for grazing in forests. These settlers were engaged in the producing economy – livestock breeding (at later stages – agriculture). They achieved high perfection in stone processing and were familiar with metal (Krainov 1972; Krasnov 1971).

Archaeological data such as finds of a stone axe and debris of three other axes of the Fatyanovo culture near the Spassky Gate and the Archangel Cathedral allow us to confidently assume that the population of the Bronze Age used this territory in economy. In the basin of the Moscow River, 22 settlements and locations with pottery of the Fatyanovo type, about 15 burial grounds and at least 60 places of finds of stone axes and other tools were detected (Krenke 2007). There are much less monuments of the Bronze Age in the Moscow region than those of the Neolithic, though they are densely grouped. Apparently, the Fatyanovo tribes, who practiced producing economy, were less associated with certain NTCs, the resource base of which played a crucial role for the cultures of the appropriating type of economy (Mesolithic, Neolithic). Modern studies of monuments of the Bronze Age on the floodplain of the Moscow river (RANIS-floodplain) and Tsaritsyno-1 (Cerera) conducted by N.A. Krenke (2007) confirmed the long existence of settlements and the established system of nature use in specific landscape conditions. An analysis of the landscape structure of the Moscow Kremlin territory and the above-mentioned settlements shows the similarity of landscape conditions and the absence of obstacles for conducting similar economy.

Forest livestock production was the main branch of the economy of the Fatyanovo culture tribes, while hunting and fishing were of a subordinate nature. At the first stage, the Fatyanovo tribes were specialized in pig breeding, which is, in general, characteristic of the initial period of forest livestock production (Tsalkin 1962; Krasnov 1971). The development of pig breeding was favored by the abundance of certain plant foods in broad-leaved and alder forests, especially in floodplains of rivers and ancient lake basins. The Fatyanovian pigs were not very different from wild pigs, which were abundant in these places, and could live on pasture forage not in summer only but in winter as well. Long grazing of pigs at one place led to the complete destruction of vegetation. Forced permanent transitions of Fatyanovo tribes to new places in search of pastures encouraged the involvement of new sites into the economy. Since this time, the renewal of indigenous forests, especially in the most dynamic floodplains, has become problematic (Nizovtsev 1990).

At the next stage of the development, the Fatyanovites appear to have cattle, sheep, goats, and even horses. Thus, the form and character of livestock breeding changes and shifts to stall livestock production (Krainov 1972). The pastures were again localized in the valley complexes, mainly on river floodplains, which, as a rule, had more open spaces such as meadow clearings and glades. Apparently, Fatyanovo tribes, using the fire or an axe cleared the forest for new pastures. The presence of traces of cleavages and the finds of cleaved axes themselves (Tsaritsyno 1 – the settlement of Cerera) indicate the forest cutting and the emergence of open spaces (Krenke 2007). Cattle breeding provided the main food, which was meat and dairy products (there was a special dish for storing milk, cooking cheese, butter, and cottage cheese). Leather and animal fur were used for making clothes and footwear. Bones were used to make tools and crafts. Rudiments of slash-and-burn farming were possible as well (Bader 1939).

N.A. Krenke (2014) believes that the valley of the Moscow River was one of the basic attractors for settling and economic development. It is possible that the changes in the water regime of rivers induced by increased cultivated area resulted in higher floods. This was the reason for developing higher hypsometric levels in the area of the modern Kremlin, similar to the Tsaritsino 1 settlement. To some extent, this may be evidenced by the topography of the stone axes finds.

The continuing intensification of economy caused sharp growth of anthropogenic press on the surrounding nature, which involved increase in area and degree of impact. The producing economy from the ecological point of view leads to deeper, sometimes irreversible changes in nature. This is due to the fact that people seem to "break away" from the ecologically balanced nature, since no other living creature conducts such an economy, but rather are engaged in hunting and gathering. R.K. Balandin and L.G. Bondarev (1984) showed that livestock breeding can feed 20 times as much people as hunting does, and agriculture can feed 20–30 times more people compared to livestock breeding. It is probable that anthropogenic load on nature increased accordingly. The producing economy is associated with the replacement of certain components of NTCs with the other ones. The extensive economy of the Fatyanovo tribes is linked with the beginning of deforestation on the floodplains of rivers and lakes. Forest regeneration on many floodplain NTC became practically impossible due to constant human influence. In the morphological structure of landscapes, in floodplain *urochishches* and *mestnosts*[2] the first stable elements of anthropogenic origin appear, which were floodplain meadows. Apparently, the Moscow River meadows in the landscapes of the modern Moscow have become a stable element since the Bronze Age (Nizovtsev 2010).

24.4 Reconstruction of Landscape Features of Nature Use in the Area of the Kremlin in the Iron Age

The invention of methods for obtaining iron and its use for the manufacture of tools and weapons made a real revolution in the history of human society, when people obtained the opportunity to master vast areas for arable land and settlements. Iron greatly increased the productive capacities of people. The development of iron contributed to changes in the social structure of society. Crafts were beginning to develop in parallel with metallurgy, blacksmithing, etc. The role of exchange of goods between tribes and peoples was growing. The spread of iron was also facilitated by the fact that deposits of iron ores, unlike copper and tin, could be found almost everywhere, especially so-called "bog ores". In line with the peculiarities of the landscape structure, the deposits of iron bog ores were sufficient not only within the territory of contemporary Moscow, but in the immediate vicinity of the future Moscow Kremlin as well. Two centers of metalworking were being formed: the Troitskoe settlement in the upper reaches of the Moscow River and in its middle reaches at the mouth of the Pakhra river.

The replacement of the material cultures of the Bronze Age by early iron cultures in the Moscow region occurred at the beginning of the sub-Atlantic climatic period. Scarce radiocarbon dating testifies that a new stage in the settlement of the Moscow region dates back to the turn of the ninth-eighth centuries BC. Artificial fortifications erected around the settlements were constituted by bulk ramparts with a palisade and deep moats from unprotected sides. Such fortified settlements around ancient towns and later unfortified ones (villages) are associated with the Diakovo culture. To date, more than 20 stationary settlements of this time are known, among which the most studied are the settlements of Mamonovskoye and Seletskoye, the village of Dubrovitsy-2 on the Desna river, the ancient settlement of Diakovo, the village of Chertov and Kolomenskoye (Gonyaniy and Krenke 1988; Krenke 2007). The fact that these settlements were located in the lower and middle reaches of the Moscow River and in the basin of the Pakhra river suggests that the population colonizing the Moscow river basin was moving upstream. Almost all the large settlements that arose during this period became the basis of the settlement network, which existed for about 1.5 thousand years. Most of the early Diakovo monuments occupy

---

[2] *Urochishche* and *mestnost* are the terms for hierarchical levels of landscape morphological units used in Russian landscape science. See Glossary and Chap. 1 for details

a border position between the lowlands of the river valley and the strip of adjacent slopes. Features of the Iron Age economy, apparently, required the mandatory combination of a variety of landscape components that existed only in the valleys of relatively large rivers. Cases when settlements were remote from the wide valley of a large river are rare (Barvikhinskoye, Korobovskoye settlements) (Gonyaniy and Krenke 1988; Krenke and Nizovtsev 1997). The peculiarities of the landscape structure of the future Kremlin territory fully met such requirements.

Within the lowlands, including floodplains and low alluvial terraces, only small monuments have been found in suitable areas that were not flooded in high water, attracting people in different historical periods. For example, on the floodplain near Dyakovo village, at least 5 different cultural complexes were detected. Most small settlements were located at the tip of the capes and at the marginal parts of the watershed slopes with relatively high elevations (up to 30 m) above the river level. This finding undoubtedly indicates suitability of the marginal areas of interfluvial plains for exploitation (apparently, agriculture) close to the central settlement (Krenke 2011). According to modern conceptions, the peculiarity of the settlement system in the Moscow river basin in the Iron Age was that in certain landscape conditions, a single system of interconnected villages and settlements ("a residential and economic complex") was formed in a small area. It is highly probable that such a residential and economic complex could also exist on the territory of the Moscow Kremlin.

Much more data on the economy of the Iron Age population are available than that for the previous period. Obvious development of agriculture involved cultivation of millet, barley, and wheat. Livestock production was specialized on the breeding of pigs and horses. River and lake fish resources were intensively used; beaver, moose, bear, and upland fowl were used as the main game resources (Krenke 2007).

Since the second century BC, change in cultural traditions in the Moscow region was associated with the arrival of the Baltic tribes, who possessed more advanced techniques of processing iron and were familiar with the use of harness. The spread of plowing practice is well illustrated by the data on occurrence of pollen of cultivated cereals in the cultural layers of the Diakovo settlement in the vicinity of the Moscow River. This encouraged the shift to a new qualitative level of farming as well as stronger impact on nature. The use of cattle as a draft force provided opportunity to increase the area of cultivated land significantly. But, at the same time, the thoroughness of its cultivation was reduced resulting in the growth of the extensive character of farming (Krasnov 1971). Our findings showed evidence that, along with the slash-and-burn system of agriculture, land rotation system (cultivated fields alternating with short-term coppice stage) and, possibly, a fallow system (cultivated fields alternating with meadows without coppice) had been practiced (Nizovtsev 1990). The settlements were surrounded by vast areas of felling and burning, as evidenced, for example, by the findings of a large amount of pyrophyte *Chamerion angustifolium* pollen during the excavations of the Akatov fortification in the neighboring Smolensk region.

Agriculture became the main type of economy with the cultivation of spring grain crops: wheat, barley, and millet (Krenke 2007). Apparently, simultaneously, several forms of agriculture were being applied as follows: slash-and-burn hoe-farming; crop rotation with short-term rewinds; plowing with two-field rotation (spring crops – fallow) using horse harnesses. Similar forms are characteristic of early farming in the forest zone (Krasnov 1971).

Settlements of the Diakovo culture were located, as a rule, in the sections of river valleys which had a very complex landscape structure, consisting of a large number of diverse, and, sometimes, contrasting natural territorial complexes. This allowed the Diakovo settlers to lead a "flexible" integrated economy. The late Diakovo tribes were again engaged in hunting that played an important role, especially furbearer hunting (mostly for a beaver). This was due to the development of trade relations (Krenke 2011). All types of agricultural land (fields, forests, pastures, etc.) were located within the river valleys, while on the interfluves intact forests were preserved. Since this time, the human impact on nature has become permanent over significant areas and comprised a variety of NTC types.

We distinguish three main types of landscape-economic systems (LESs) (Nizovtsev 2010) that were developing at that time. Firstly, there were small residential LESs: villages and settlements with adjoining permanent miniature arable plots (arable agro-geosystems at the level of *facies* and *podurochishches*[3]), located on capes and narrow interfluves between river valley slopes and erosional landforms. The Borovitsky hill of the Moscow Kremlin was, perhaps, the example for this kind of sites. Secondly, there were pastoral agro-geosystems, occupying floodplain, valley, and *balka* NTCs. Finally, LESs with slash-and-burn agriculture comprised the vastest area. These LESs were very unstable, and to a greater extent, they represented anthropogenic modifications of the NTCs at the level of *urochishche* as the depth of human impact on nature was still very low.

The highest suitability for cutting was inherent for NTCs of sandy-loam and sandy alluvial terraces, relatively well drained glaciofluvial plains, and valley outwash plains with sandy, sandy-loam and, less often, loamy soils. Note that for such kind of farming, natural fertility of soils was less important than good drainage and the possibility of cultivation without the use of arable tools, since ash fertilized the soil. Since growing period in the Moscow region is typically short and labor-consuming, rather high duration of the spring drying was favorable for the early readiness of soil for sowing works. In contrast, poorly drained flat surfaces of morainic and glaciofluvial plains with soil moisture surplus and insufficient heat supply ("cold soils") have not been used in agriculture for a long time. The light soils were of great importance, because they required less physical processing costs, and this was one of the decisive factors of that time.

Anthropogenic impact on these lands was periodical, since soils at the plots cleared for plowing were exhausted and abandoned after 2–3 years of exploitation. Consequently, these LESs, for 3–4-year long periods, were converted to nonagricultural agro-geosystems used as forest pastures. In 50–60 years, the forests had to be

---

[3] *Facies* and *Podurochishche* are the terms for hierarchical levels of landscape morphological units used in Russian landscape science. See Glossary and Chap. 1 for details

burned again. This period is not sufficient for restoring the native forests with dominance of spruce, pine, oak, and lime. Therefore, the ecological consequences of slash-and-burn farming were becoming noticeable as secondary small-leaved birch and aspen forests became widespread.

Arable LESs were located in close proximity to the settlements and were comparable with them in area. They occupied small areas on valley slopes (with gradient ranging from 2° to 8°), floodplains, slopes of low near-valley outwash plains, and the areas in the vicinity of water divides. Such a location was not accidental, since it represented the optimal combination of natural factors favorable for arable farming of that time. Gentle slopes, proximity to valleys, and light soil-forming sediments (loamy sand and sandy loam) provide good drainage and an excellent water-holding characteristic and aeration in the dominant Umbric Albeluvisols (sod podzolic soils) and Albeluvisols (podzolic soils). Despite the moderately long snowmelt period, in spring, these soils dry quickly and are characterized by sufficient heat supple ("warm soils") and early readiness for plowing. At the same time, during the dry periods, the loamy soils of these habitats could retain a certain amount of moisture for a long time. This allowed farmers to receive guaranteed yields on these lands even in extreme weather conditions.

More remote and slightly inclined (2–4°) surfaces of alluvial terraces, near-valley outwash plains, and interfluvial morainic plains could be used for plowing using short-term (with the duration of fallow period up to 4–5 years) and medium-term (approximately 10–20 years of "turnover") rotation. These plots were deeply dissected by short coastal ravines and gullies, well drained and also had relatively fertile soils with early readiness in spring. However, the rapid soil depletion and the small productive capacities of the Diakovo tribes did not allow them to plow them longer than 3–4 years. Subsequently, these plots were abandoned and used for grazing cattle in small forests (Gunova et al. 1996).

24.5   Conclusion

Thus, the development of the iron processing technology and the production of iron tools, encouraged dramatic enlargement of the productive capacities of people. The Diakovo tribes introduced relatively productive economy based on agriculture and livestock production. The deforestation began to spread first due to clearing for settlements, and, later, for arable lands. Slash-and-burn farming induced a new kind of landscape disturbance due to fire processing. The highest degree of disturbance occurred in NTCs of alluvial terraces and the areas of well drained near-valley outwash plains, and inclined moraine plains. The impact on these lands was periodical, since the plots cleared for plowing were depleted and abandoned in 3–4 years and the forests had to be burned again. There were practically no permanent fields, so the human impact was reversible and limited to the biota. Since that time, huge secondary forests have become a constant element of landscapes. Despite the high density of the Iron Age settlements in a number of districts of the Moscow region and the intensive economic activity, the ecological balance was not yet destroyed and settlements existed for hundreds of years; game resources were not depleted; the Moscow river retained its exceptional purity, as evidenced by finds of *Hucho taimen* bones in the late Diakovo layer (Krenke and Tsepkin 1991).

Acknowledgments The research was financially supported by Russian Foundation for Basic Research (project No. 19-05-00233).

References

Bader, O. N. (1939). To the history of the primitive economy on the Oka river and in the Upper Volga region in the era of early metal. *Bulletin of Ancient History, 3*, 113. (in Russian).

Balandin, R. K., & Bondarev, L. G. (1984). *Nature and civilization*. Moscow: Mysl. (in Russian).

Gonyaniy, M. I., & Krenke, N. A. (1988). The structure of the settlement of the Diakov tribes in the basin of the Pakhra river. *Soviet Archeology, 3*, 54–62. (in Russian).

Gunova, V. S., Kiryanova, N. A., Krenke, N. A., Nizovtsev, V. A., & Spiridonova, E. A. (1996). Agriculture and the system of land use in the valley of the Moscow river in the Iron Age. *Russian Archeology, 4*, 93–120. (in Russian).

Krainov, D. A. (1972). *The oldest history of the Volga-Oka interfluve. Fatyanovo culture*. Moscow: Nauka. (in Russian).

Krasnov, Y. A. (1971). *Early farming and cattle breeding in the forest belt of Eastern Europe*. Moscow: Nauka. (in Russian).

Krenke, N. A. (2007). Formation of the cultural landscape in the Moscow river basin from the Bronze Age to the Middle Ages. *Russian Archeology, 1*, 64–78. (in Russian).

Krenke, N. A. (2014). Moskvoretsky monuments of the Fatyanovo culture. *Russian Archeology, 4*, 5–18. (in Russian).

Krenke, N. A. (2011). *Diakovo settlement: The culture of the population of the Moscow river basin in the I millennium B.C. – I thousand A.D.* Moscow: Institute of Archeology RAS. (in Russian).

Krenke, N. A., & Tsepkin, E. A. (1991). Fisheries on the Moscow river from the V century B.C. to the VII century A.D. *Soviet Archeology, 1*, 104–111. (in Russian).

Krenke, N. A., & Nizovtsev, V. A. (1997). Man and landscape on the territory of the Moscow region in the Iron Age. In *The history of studying, using and protecting the natural resources of Moscow and the Moscow region* (pp. 81–91). Moscow: Yanus-K. (in Russian).

Nizovtsev, V. A. (1990). The history of economic development of the landscapes of the south-west of Moscow region (the pre-Mongol period). In *Landscapes of the Moscow region and Moscow, their use and protection* (pp. 18–29). Moscow: Moscow Division of Geographical Society USSR. (in Russian).

Nizovtsev, V. A. (2010). Landscape conditions of the early stages of the settlement of Borovitsky Hill. In *Monuments of material culture of the IV millennium B.C. – the first half of the first millennium A.D* (pp. 23–43). Moscow: Golden-Bi. (in Russian).

Sakharov, A. N. (Ed.). (1997). *The history of Moscow from ancient times to the present days. Vol 1*. Mosgorarkhiv: Moscow. (in Russian).

Spiridonova, E. A. (1991). *Evolution of the vegetation cover of the Don basin in the Upper Pleistocene-Holocene (Upper Paleolith-Bronze)*. Moscow: Nauka. (in Russian).

Tsalkin, V. I. (1962). Cattle breeding and hunting in the forest belt of Eastern Europe in the early Iron Age. *Materials and Research on Archeology of the USSR, 107*, 140–142. (in Russian).

Chapter 25
GIS-Based Study of Landscape Structure and Land Use Within the River Valleys in the Southern Tomsk Region: Spatial-Temporal Aspects

Vadim V. Khromykh and Oksana V. Khromykh

Abstract The chapter is devoted to the study of landscape structure and its natural and anthropogenic changes within the river valleys in the southern Tomsk region. The study is based on field observations, remote sensing data, and GIS analysis. Geodatabase and digital elevation models were created. Changes in landscape structure as a result of different factors were considered. The process of mire geosystems formation on terraces has been studied. Anthropogenic changes in landscape structure in the vicinity of Tomsk were revealed. The key factors of the anthropogenic modification of geosystems in the study area are the increased industrial and agricultural activities since the middle twentieth century, such as sand and gravel extraction from the riverbed, draining, land reclamation, and plowing. Also, important factors of the anthropogenic modification of geosystems are transport construction and expansion of the urban and rural areas. Now, almost all geosystems within the river valleys near Tomsk are exposed to anthropogenic modifications to various degrees. The vector of the changes is directed towards considerable desiccation due to the lowering of the groundwater level, which occurred due to the overlapping of various anthropogenic factors.

Keywords River valley · Landscape mapping · Landscape metrics · GIS · DEM · Spatial analysis

V. V. Khromykh (✉) · O. V. Khromykh
Tomsk State University, Tomsk, Russia
e-mail: geo@mail.tomsknet.ru

© Springer Nature Switzerland AG 2020
A. V. Khoroshev, K. N. Dyakonov (eds.), *Landscape Patterns in a Range of Spatio-Temporal Scales*, Landscape Series 26,
https://doi.org/10.1007/978-3-030-31185-8_25

25.1 Introduction

The study of landscape structure and land use features within the river valleys in the southern Tomsk region has both theoretical and practical importance. These valleys are located in a transition zone from the Altay mountain region to the West Siberian Platform. They have complex landscape structure, which is conditioned by location. Valley geosystems have a high rate of the natural transformations. Besides, many geosystems have undergone serious anthropogenic modification. Intensive changes of the Tom valley geosystems have been observed since the middle twentieth century due to increased industrial and agricultural activities, such as sand and gravel extraction from the river bed, land reclamation, as well as plowing, transport construction, and expansion of urban and rural areas.

The analysis of the literature sources showed a certain lack of works containing an integrated description of the valley natural systems within the study area and a description of their evolution. The available publications are based mainly on medium-scale landscape mapping performed more than 30 years ago. At the same time, the development of GIS-based mapping and spatial analysis technologies in recent years provides the opportunity to bring such studies to a new level.

The goal of this research is to characterize the landscape structure of the river valleys in the southern Tomsk region, as well as the natural and anthropogenic changes in valley geosystems since the late nineteenth century.

The objects of the study are the valleys of the Tom and the Ob rivers and their large tributaries in the southern Tomsk region (southeastern West Siberia). The research focused on the intra-valley differentiation of geosystems (at the *urochishche* and *mestnost* hierarchical levels[1]), their current state, natural dynamics, evolution, and anthropogenic modification.

The research was based on the theoretical and methodological ideas and works in the field of geography, basic and applied landscape studies by A.G. Isachenko, F.N. Milkov, N.L. Beruchashvili, V.B. Sochava, A.A. Krauklis, V.S. Mikheev, V.V. Kozin, I.I. Mamay, K.N. Dyakonov, A.V. Khoroshev. We took into account the experience of regional studies by A.A. Zemtsov, V.I. Bulatov, V.S. Khromykh, N.S. Evseeva, P.N. Ryazanov, V.V. Surkov, V.V. Khakhalkin, O.N. Baryshnikova, L.V. Sherstobitova, and others. To develop a methodology for large-scale GIS-mapping of valley geosystems, the authors used theoretical concepts and practical recommendations in the field of cartography, GIS-mapping, and geosystems modeling by R.F. Tomlinson, P.A. Burrough, M.N. DeMers, J.P. Wilson, K. McGarigal, V.I. Kravtsova, V.S. Tikunov, A.V. Koshkarev, I.K. Lurie, V.V. Sysuev, A.K. Cherkashin, A.D. Kitov, and others.

In this study, we applied methods of integrated physical-geographical research, including field methods and remote sensing. The newest methods of GIS-mapping

---

[1] *Urochishche* is the terms for hierarchical level of landscape morphological units (natural territorial complexes, NTC) higher than *facies* and lower than *mestnost* used in Russian landscape science. See Glossary and Chap. 1 for details.

and integrated spatial analysis were widely used as well. Comparative-cartographic and historical-cartographic methods were helpful in studying the geosystems dynamics and evolution. Geo Statistical methods were applied using GIS software: ArcGIS 10.3 (ESRI Inc.), ERDAS Imagine (Intergraph), Agisoft PhotoScan (GeoScan), Easy Trace 8.65 (EasyTrace Group).

25.2 Approaches and Terminology

In this study, we use the definition of the term "landscape" as an open dynamic spatial-temporal system. V.B. Sochava (1963, 1974, 1978) introduced the system approach into Russian-language physical geography in the 1960s–1970s. Cherkashin (2005) argues that the advantage of the geosystem approach is the opportunity to consider the patterns of structure and development of landscapes through the expression of the natural technology of elements transformation in the transition from state to state under the influence of environmental factors. Bastian et al. (2015) emphasize some differences in understanding the term "geosystem" by different researchers ranging from narrow biologically oriented "ecosystem" point of view to wider physical-geographical conception (Christopherson 2014).

Valley geosystems are considered within the frames of structural-genetic or genetic-morphological conception formulated by N.A. Solnetsev (1948, 1960). He elaborated rationales for five hierarchical levels of subordinate morphological geosystem units: *facies, podurochishche, urochishche, mestnost*, and *landscape* (as unit of hierarchical level) (Dyakonov 2005).[2]

It should be noted that landscape is characterized by both inter-component (vertical) and inter-complex (horizontal) interrelations. To study the inter-complex relationships, F.N. Milkov (1966) proposed the term *"paradynamic landscape complex"*, which was defined as a system of spatially adjacent regional or typological units characterized by the presence of interchange between matter and energy between them. A particular case of paradynamic complexes are *paragenetic landscape complexes* defined as dynamic systems of conjugate landscape complexes, with the holistic features determined both by genetic unity and genetic conjugation. Their functioning is determined by integration or disintegration of system-forming matter or energy flows in the gradient zone. River valleys are a vivid example of holistic paragenetic landscape systems (Kozin 1979).

Landscape change is the acquisition of new properties or the loss of the previous ones because of external impact or the self-development process. In the 1980s, in the works by V.B. Sochava (1978) and his followers (Krauklis 1979; Isachenko 2004; Kozin and Petrovskiy 2005), the conception of "landscape change" was recognized as generic and divided into three kinds: functioning, dynamics, and evolution.

---

[2] See Glossary for details.

The *functioning* of a landscape is a permanent sequence of constantly operating processes of exchange and transformation of matter, energy, and information, which ensures the preservation of the landscape state for a considerable period of time (Preobrazhenskiy 1982). Functioning often has a rhythmic (diurnal and annual) character and is not accompanied by a transition of the landscape from one serial state to another, which distinguishes it from dynamics (Kozin and Petrovskiy 2005).

Dynamics is the landscape changes not accompanied by changes in its structure, i.e. the variation of state attributes within a single invariant. The concept of an invariant was introduced into geography by V.B. Sochava to denote "a set of inherent properties of the geosystem, which remain unchanged during the transformation of geosystems" (Sochava 1978, p. 293). The notion on landscape invariant involves its vertical, horizontal, and temporal structure that is a relatively stable unity of elements, subsystems of the landscape (Aleksandrova 1986). Examples of dynamic changes are the serial chronosequences of facies, succession shifts, and changes in landscape conditions.

The landscape *evolution* is an irreversible directed change, leading to a radical transformation of landscape structure, to the replacement of one invariant by another, i.e. to the emergence of a new geosystem. The mechanism of landscape evolution consists in the gradual quantitative accumulation of the new structure elements and the replacement of the old structure elements.

25.3 Study Area

The study area comprises 1800 km^2 within the valleys of the Tom and Ob rivers and their large tributaries within Tomsk and Kozhevnikovo districts in the southern Tomsk region. The range of geographical coordinates is 55°40′–56°53′N 83°34′–84°59′E. The nature peculiarity of this territory is determined by transition from the Altai-Sayan Mountains to the West Siberian Plain.

Economically, the territory is well developed and has a high population density. There are big cities: Tomsk (population 574,000) and Seversk (107,000). In the south, the territory borders the industrial region of Kuzbass (the upper Tom basin with big cities of Kemerovo and Novokuznetsk). Thus, both natural and anthropogenic factors influence the formation of geosystems.

25.4 Spatial Data

Our study was based on field materials, various map sources and remote sensing data (RSD). Field research was performed by authors in expeditions from 2001 to 2018. Integrated landscape observations (more than 600 test sites) included geobotanical test areas and soil descriptions and were geo-referenced with the GPS. For GIS-mapping, we used topographic maps 1:25,000 dated 1977–1981 (31 maps), topographic plans 1:5000 dated 2000 (80 sheets), topographic cadastral plans 1:10,000 dated 1998 (77 sheets), topographic plans 1:10,000 dated 1896 (22 sheets), topographic plan of Tomsk 1:10,000 dated 1930–1932, soil and vegetation maps of Tomsk Region dated 1928–1929, maps of forest survey 1:50,000, soil map of Tomsk Region 1:100,000, maps of geological survey 1:200,000, aerial photos dated 1944–1973s (158 images), space images from Terra (Aster) dated 2002–2006 with spatial resolution 15 m (7 images), space images from Quick Bird II dated 2005–2009 with spatial resolution 2.4 m (37 images) and data from drone SuperCam S250 dated 2015–2017 (680 images).

25.5   Spatial Analysis: Methods and Results

Appearance of space images with very high spatial resolution, availability of data from drones, development of computerized integrated spatial analysis methods (Burrough 1996; Mitchell 1999; Wilson and Gallant 2000; Sysuev 2003; Puzachenko and Kozlov 2006; McGarigal et al. 2009) allowed the authors to improve the GIS-mapping methodology and to adapt it for the fine-scale landscape research of river valleys (Khromykh and Khromykh 2007a, b; Khromykh et al. 2013). Our approach to GIS-based landscape mapping relies upon the use of maximum number of available different dated large-scale maps, RSD, and georeferenced field materials. Then the results are integrated into the common dataset in the same cartographic projection and coordinate system and analyzed with the newest GIS-techniques. The approach was applied to the fine-scale GIS-mapping of river valleys carried out in a few stages (Khromykh and Khromykh 2016).

On the first stage we built the common personal geodatabase (GDB) in ArcGIS MDB format uniting graphic and attribute information (including field materials).

Then the high-detail digital elevation models (DEM) were created from the topographic maps 1:5000–1:25,000 and orthophotos from drone. To create models, we used ArcGIS 3D Analyst (ESRI Inc.) in triangulated irregular network (TIN) format based on Delaunay triangulation method. Some models consist of more than million triangles (Khromykh and Khromykh 2007a, b). In our opinion, it is more reasonable to use TIN instead of regular elevation network (GRID) for modeling relief of large plain river valleys because wide flat areas alternate with the terrace slopes and steep valley slopes. In this case, most of the simulated territory will provide redundant information, because grid cells on the flat areas will have the same elevation values. Nevertheless, on the sections of steep ledges of the relief, the size of the grid cell may be too large, and, accordingly, the spatial resolution of the model is insufficient to describe topography in necessary details. Since an irregular network of triangles is used, flat sections are modeled by a small number of huge triangles. In areas of steep ledges, where all the relief faces need to be shown in detail, the surface is displayed by numerous small triangles. This allows more efficient use of computer memory resources to store the model. However, the TIN model can hardly be analyzed on the basis of a raster map algebra, which complicates, for example, calculating zonal statistics. Therefore, we preferred to build a primary DEM in TIN format, and then to convert TIN to GRID for the purposes of proper geostatistical analysis.

Regular GRID elevation models for river valleys were converted from TIN with the cell size 10 m. Raster elevation models, such as GRID, are very important for landscape analysis because they allow creating common continual "geo-field" (surface) with the possibilities to calculate morphometric indices for each cell of model (and on its base – indices of humidification, drainage, etc.) and to combine them with the other kinds of spatial information. By so doing, we created the digital "landscape-gradient" models with the simultaneous use of morphometric indices calculated from DEM and vegetation indices (NDVI etc.) calculated from RSD. This approach nowadays is widely applied (McGarigal et al. 2009). Raster approach to landscape mapping allows avoiding sharp line borders of geosystems, which are characteristic of vector maps. In our opinion, it is the appropriate technique for mapping continual features of landscapes – one of the critical issues in the modern landscape science (Khoroshev 2016).

The delineation of large geomorphologic units within the Tom and the Ob valleys (floodplain and terraces) was performed based on results of morphometric analysis using DEM, geological maps, space images Terra (Aster), and field materials. Landscape mapping at the level of *mestnost* was made based on large geomorphologic units within river valleys (Khromykh and Khromykh 2018a). Our investigation showed that floodplain occupies the largest area up to 40% of valleys. The width of the Tom and the Ob floodplain reaches 10 km. The relative elevation of the floodplain surface above the water edge does not exceed 7–8 m.

A large-scale landscape mapping at the level of *urochishche* was built based on differentiation of vegetation and soil cover within the types of *mestnost* with the use of topographic maps 1:5000–1:25,000, RSD from Terra and Quick Bird II (Fig. 25.1), aerial photos, data from drone, and field materials. The processing of space images Terra (Aster) was conducted in ERDAS Imagine, with the use of supervised and nonsupervised classifications (ISODATA algorithm) and NDVI calculation. Images from drone were used for creation of orthophoto in Agisoft PhotoScan software. As a result of large-scale landscape mapping more than 7000 polygons of geosystems were delimited, classified into 120 types of *urochishches*.

The zonal statistics for geosystems were calculated based on map of the slope gradient using ArcGIS Spatial Analyst module. The average slope gradient for each geosystem was determined. This allowed us to estimate the degree of drainage for geosystems. We came to a conclusion that the best drained geosystems are located in the upper part of the valley, where the average slope was 0.92° versus 0.58° in geosystems of the lower part. On the basis of integrated spatial analysis in GIS, we identified the most frequent neighbors for each geosystem. We also studied geosystem position to simplify in some cases (especially in the riverbed floodplain) identification of paragenetic series of geosystems, considering common genesis and adjacency.

The geosystems have various degrees of anthropogenic modification. A.G. Isachenko (2003) proposed to evaluate it using the coefficients (ranging from

Fig. 25.1 Landscape mapping of Ushayka valley using Quick Bird II image

0.1 to 1.0) that reflect the degree of disturbance of elementary geosystems. By these coefficients, all geosystems were divided into three groups: geosystems with low (0.1–0.5), high (0.6–0.7) and very high (0.8–1) degrees of anthropogenic modification. Geosystems with a low degree of anthropogenic modification experience a weak anthropogenic load or recover rapidly after a small impact. Geosystems with high degree of modification are subject to significant change in some component (usually vegetation) and require constant anthropogenic maintenance but can recover after its termination. In geosystems with a very high degree of anthropogenic modification, almost all components (first of all, the abiotic template) have been seriously changed. They are characterized by a shift to another invariant. This approach allows us to characterize modern geosystems in a proper way, avoiding the implicit classification into natural and anthropogenic ones.

According to our results, floodplain type of *mestnost* has the most complex landscape structure: more than 30 types of *urochishches*. It is explained by specific natural conditions of the floodplain. Floodplain landscapes are characterized by the greatest dynamism and the youngest age in comparison with any watershed landscapes. The reason is the variability of the hydrological regime, which is the main control over the formation of floodplains. In general, hydrodynamic factors (erosion-accumulative activity of the river, "alluviality") play an important role in the development and formation of floodplain landscapes and their dynamics, as well as zonal and provincial features of the territory (climate, tectonics, the extra-flooded surroundings). All other components are influenced by these factors and play a subordinate role in comparison with them. No less important for the change of the floodplain geosystems is their self-development from the moment of the emergence of the newly formed riverbank above the water level until the sub-climax state, i.e. stage, close to the typical zonal complexes. These geosystems form unique paragenetic series, which, according to Kozin (1979), can be represented as a sequence "the riverbed – tributary channel – oxbow – fen – wet meadow – lowland meadow – plain meadow".

The emerging low floodplain areas near the riverbed represent the first element in the series of geosystems of the riverside elementary paragenetic complex (EPGC), which is associated with maximum sedimentation of alluvium (Kozin 1979). The riverbank *urochishches* formed of gravel, pebble, and sand with pioneer vegetation on Arenic Fluvisols occur widely in this area (Khromykh et al. 2018b). The average size of these geosystems is small, less than 0.11 km². The largest number of their neighbors (27% of the length of borders) belong to the geosystems of the levees, with willow on Umbric Fluvisols, which are the next element in riverside EPGK dominating in the near-riverbed floodplain. The runnels are occupied by the geosystems of *Equisetum sp.* and *Carex* sp. dominated meadows on Umbri-Gleyic Fluvisols. The forb-grass meadows predominate on the levees.

The central floodplain is remarkable for alternation of alluvial ridges and runnels. The geosystems of the low central floodplain form the oxbow-runnel EPGK, continuing evolutionary series of the riverside EPGC (Kozin 1979). The main physicogeographic process here is filling low-lying fluvial landforms with lithogenic and biogenic sediments. The willow shrubbery on silty Gleyic Fluvisols dominates in depressions on the periphery of dead channels and oxbow wetlands. In the depressions, there are the sedge meadows on Umbri-Gleyic Fluvisols. As the distance from the river and channels increases, the oxbow patches are replaced by forest-meadow ones (Khromykh and Khromykh 2016). Young central floodplain is occupied mainly by willow and willow-birch nettle-grass forests on Umbric Fluvisols. The highest ridges are occupied by forb-grass meadows with predominance of *Poa pratensis*, *Dactylis glomerata*, *Alopecurus pratensis*, *Agropyron repens*, and pine-birch-aspen forest with shrubs and forbs on Umbric Fluvisols. At the same time, due to the more pronounced influence of zonal factors, high floodplain meadows are more prevalent in the upper portions of Tom and Ob valleys, while forests dominate the lower parts of the valleys.

Flat sites on near-terrace floodplain are usually occupied by pine and pine-birch forests with sedge on Histic Gleysols or birch and birch-aspen wet grass forests on Histic Gleysols or Histosols.

In the floodplain of the large tributaries of the Tom and the Ob, gently sloping areas with aspen-pine-spruce and birch-spruce shrubbery forests on base-saturated Umbric Fluvisols prevail. There are also alluvial ridges with pine-aspen-birch forests on Umbric Fluvisols.

Within the first terraces of the Tom and the Ob, the most widespread *urochishches* are represented by flat and sloping areas with *Pinetum herbosum* forests on Umbric Albeluvisols or birch and pine-birch-aspen forests with shrub layer on

Umbric Albeluvisols. The flat poorly-drained areas are usually occupied by sedge-hypnum and birch-pine sedge-moss bogs on Histosols.

On the second terrace, hummocky terrain geosystems are typical. Their origin is related to the system-forming activity of wind. Therefore, hummocky areas with pine and birch-cedar-pine forests on sandy loamy Umbric Albeluvisols predominate. Within the lower part of the valleys, there are depressions with wet pine sedge forests on Umbri-Gleyic Albeluvisols. Pine forests on sandy Podzol are characteristic of the high hummocks. Flat poorly-drained areas are occupied by sedge-hypnum and birch, birch-pine sedge-sphagnum bogs on Histosols and Histic Gleysols.

The analysis of valley geosystems dynamics and evolution was based on the comparison of the maps from various dates, RSD, and field research materials. According to our findings, natural changes of valley geosystems are driven mainly by the erosion-accumulative activity of the rivers and the mires formation.

Most changes of geosystems caused by the erosion-accumulative activity of the rivers are characteristic of floodplain geosystems, which evolve from sandy side bars with pioneer vegetation on Arenic Fluvisols to well-drained alluvial ridges of central floodplain with the forests and meadows on Umbric Fluvisols (Khromykh and Khromykh 2016).

The processes of the mire formation mainly develop on flat slightly drained areas of the terraces. On the terraces, sedge and sphagnum-dominated bogs are widespread. In the depressions of the central floodplain, lowland sedge fens predominate. The evolution of bog geosystems includes several stages, which are characterized by a successive change of the dynamic states of geosystems, up to a change in the invariant.

By comparing the different dated RSD and maps, we thoroughly examined the process of lake disappearing and the subsequent development of the mesotrophic mires on the example of disappearance of Lake Strashnoe on the second terrace in the vicinity of the village Timiryazevskoe in the twentieth century (Fig. 25.2).

Lake Strashnoe was located 400 m to the southeast from Lake Peschanoe and was studied in the 1930s by B.G. Ioganzen, who wrote: "The Lake Strashnoe has been known to us since 1920–1921 ... The shallowing of the lake is striking ... The current state of the lake is such that it is not difficult to predict the rather rapid decrease of the existing free water space and the closure of this site with the rest of the peat bog" (Ioganzen et al. 1951, p. 183). In our opinion, Ioganzen, in 1938, described the first stage of mire formation, when the lake depression was being filled with dying plants and the bottom level was rising (Fig. 25.2a). In the opinion of Mirkin et al. (2000), at this stage, the conditions for plant growth can improve, and the floating macrophytes (*Potamogeton* sp., *Stratiotes* sp., *Lemna* sp., etc.) are replaced by tall semisubmerged plants – *Phragmites* sp., *Typha* sp., *Equisetum palustre*. Gradually, the lake eventually grows and turns into a fen with progressing peat accumulation; sedges, *Salix cinerea*, and *Alnus fruticosa* appear. This is how the lake geosystem with silty Gleysols in coastal line evolves into a natural complex of the fen with Histosols (Fig. 25.2b). Year by year, the layer of peat increases, the roots of the plants no longer reach the soil and start to feed at the expense of peat. Mineralization of peat is slow and not complete at this stage. Therefore nutrient-sensitive plants are gradually replaced by species that are adapted to limited nutri-

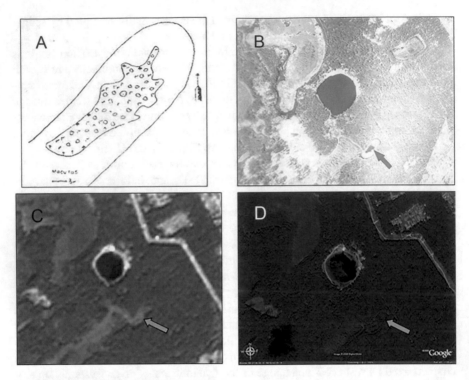

Fig. 25.2 Disappearance of Lake Strashnoe
(**a**) – Scheme of lake made by B. G. Ioganzen in 1938 [8]
(**b**) – Lake on aerial photo dated 1949 (indicated by arrow)
(**c**) – Young sedge-sphagnum bog instead of lake on space image dated 2003
(**d**) – Pine-birch sphagnum bog on space image dated 2018

tion (*Betula pubescens, Menianthes trifoliata, Calla palustris* etc.). Thus, the invariant changes again, and a new geosystem of the mesotrophic sedge-sphagnum mire is formed (Fig. 25.2c). Now the edges of the former Strashnoye lake are overgrown with a young birch and are virtually indistinguishable from neighboring forest areas on contemporary space images (Fig. 25.2d). On the high terraces, the geosystems of mesotrophic sedge-sphagnum mire under certain conditions can continue to evolve towards the sphagnum peat bogs. In this case, the further growth of the peat layer leads to an even poorer mineral nutrition, a complete isolation of the bog surface from the mineralized groundwater occurs, and the entire bog begins to become covered with sphagnum mosses. A boggy form of pine appears, as well as *Vaccinium oxycoccus* and species from the *Ericaceae* family (*Ledum palustre, Chamaedaphne calyculata, Andromeda polifolia*, etc.).

In order to identify land use changes and anthropogenic transformation of the geosystems within the Tom River valley in the vicinity of Tomsk, we compared a digitized vegetation map dated 1929 (Spring 1929) and an up-to-date landscape map (Fig. 25.3). The result showed obvious cardinal anthropogenic transformation of natural complexes during past 80 years.

The greatest changes of geosystems were observed in the floodplain. The most dramatic process is anthropogenic desiccation due to the lowering of the groundwater level, which occurred due to a set of superimposed factors: lowering of the water level due to the extraction of gravel, the activity of water intake for needs of Tomsk, land reclamation, deforestation, increasing of arable lands, and horticultural areas (Khromykh et al. 2015). Disturbance of the flood regimes resulted in a shift of almost all the geosystems one step higher in the paragenetic sequence of floodplain geosystems. Thus, many riverbed geosystems have become riparian gravel and pebble side bars, which are well visible on the site near the Lagerny Garden, where the river Tom has narrowed from 514 m to 308 m, and near the bridge from 770 m to 180 m. Geosystems of the riverside floodplain evolve rapidly into the category of the central floodplain complexes, which have been noted by researchers since the 1980s (Ryazanov and Surkov 1986). This is evidenced by the displacement of willow shrubberies to the riverbed and their replacement by meadows. High floodplain, not flooded for several decades, evolved into an anthropogenic terrace. Here there is a rapid transformation of Umbri-Gleyic Fluvisols to Umbrisols, which was confirmed by our field observations. At the same time, the extraction of gravel from Tom riverbed resulted in destruction of the riverside geosystems on the island Basandayskiy, which area was decreased almost twice – from 0.176 km^2 in 1929 to 0.094 km^2 now (Fig. 25.3).

The connection between the Tom River and Lake Sennaya Kuriya was destroyed after the construction of the bridge. The big reservoir for agricultural use was built instead of Lake Kalmatskoe on the floodplain. Dramatic changes took place in land use: almost all the geosystems of floodplain meadows were transformed to geosystems of plowed areas or gardens.

Lowering the groundwater level affected the geosystems of the terraces. The size and depth of Peschanoe Lake decreased; the shape of the lake was transformed from round to a horseshoe one. Many sphagnum bogs on Histosols were modified into dried pine-birch bogs on drained Histosols (shown by arrows in Fig. 25.3). In the same place, where large bog expanses still remain, intensive land reclamations are being carried out (Fig. 25.4). Our analysis based on GIS-mapping showed a reduction of the area of bogged forests four times, from 8.02 km^2 in 1929 to 1.99 km^2 now. The area of residential rural zones increased: the area of Kislovka increased three times (from 0.42 km^2 in 1929 to 1.41 km^2), the area of Timiryazevskoe increased eight times (from 0.36 km^2 in 1929 to 2.83 km^2).

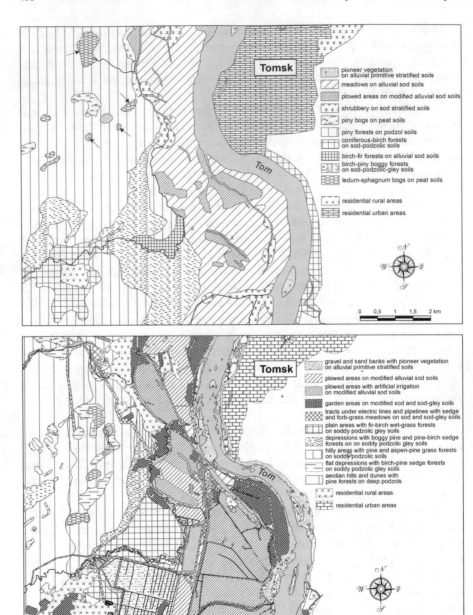

Fig. 25.3 Comparison of the vegetation map in the vicinities of Tomsk dated 1929 (at the top) with the up-to-date landscape map (below). On the map dated 1929 arrows indicate bogs, which were drained

Fig. 25.4 The fragment of the Tom valley 3D-model with the imposed aerial photo dated 1954 Vertical scale five times larger than horizontal scale. 1 – Modern bank of the Tom River, 2 – bog expanse drained now, 3 – modern land reclamation channels

25.6 Conclusions

The large-scale GIS-mapping of landscape structure and land use within the river valleys in the southern Tomsk region was performed in this study. We created high-detail DEMs and "landscape-gradient" models with the simultaneous use of morphometric indexes calculated from DEM and vegetation indexes (NDVI etc.) calculated from RSD. More than 7000 polygons of geosystems were mapped and classified into 120 types of *urochishches*.

The main factors of natural changes of valley geosystems are the erosion accumulative activity of the rivers and the mire formation. According to our analysis, the processes of natural landscape dynamics prevail only in some parts of Ob valley and in lower part of the Tom valley.

At the present time, almost all geosystems within the Tom River valley near Tomsk are exposed to anthropogenic modifications to various degrees (Fig. 25.5). The vector of the landscape dynamics is directed towards considerable desiccation due to the lowering of the groundwater level, which occurred due to the overlapping of various anthropogenic factors.

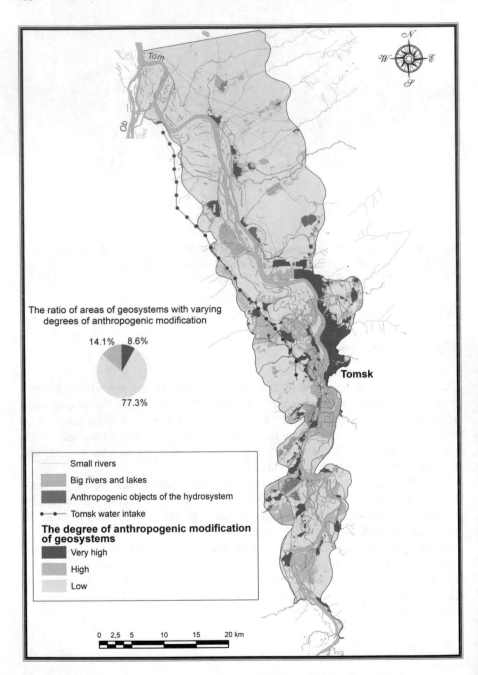

Fig. 25.5 Degree of geosystems anthropogenic modification within the Tom valley

Acknowledgements   This study was supported by the Tomsk State University Competitiveness Improvement Programme.

References

Aleksandrova, T. D. (1986). *Conceptions and terms in landscape science.* Moscow. (in Russian).

Bastian, O., Grunewald, K., & Khoroshev, A. V. (2015). The significance of geosystem and land-scape concepts for the assessment of ecosystem services: Exemplified in a case study in Russia. *Landscape Ecology, 30,* 1145–1164.

Burrough, P. A. (1996). *Principles of Geographical Information Systems for land resources assess-ment.* Oxford.

Cherkashin, A. K. (Ed.). (2005). *Landscape-interpretation mapping.* Novosibirsk: Science. (in Russian).

Christopherson, R. W. (2014). *Geosystems: An introduction to physical geography* (9th ed.). Upper Saddle River: Prentice Hall.

Dyakonov, K. N. (2005). Basic concepts of landscape science and their development. *Proceedings of Moscow University, Series 5 Geography, 1,* 4–12. (in Russian).

Ioganzen, B. G., Popova, M. A., & Yakubova, A. I. (1951). Reservoirs of the surroundings of Tomsk. *Proceedings of Tomsk University, 115,* 121–190. (in Russian).

Isachenko, A. G. (2003). *Introduction to ecological geography.* St. Petersburg: St. Petersburg University Publishing House. (in Russian).

Isachenko, A. G. (2004). *Theory and methodology of geographical science.* Moscow: Academia. (in Russian).

Khoroshev, A. V. (2016). Modern lines in structural landscape science. *Proceedings of Russian Academy of Sciences, Geographical Series, 3,* 7–15. (in Russian).

Khromykh, V. V., & Khromykh, O. V. (2007a). Experience of automized morphometric analysis of Tom valley geosystems based on DEM. *Proceedings of Tomsk University, 298,* 208–210. (in Russian).

Khromykh, V. V., & Khromykh, O. V. (2007b). The use of GIS for the study of valley landscapes dynamics (on the example of Tom valley). *Proceedings of Tomsk University, 300*(1), 230–233. (in Russian).

Khromykh, V., & Khromykh, O. (2016). Spatial structure and dynamics of Tom river floodplain landscapes based on GIS, digital elevation model and remote sensing. In O. S. Pokrovsky (Ed.), *Riparian zones: Characteristics, management practices and ecological impacts* (pp. 289–309). New York: Nova Science Publishers.

Khromykh, V. V., Khromykh, O. V., & Erofeev, A. A. (2013). The landscape approach to detach-ment of water-securing zone of river Ushaika based on GIS-mapping. *Proceedings of Tomsk University, 370,* 175–178. (in Russian).

Khromykh, O. V., Khromykh, V. V., & Khromykh, V. S. (2015). Natural and anthropogenic dynam-ics of the floodplain landscapes near Tomsk. *Proceedings of Tomsk University, 400,* 426–433. https://doi.org/10.17223/15617793/400/64. (in Russian).

Khromykh, V. V., & Khromykh, O. V. (2018a). A study of natural and anthropogenic changes of soils within the Tom river valley based on GIS, remote sensing and field observations. *IOP Conference Series: Earth and Environmental Science 201:012006.*

Khromykh, V. V., Khromykh, V. S., & Khromykh, O. V. (2018b). Features of soils in the flood-plain landscapes of Siberian Rivers. *IOP Conference Series: Earth and Environmental Science, 201:012007.*

Kozin, V. V. (1979). *Paragenetic landscape analysis of river valleys.* Tyumen: Tyumen State University Press. (in Russian).

Kozin, V. V., & Petrovskiy, V. A. (2005). *Geoecology and environmental management: Terminological dictionary*. Smolensk: Oykumena. (in Russian).

Krauklis, A. A. (1979). *Problems of experimental landscape science*. Novosibirsk: Nauka. (in Russian).

McGarigal, K., Tagil, S., & Cushman, S. A. (2009). Surface metrics: An alternative to patch metrics for the quantification of landscape structure. *Landscape Ecology, 24*, 433–450.

Milkov, F. N. (1966). Paragenetic landscape complexes. *Proceedings of Voronezh Branch of the All-Union Geographical Society*, 7–35. (in Russian).

Mirkin, B. M., Naumova, A. G., & Solomesh, A. I. (2000). *The modern science of vegetation*. Moscow: Logos. (in Russian).

Mitchell, A. (1999). *The ESRI guide to GIS analysis*. Redlands: ESRI Press.

Preobrazhenskiy, V. S. (Ed.). (1982). *Protection of landscapes (Dictionary)*. Moscow: Progress. (in Russian).

Puzachenko, Y. G., & Kozlov, D. N. (2006). Versions of landscape mapping. In K. N. Dyakonov (Ed.), *Landscape science: Theory, methods, regional studies, practice. Proceedings Int. landscape conference* (pp. 123–125). Moscow: MSU Publishing House. (in Russian).

Ryazanov, P. N., & Surkov, V. V. (1986). Floodplain natural territorial complexes of the downstream of Tom river and some of the trends of their change. *Geography and Natural Resources, 1*, 59–65. (in Russian).

Sochava, V. B. (1963). Definition of some notions and terms in physical geography. *Proceedings of the Institute of Geography of Siberia and Far East, 3*, 50–59. (in Russian).

Sochava, V. B. (1974). Geotopology as a division of geosystem concept. In V. B. Sochava (Ed.), *Topological aspects of geosystem concept* (pp. 17–35). Novosibirsk: Nauka. (in Russian).

Sochava, V. B. (1978). *Introduction to the theory of geosystems*. Novosibirsk: Nauka. (in Russian).

Solnetsev, N. A. (1948). The natural geographic landscape and some of its general rules. In *Proceedings of the Second All-Union Geographical Congress* (Vol. 1, pp. 258–269). Moscow: State Publishing House for Geographic Literature (in Russian). In J. A. Wiens, M. R. Moss, M. G. Turner, & D. J. Mladenoff (Eds.), (2006). *Foundation papers in landscape ecology* (pp. 19–27). New York: Columbia University Press.

Solnetsev, N. A. (1960). About relations between "alive" and "dead" nature. *Proceedings of Moscow University, Series 5 Geography, 6*, 10–17. (in Russian).

Spring botanical excursions in the vicinity of Tomsk. (1929). Tomsk. (in Russian).

Sysuev, V. V. (2003). *The physical-mathematical foundations of landscape science*. Moscow: MSU Publishing House. (in Russian).

Wilson, J. P., & Gallant, J. C. (2000). *Terrain analysis: Principles and applications*. New York: Wiley.

Part VIII
How Patterns Determine Actual Land Use

Chapter 26
The Development of the Territorial Planning and Agrolandscapes Projecting in Russia

Valery I. Kiryushin

Abstract A new ideology of the territorial planning is forming on the basis of landscape-ecological knowledge instead of the traditional land management projecting. Methods for the agrolandscape projecting are proposed for the local level. It includes the system of agroecological assessment of lands, their typology, landscape and ecological classification and the procedure of projecting based on AgroGIS.

Keywords Landscape · AgroGIS · Planning · Ecological classification

The transition to sustainable development, in correspondence with the biospheric paradigm, is closely related to the improvement of territorial planning as an obligatory condition for nature management optimization. The ecologically oriented planning, especially, landscape planning, becomes the more significant among the various types of planning. The aim of landscape planning is to promote sustainable development and preservation of landscape functions as a life-support system. Such a planning is actively applied in world practice. Territorial planning on ecological and landscape bases is developing in European countries in various forms. The role of the state in the process is growing as well as the role of society. The fundamental understanding of the problem significance also increases. This is stipulated by European Landscape Convention adopted in 2004, which is based on the recent progress in landscape ecology.

At the beginning of the twenty-first century, the priorities in the world practice of landscape planning are seen as follows: optimizing land-use technologies aimed at preserving landscape and biological diversity; consideration of anthropogenic burden impact on biological diversity; establishing optimal proportions of land use types in landscapes; adaptation of spatial solutions to the landscape hierarchy;

V. I. Kiryushin (✉)
Dokuchaev Soil Science Institute, Moscow, Russia

© Springer Nature Switzerland AG 2020
A. V. Khoroshev, K. N. Dyakonov (eds.), *Landscape Patterns in a Range of Spatio-Temporal Scales*, Landscape Series 26,
https://doi.org/10.1007/978-3-030-31185-8_26

projecting the landscape-ecological frameworks; determination of land use priorities and intensity of anthropogenic burden based on the forecast of chain reactions between landscape components; comparison of alternative scenarios and search for compromise solutions for multifunctional land use, based on the analysis of the contradictions between ecological, economic, social, technological conditions and interests of land users (Dyakonov and Khoroshev 2011; Kiryushin 2018).

Russia has a very particular situation because the past system of territory planning was destroyed, and the new one has not been developed yet. In the Soviet period, territorial planning in agriculture was developed as a powerful land-management system. It involved the development of projects of intra-farm land management at the local level, inter-farm land management projects at the level of administrative district, general land management schemes at the higher regional levels. The projecting was based on materials of large-scale soil mapping that covered the country's agricultural lands, with due regards to geobotanical data and other types of field investigation. Schemes of natural and agricultural zoning and agro-production groups of soils provided opportunities to design relatively differentiated systems of agriculture. However, these opportunities for a differentiated approach to agricultural nature management had a rather narrow area of application, as the priorities of land management were determined by strict planning tasks for the structure of crop areas, livestock and other indicators that did not have agro-ecological substantiation. Quite often, the projects of intra-farm land management simply replicated standards, which resulted in significant economic and ecologic expenses. Irrational distribution of crops, standardized agrotechnologies, discrepancy of technological activities and agroecological conditions, the wide involvement of marginal lands into rotation, inadequate application of chemicals frequently led to soil degradation. Mistakes in windbelt projecting lead to the accumulation of snow, and, hence, to increased erosion during snowmelt, e.g., if windbelts were constructed along the slopes to hold the prevailing winds. Mistakes in the projection and placement of drainage and land reclamations for irrigation were followed by environmental and economic damage. In the 1970s, the expansion of land reclamation and unsystematic activity discredited and slowed its development in the country. These years showed a beginning of awareness of regional environmental disasters in the USSR, like in the Aral Sea region, and the impending global environmental crisis. In the 1980s–1990s, this awareness grew into a new ideology of nature management, which was declared at the UN conference in Rio de Janeiro in 1992 in terms of "sustainable development". In the same year, the session of the Russian Agricultural Academy (RAA) was held in the Talovsky district of the Voronezh region. It was dedicated to the centenary of the publication of Dokuchaev's famous book "Our steppes before and now" (Dokuchaev 1892). This book anticipated the biospheric paradigm of nature management and formulated, in terms of modern science, the foundations for landscape adaptive agriculture. The development of this fruitful idea was interrupted by hundred years of euphoria related to the ideology of nature conquest, complicated by the soviet command and planning system. The resolution of the Russian Agricultural Academy session stated: "The most important issue faced by agrarian scientists is the development and popularization of Dokuchaev's scientific heritage, the necessity of the complex investigation and creation (construction) of the ecologically and economically balanced highly productive and sustainable agrolandscapes, which are maximally adapted to the local natural conditions".

The ecology-oriented approaches to the territorial planning were developed during the *Perestroika*. These approaches included the regional-specific plans, proposing ecologically justified placement of settlements, industrial objects, recreation facilities, etc. Also, they included territorial integrated schemes for nature protection (*TerKSOP*), which proposed the norming of the anthropogenic loads, the restriction of industrial objects placement in order to preserve the natural balance, as well as the integrated measures on environmental protection and nature management.

The period of the most intensive practical activity on land management projecting dates back to the middle 1980s when the zonal agricultural systems were being implemented. The projecting *"Giprozem"* institutes elaborated numerous intra-farmland management projects for most agricultural enterprises in the country. Despite their ecological insufficiency and the rigid commitments to the state planning imperatives, they played an important role in the differentiation of agriculture due to the diverse ecological conditions.

In the 1990s, as a result of land reform, these works were ceased, and a powerful land management service was destroyed. The reforming of the agricultural enterprises that occurred in 1992–1993 was too fast and unprepared. The state stopped fulfilling its most important function, which is land management and which is the major mechanism for restoration of land, solving of ecological, normative, socioeconomic, organizational and territorial issues. The state mostly focused on the fiscal and political goals, on the re-distribution of the property and land-cadaster actions.

Despite the termination, territorial planning works in agriculture, scientific research in this area, were carried on most intensively within landscape science and other geographical sciences. The largest investigation centers are the Institute of Geography of Russian Academy of Sciences, the Department of Physical Geography and Landscape Science of Lomonosov, Moscow State University, and the group of classic universities. The review of Russian and European experience in the landscape planning (Drozdov 2006), shows that it is aimed at the development of a sustainable production organization for rational nature management and nature protection in particular areas in accordance with the long-term goals of society. The main purposes of landscape planning include the preservation of major landscape functions as a life support system; the identification of stakeholders' interests and the analysis of emerging conflicts; developing action plans to resolve conflicts and achieve common goals; cooperation for the sustainable development of the territory.

To solve these issues, it is necessary to determine: the landscape functions, its resource potential; its sensitivity, buffer capacity, sustainability limits; current and planned loads with pointing out of their sources; environmental risks and possible consequences of existing and planned types of economic activities; contradictions between the needs of the landscape protection and its use.

The landscape planning is realized as the hierarchical system at various landscape levels. At the regional level, landscape programs that define the main directions of nature management are developed. At the local level, landscape plans designed for coordinated solution of nature protection and land-use issues at the lowest (municipal) administrative-territorial level are created.

Nowadays, we have a lot of examples of successful landscape projecting in different regions for various purposes. However, it has not yet become an ideology, accepted by the state and society as a necessary condition for optimizing the use of natural resources in accordance with the declaration of sustainable development.

Despite the vitality of the problem, the opportunities for implementation are still quite constrained. Therefore, the most realistic perspective is the integration of landscape planning methodology into the legal territorial planning procedures with focus on improving their scientific rationales and increasing efficiency.

Along with the development of landscape planning studies in the last 20 years, approaches have been developed for the agrolandscapes projecting, which was encouraged by the resolution of the abovementioned RAA session. The first attempts for the land differentiation on the landscape basis applied the concept of genetic-morphologic landscape structure[1] and other categories of the geographic landscape science. However, the genesis-based classifications of landscapes turned out to be insufficient for the ecologically-oriented differentiation of agriculture. We need specific landscapes classification focused on agroecological assessment. For this purpose, we have elaborated the agroecological typology of lands (landscapes) (Kiryushin 1993, 2011), which provided definitions for the basic terms, such as "agricultural landscape", "agrolandscape", "elementary agrolandscape area". The definition of the natural landscape served as a reference point. The definition by V.B. Sochava (1978) seems to be the most suitable: "Natural landscape is the geosystem of the least regional dimension, consisting of genetically and functionally interrelated local geosystems, formed on a single morphostructure in the conditions of the local climate (*urochishche, podurochishche, facies[2]*)".

According to GOST[3] 17.87.1.02.88, the agricultural landscape is a landscape used for the purposes of agricultural production, and which was formed under its impact. We added details and defined it as *anthropogenic-natural landscape, formed under the impact of agricultural activity, where the natural basis is coupled with the producing and social structure*. To examine agricultural landscape from the viewpoint of requirements of agricultural crops for cultivation conditions, we defined the *agrolandscape as a geosystem with a certain set of binding agroecological factors (which determine the choice for certain agricultural system), the functioning of which occurs within single network of matter and energy migration*. Therefore,

---

[1] Genetic-morphologic landscape structure is one of the concepts of landscape structure developed in Russia. It is based on the understanding of strong interdependencies among physical environment, genesis of geologic substrate, soils and plant cover. See Chap. 1 for details.

[2] *Facies, podurochishche, urochishche, mestnost* are the Russian terms for the hierarchical levels of landscape morphological units. See Chap. 1 and Glossary for details.

[3] GOST is State Standard of Russian Federation.

within the agricultural landscape, along with the urban, technogenic and other landscapes, we distinguish the agrolandscape, within which the crop cultivation infrastructure is formed (field, pasture, orchard, etc.). From the viewpoint of the genetic-morphological structure, the agrolandscape may correspond to the natural landscape, *mestnost*, or *urochishche*.

The elementary structural unit of agrolandscape consists of one or several *facies*, which are uniform from the viewpoint of agriculture. The elementary agrolandscape area is defined as an elementary structural unit, which is related to a mesorelief element with elementary soil structure (by the elementary soil area, in more rare cases) in homogeneous geologic and microclimatic conditions.

These concepts provide the basis for the agrolandscape-ecological land typology needed for the projecting of adaptive-landscape systems of agriculture (ALSA). The upper level of its hierarchy is represented by the *agroecological groups of lands*, which differ in the binding agroecological factor (upland lands, erosional lands, over-moistened lands, salinized lands, solonetz lands, lithogenic lands, etc.). The adaptive-landscape agricultural systems are developed for the implementation on agroecological groups of lands (in other words, on agrolandscapes). The groups include *agroecological land types* (ecologically homogeneous areas for a crop, or a group of crops, cultivation). Crop rotations, mowing rotations, pastures rotations are formed within the groups of lands. Agroecological groups of lands consist of the elementary areas of the agrolandscape (in other words, land types), with respect to which the elements of agrotechnology are differentiated.

Identification of species and groups of lands is conducted in GIS based on agro-ecological assessment of lands by integrating the electronic map layers, which depict the forms and elements of mesorelief, soil-forming rocks, microclimate, hydrogeological conditions, microstructure of soil cover, soil properties, etc. The GIS were developed on the basis of the materials of soil and landscape mapping at the scale of 1:10000, on the basis of the landscape-ecologic lands classification (Kiryushin 2011).

The projecting of agrolandscapes begins with the distribution of agricultural crops. To do this, we create digital maps of land-type suitability for cultivation of crops for sale. The procedure is conducted by comparing the database of land agro-ecological parameters (for each unit) with agro-ecological requirements of crops and species. As a result of map layers overlapping, agroecological types of land are identified and form the field infrastructure; crop rotation fields and production areas are delineated. Taking into account the soil-landscape relationships and the flows of energy and matter, we project the anti-erosion territorial organization and develop measures to prevent the ecological conflicts and erosion sources. Then we delimit the elements of the ecologic framework at ecotones. We differentiate the location of crops, considering the habitats of birds and insects beneficial for pest control or for pollination of plants. Microreserves are projected. The system of protective forests is developed, and the necessity or usefulness of different land reclamations is substantiated. The whole procedure is referred to as the *landscape-adaptive agricultural systems*, including the system of soil cultivation, fertilizing and crop protection. Experience of such projects was obtained in various nature-agricultural conditions and may be applied in other areas.

Taking into account the methodology of landscape planning and the experience of land management with various degrees of approximations to landscape-ecological planning, we have the opportunity to create the models of integrated territory planning at the local and regional levels.

It is obvious that at the local level, the projecting of the landscapes should be enhanced by projecting of the productive infrastructure (animal farms, enterprises on production recycling, etc.) and the social infrastructure (settlements, roads, recreation facilities, etc.). Such an integrated approach may be called a project of agro-industrial production of an agricultural enterprise, which, essentially, optimizes the agricultural landscape.

At the regional level, territorial planning was conducted in various forms as follows: inter-farm land management, district planning schemes, the development of new lands, the withdrawal and allotment of land tenures. As the land management is developed, the abovementioned *TerKSOP*s and integrated schemes for nature protection and environmental management became closer to landscape planning. It is obvious that regional (municipal) territorial planning should be developed on the basis of the existing experience with the application of landscape planning methods. Probably, it may be implemented as a framework plan, in which it is necessary to determine the purpose and the appropriate land use type: recreational, residential, silvicultural, industrial, for energy production purposes, for mining, etc. Besides, an The alternative planning options are needed to provide better rationales for choice of decision. For example, before conducting oil extraction or other mineral mining in areas with nutrient-rich Chernozem soil, the expected effect should be compared with that of agricultural use of soils. The same holds true for the various alternatives of agricultural use of floodplain lands (inundations during the construction of hydroelectric power stations, the destruction by floating mines during gold mining works, etc.).

At the regional level, manuals on construction and development of adaptive landscape systems of agriculture for the administrative area were created following the innovative book "Adaptive landscape management systems of the Novosibirsk Region" (Kiryushin and Vlasenko 2002). The upcoming stage involves projecting of landscape-adaptive agricultural systems and agrolandscapes. Regional geoinformation systems are crucially needed for the purpose of agroecological assessment of lands and agricultural landscapes projecting. Such GIS are based on the following set of digital maps:

- Nature-agricultural zoning of the area
- Geomorphologic conditions
- Agroclimatic resources
- Hydrogeologic and hydrologic conditions
- Soil cover structures
- Agroecologic group of lands
- Crop yields at the extensive agrotechnologies (natural fertility potential)

- Crop yields at the intensive agrotechnologies (the productive potential, given that modern achievements of scientific and technical progress are implemented)
- Qualitative production index
- Economic indicators of crop production

Digital maps include databases on all of the abovementioned items.

On the basis of regional AgroGIS, models concerning adaptive-landscape agriculture can be developed for various agro-ecological groups of lands. Within a framework of ALSA, the packages of agrotechnologies were developed for the different levels of intensification.

Creation of a regional electronic AgroGIS, in addition to its purpose for the projecting of ALSA, is important for the forming of regional agro-technological policy, the planning of production by agro-industrial enterprises, the agro-industrial orientation of commodity producers, the choice for placing crops and technologies for their cultivation, assessing potential yield, opportunities for investments on different lands and the need for production resources, etc.

Availability of modern information technologies provides the opportunity to ensure open access to regularly updated websites containing methodical support for adaptive-landscape agriculture in Russia. This support is required by agricultural land users and engineers. The web resources should contain basic official documents, methodical and normative materials concerning the issues of agroecological assessment, lands typification, zoning and monitoring, projecting of landscape adaptive agriculture systems and modern agrotechnologies.

References

Dokuchaev, V. V. (1892). *Our steppes in the past and at the present time*. St. Petersburg. (in Russian).

Drozdov, A. V. (Ed.). (2006). *Landscape planning with elements of engineering biology*. Moscow: KMK. (in Russian).

Kiryushin, V. I. (1993). *The concept of adaptive-landscape agriculture*. Pushchino: Pushchino Scientific Center. (in Russian).

Kiryushin, V. I. (2011). *Theory of adaptive-landscape agriculture and the projecting of agrolandscapes*. Moscow: KolosS. (in Russian).

Kiryushin, V. I., & Vlasenko, A. N. (Eds.). (2002). *Adaptive-landscape systems of agriculture in the Novosibirsk region*. Novosibirsk: SibNIISKhim SB RAS. (in Russian).

Sochava, V. B. (1978). *Introduction to the theory of geosystems*. Novosibirsk: Nauka. (in Russian).

Dyakonov, K. N., & Khoroshev, A. V. (2011). Challenges and objectives for landscape planning. In K. N. Dyakonov (Ed.), *Actual problems of landscape planning* (pp. 8–13). Moscow: Moscow State University Press. (in Russian).

Kiryushin, V. I. (2018). *Ecological foundations for projecting agricultural landscapes*. Sankt-Petersburgh: KVADRO. (in Russian).

Glossary

The Glossary contains definition of basic terms that are widely used in Russian-language landscape science and have no direct analogs in international literature. Each term is supplied by the reference to the chapters that provide detailed explanations and examples. The list of the references includes the foundation publications in which the terms were proposed and explained.

Adaptive-landscape systems of agriculture (ALSA) a tool to adapt agricultural land use and technologies to landscape pattern (Kiryushin 2011) (Chap. 26).

Altitudinal landscape complexes (ALC) original paradynamic systems of landscapes resulting from vertical transformation of relief. Their development is closely related to an interaction of contrast environments subject to both endogenous and exogenous factors (Chap. 14).

Autonomous landscape landscape-geochemical unit, part of catena, which receives matter from atmospherc only and has no connections with groundwater (e.g., on flat interfluve or terrace) (Glazovskaya 1964, 2002; Perelman 1972; Fortescue 1992) (Chaps. 9 and 10).

Azonal features of landscape attributes of a landscape determined by abiotic template (topography and geology) being in some contradiction with hydroclimatic conditions of the landscape zone (Isachenko 1973).

Balka the Russian term for flat-bottom erosional landforms sometimes with ephemeral water courses (Chaps. 13 and 24).

Basin configuration (structure) a set of landscape units within a basin integrated by lateral flows; the spatial relationships are determined by "similarity of geotopes with respect to their hydrofunctioning and their relationship to the basins of surface flows" (Grodzinsky 2005) (Chap. 1).

Biocentric network configuration (structure) a set of landscape units integrated by "the biotic migrations in a landscape" (Grodzinsky 2005). Similar to *patch-matrix-corridor concept* of landscape structure (Chap. 1).

Biocirculation structure a set of landscape units shaped by solar radiation input depending on elevation and slope aspect (Solntsev 1997) (Chap. 1).

Cascade landscape-geochemical system a *paragenetic* set of elementary landscape units integrated by unidirectional lateral matter flows. Catena and basin are the particular cases (Glazovskaya 1988) (Chap. 9).

Characteristic time scale time of full change of object or time of one full cycle at cyclic character of changes (Armand and Targul'yan 1976) (Chaps. 1, 16, 17 and 19).

Chorion geosystem consisting of the core and the adjusting area which is subject to material and energetic influence of the core (Reteyum 1988) (Chaps. 1 and 2).

Coefficient of lateral (L) migration ratio between element content in autonomous and subordinate elementary landscape units (Perelman 1972) (Chap. 9).

Coefficients of radial (R) migration ratio between element content in soil horizon and in parent rock (Perelman 1972) (Chap. 9).

Derivative facies state of elementary geosystem induced by human or natural disturbances (Sochava 1978) (Chap. 2).

Ecotope site with certain combination of conditions of nutrients and water supply (Chaps. 6, 8 and 13).

Elementary landscape unit (ELU) the term in landscape geochemistry for the smallest uniform unit with homogeneous bedrocks, soil, and moisture. Catena consists of a toposequence of elementary landscape units (autonomous, trans-eluvial, trans-accumulative, super-aqual, sub-aqual) (Perelman 1972) (Chap. 9).

Eluvial landscape see *Autonomous landscape.*

Epifacies a whole set of variable and final (climax) states of a geosystem (Sochava 1978) (Chap. 2).

Extrazonal landscape a landscape with vegetation and corresponding soils which are not typical for the zone where it is located but inherent for another zone (e.g., island of broad-leaved forests located within the taiga zone) (Isachenko 1973) (Chap. 20).

Facies a morphological unit of a landscape; the smallest natural unit with uniform lithology of surficial sediments, character of relief, humidity, microclimate, soil, and biocoenosis occupying usually part of relief microform. Facies is a constituent of *podurochishche* and *urochishche* (Solntsev 1948) (Chap. 1, 2).

Factoral-dynamical series a sequence of landscape units along an ecological gradient (water supply, nutrients supply, permafrost features, etc.) (Krauklis 1979) (Chap. 8).

Fibration a tool to stratify complex geographical objects (polygeosystem) into a set of possible system-based interpretations (Cherkashin 2005) (Chap. 2).

Genetic-morphological configuration (structure) a set of landscape units shaped by the relationships between geotopes in common characteristics of their origin and evolution form (Grodzinsky 2005). The most well-developed methodology for description and investigation was elaborated by N.A. Solntsev and his team (Solntsev 1948, Dyakonov et al. 2007) (Chap. 1).

Geochemical barrier a site in a landscape where matter loses mobility due to abrupt change in geochemical conditions (e.g., accumulation of iron oxides at the border of bog and drained area) (Perelman 1972) (Chap. 9).

Geochemical structure a set of interacting sites with contrasting conditions of matter migration (Chap. 9).

Geochore heterogeneous geosystem formed by adjacent geomers which constitute structural, functional, and dynamical whole. Geochores are organized in a hierarchy (micro-, meso-, topo- macrogeochores) (Sochava 1963, 1978). Close to the term proposed by Neef (1963) (Chap. 2, 8, 10).

Geocirculation structures emerge as a result of various intensity and directions of lateral flows depending on relief dissection (Solntsev 1997) (Chap. 1).

Geocomponent, landscape component constituent of *natural territorial complex* (NTC) (parent rock, soil, water, air, plants, and animals, sometimes also litter) (Chaps. 1, 6 and 16).

Geomer homogeneous geosystem, the minimum area representative of all the components (Sochava 1978). The *facies* is elementary geomer (Chaps. 2, 8 and 10).

Geostationary structures a set of landscape units shaped by radial (vertical) connections between *geocomponents* under the control of abiotic template which is responsible for nutrient supply, water percolation, air regime in soil, etc. (Solntsev 1997) (Chap. 1).

Geosystem an open system formed by interactions of natural and cultural components generating stable patterns (invariants) at various spatiotemporal scales (Sochava 1978) (Chap. 1, 2).

Goltsy zone of the high mountain tundra with dominance of stony habitats above the upper boundary of the forest in Siberia (Chap. 8)

Heterolithic catena catena with heterogeneous geological substrate (Kasimov et al. 2012) (Chap. 9).

Horizontal (spatial) structure spatial pattern formed by laterally interacting elements (Chap. 1).

Hydrocirculation the lateral water flowing downward under force of gravity. Hydrocirculation indicators show the distribution of water under force of gravity (Chap. 11).

Hygrotope water supply level in a landscape unit (e.g., dry, wet) (Chap. 23).

Intrazonal landscape landscape differing much from the dominant zonal landscape due to strong influence of a particular factor. Intrazonal landscape may occur within various zones but do not form their own zone (e.g., bogs, floodplain meadows) (Isachenko 1973) (Chaps. 2, 13, 20).

Invariant a set of landscape properties that are preserved under external influences and dynamic changes. At a certain hierarchical level of geosystem, invariant properties are inherent for the lower-level geosystems. Invariant is transformed in a course of evolution (Sochava 1978) (Chaps. 2, 3, 6, 8, 15 and 21).

Khasyrei a drained lake in thermokarst plain in tundra (Chap. 4).

Kind of landscape a set of individual landscapes united by common genesis of abiotic template (e.g., morainic, karst, erosional).

Landscape (Geographical landscape) genetically uniform territory with regular and typical occurrence of interrelated combinations of geological composition, landforms, surface and ground waters, microclimates, soil types, phytocoenoses, and zoocoenoses (Solntsev 1948).

Landscape component see Geocomponent.

Landscape envelope a sphere of strong interactions between the atmosphere, the hydrosphere, the lithosphere, and the biosphere within the layer 10n meters above and below the Earth's surface. Weathering crust, soils, clay minerals, and chemical composition of waters are the examples of specific matter generated by this interaction (Chap. 10).

Landscape geochemistry history of atoms in a landscape (Perelman 1972) (Chaps. 5, 9 and 10).

Landscape geophysics investigation of physical fields and energy cycles in a landscape (Chaps. 3 and 19).

Landscape site the stable part of the landscape described by the landform, upper layer of soil-forming bedrock, and moistening regime; abiotic template of a landscape (Chap. 17).

Landscape-areal representation of a landscape as a product of spatial differentiation with focus on internal homogeneity (Reteyum 2006) (Chap. 13).

Landscape-system representation of a landscape as a product of spatial integration with focus on internal heterogeneity and uniting processes (Reteyum 2006) (Chap. 13).

Mathematical morphology of landscape a branch of science studying quantitative regularities (laws) of mosaics formed by land units on the Earth's surface (landscape patterns) and techniques of their mathematical analysis; "the landscape geometry" (Victorov 2006).

Mestnost a morphological unit of a landscape, a territory within *landscape* with particular combination of main *urochishches* (Solntsev 1948).

Microlandscape level spatial variability of certain attributes (e.g., chemical elements or isotope content) within the elementary landscape units (*facies*).

Migration structure a set of matter migration conditions in a landscape; characterized by lateral and radial migration (Chap. 9).

Monolithic catena catena with uniform geological substrate (Kasimov et al. 2012) (Chap. 9).

Monosystem geosystem described with focus on radial interactions between geocomponents (rocks, soils, vegetation, water, air) (Chap. 2).

Natural territorial complex (NTC) a set of interacting geocomponents and spatial units forming a whole at any hierarchical level. *Facies, urochishche, mestnost,* *landscape*, *physical-geographical province*, and *physical-geographical country* are the examples of NTC of various hierarchical levels.

Opolje a kind of landscapes within elevated, well-drained rolling plains covered by loess or loess-like loams of periglacial origin, with fertile soils, in most cases cultivated. Commonly, in the East European Plain, *opolje* is a former island of extra-zonal broad-leaved forests on Phaeozems within the taiga or mixed-forest zone (Chap. 23).

Ouval landform composed of solid bedrocks with well-manifested slopes and flat surface.

Paradynamic complexes, paradynamic systems of landscapes a set of contrasting landscape units linked by matter flow (Milkov 1981) (Chaps. 14 and 25).

Paragenetic system a set of neighboring interacting landscape units linked by common genesis (Milkov 1981) (Chap. 25).

Paragenetical configuration (structure) a set of landscape units (geotopes) along with lines of concentration of the horizontal flows (Grodzinsky 2005) (Chap. 1).

Plakor flat or slightly inclined part of the interfluve or terrace at a well-drained location with loamy soils without any influence of groundwater (Chap. 8).

Physical-geographical country the largest natural territorial complex with uniform tectonic macrostructure (platform, fold area), climate continentality, system of atmospheric circulation, and a peculiar combination landscape zones or altitudinal belts (e.g., East European Plain, West Siberia, Caucasus) (Chap. 18).

Physical-geographical province the regionalization unit in physical geography with peculiar geomorphological and geological conditions due to similar neotectonics tendency and Quaternary history within a zonal division (e.g., provinces of the Meshcherskaya lowland and Smolensko-Moskovskaya highland within the mixed-forest zone of the East European Plain) (Chap. 24).

Podurochishche a morphological unit of a landscape, a natural unit composed by group of *facies* densely linked by genesis and dynamic processes due to common position on element of relief mesoform with the same solar exposure (Solntsev 1948). *Podurochishche* is a constituent of *urochishche*.

Polessje a kind of landscapes within outwash lowland sandy plains covered mainly by pine forests and mires (e.g., Meshchera lowland) (Chap. 23).

Polystructuralism, structure multiplicity the concept in landscape science accepting the possibility of various structural projections of a landscape based on a number of system-forming relations. Each kind of geocomponent interactions generates its particular set of connections and, hence, a system which is holistic in relation to this kind of interactions (Solntsev 1997) (Chaps. 1 and 2).

Polygeosystem methodology representation of a landscape as a set of systems each of which reflects a certain part of objects in terms of the corresponding special system theory (Chap. 2).

Polysystem mapping multidimensional modeling of geographical space for the solution of various planning and forecasting problems (Cherkashin 2005) (Chap. 2).

Positional-dynamical configuration the connection between geotopes with surface flows of matter and energy and their relationship with lines of change in direction and the intensity of these flows form (Grodzinsky 2005) (Chap. 1).

R-L analysis geochemical investigation of spatial redistribution of chemical element in a landscape based on calculation of coefficients of radial (R) and lateral (L) migration (Chap. 9).

Serial geosystem short-living spontaneous geosystems replacing each other. The chronosequence of serial geosystems ends in the formation of equifinal state (Sochava 1978) (Chap. 10).

Sor solonchak with salt crust (Chap. 13).

Super-aqual landscape landscape-geochemical unit, part of catena, which receives matter from water streams and groundwater (e.g., on floodplain) (Glazovskaya 1964, 2002; Perelman 1972; Fortescue 1992) (Chaps. 9 and 10).

Trans-eluvial landscape landscape-geochemical unit, part of catena, in transitional position which receives matter from upper parts of a slope and transfers it to the lower (Glazovskaya 1964, 2002; Perelman 1972; Fortescue 1992) (Chaps. 9 and 10).

Trans-accumulative landscape landscape-geochemical unit, part of catena, at the toe slope which receives matter from a slope and is a site of matter accumulation (Glazovskaya 1964, 2002; Perelman 1972; Fortescue 1992) (Chaps. 9 and 10).

Urochishche a morphological unit of a landscape which is a system of facies interrelated by genesis, dynamic processes, and territory occupying one relief mesoform.

Vertical (component) structure composition and interactions of *geocomponents* (Solntsev 1948).

Vysotsky-Ivanov annual atmospheric humidity factor ratio of annual precipitation to maximum possible evaporation from open water surface (Chap. 20).

Zonal features of landscape attributes that are typical for the landscape zone in accordance with hydroclimatic conditions (Isachenko 1973).

References

Armand, A. D., & Targul'yan, V. O. (1976). Some fundamental limitations on experimentation and model-building in geography. *Soviet Geography, 17*(3), 197–206.

Cherkashin, A. K. (2005). *Polysystem modeling.* Novosibirsk: Nauka. (in Russian).

Dyakonov, K. N., Kasimov, N. S., Khoroshev, A. V., & Kushlin, A. V. (Eds.). (2007). *Landscape analysis for sustainable development. Theory and applications of landscape science in Russia.* Moscow: Alex Publishers.

Fortescue, J. A. C. (1992). Landscape geochemistry: Retrospect and prospect – 1990. *Applied Geochemistry, 7*, 1–53.

Glazovskaya, M. A. (1988). *Geochemistry of natural and technogenic landscapes of the USSR.* Moscow: Vysshaya shkola.

Glazovskaya, M. A. (2002). *Geochemical foundations for typology and methods of research of natural landscapes.* Smolensk: Oikumena. (in Russian). First published: Glazovskaya, M.A. (1964). *Geochemical foundations for typology and methods of research of natural landscapes.* Moscow: MSU Publishing House. (in Russian).

Grodzinsky, M. (2005). *Understanding landscape. Place and space.* In two volumes. Kiev: Kiev university. (in Ukrainian)

Isachenko, A. G. (1973). *Principles of landscape science and physical-geographic regionalization.* Carlton, VIC.: Melbourne University Press.

Kasimov, N. S., Gerasimova, M. I., Bogdanova, M. D., et al. (2012). Landscape-geochemical catenas: Concept and mapping. In N. S. Kasimov (Ed.), *Landscape geochemistry and soil geography* (pp. 59–80). Moscow: APR. (in Russian).

Kiryushin, V. I. (2011). *Theory of adaptive-landscape agriculture and the projecting of agro-landscapes.* Moscow: KolosS.

Krauklis, A. A. (1979). *Problems of experimental landscape science.* Novosibirsk: Nauka. (in Russian).

Milkov, F. N. (1981). *Physical geography: Present-day state, regularities, problems.* Voronezh: Voronezh University Press. (in Russian).

Neef, E. (1963). Topologische und chorologische arbeitsweisen in der landschaftsforschung. *Petermanns Geographische Mitteilungen, 107*(4), 249–259.

Perelman, A. I. (1972). *Landscape geochemistry.* Moscow, Vysshaya Shkola. (Translated form Russian). (Geol. Surv. Canada Trans. No. 676, Part I and II).

Reteyum, A. Y. (1988). *The terrestrial worlds (on holistic studying of geosystems).* Moscow: Mysl'. (in Russian).

Reteyum, A. Y. (2006). Research installations of landscape science. In K. N. Dyakonov (Ed.), *Landscape science: Theory, methods, regional studies, practice. Proceedings of XIth International landscape conference* (pp. 46–49). Moscow: MSU Publishing House. (in Russian).

Sochava, V. B. (1963). Definition of some notions and terms in physical geography. *Proceedings of the Institute of Geography of Siberia and Far East SB AS USSR, 3*, 50–59. (in Russian).

Sochava, V. B. (1978). *Introduction to the theory of geosystems*. Novosibirsk: Nauka.

Solntsev, N. A. (1948). The natural geographic landscape and some of its general rules. In *Proceedings of the second all-union geographical congress* (Vol. 1, pp. 258–269). Moscow: State Publishing House for Geographic Literature. (in Russian). See also in J.A. Wiens, M.R. Moss, M.G. Turner, & D.J. Mladenoff (Eds.) (2006). *Foundation papers in landscape ecology* (pp. 19–27). New York: Columbia University Press.

Solntsev, V.N. (1997). *Structural landscape science*. Moscow.

Victorov, A. S. (2006). *Basic problems of mathematical landscape morphology*. Moscow: Nauka. (in Russian).